METHODS IN MOLECULAR BIOLOGY

Series Editor
John M. Walker
School of Life and Medical Sciences
University of Hertfordshire
Hatfield, Hertfordshire, AL10 9AB, UK

For further volumes:
http://www.springer.com/series/7651

Biosensors and Biodetection

Methods and Protocols Volume 1:
Optical-Based Detectors

Second Edition

Edited by

Avraham Rasooly

National Cancer Institute
National Institutes of Health
Rockville, MD, USA

Ben Prickril

National Cancer Institute
National Institutes of Health
Rockville, MD, USA

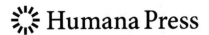 Humana Press

Editors
Avraham Rasooly
National Cancer Institute
National Institutes of Health
Rockville, MD, USA

Ben Prickril
National Cancer Institute
National Institutes of Health
Rockville, MD, USA

ISSN 1064-3745   ISSN 1940-6029 (electronic)
Methods in Molecular Biology
ISBN 978-1-4939-8316-2   ISBN 978-1-4939-6848-0 (eBook)
DOI 10.1007/978-1-4939-6848-0

Printed on acid-free paper

This Humana Press imprint is published by Springer Nature
The registered company is Springer Science+Business Media LLC
The registered company address is: 233 Spring Street, New York, NY 10013, U.S.A.

*This book is dedicated to the memory
of Jury Rasooly Ph.D., Malkah Rasooly,
and Ilan Rasooly.*

Preface

Biosensor Technologies

A biosensor is defined by the International Union of Pure and Applied Chemistry (IUPAC) as "a device that uses specific biochemical reactions mediated by isolated enzymes, immunosystems, tissues, organelles or whole cells to detect chemical compounds usually by electrical, thermal or optical signals" [1]; all biosensors are based on a two-component system:

1. Biological recognition element (ligand) that facilitates specific binding or biochemical reaction with the target analyte
2. Signal conversion unit (transducer)

Since the publication of the first edition of this book in 2009, "classical" biosensor modalities such as electrochemical or surface plasmon resonance (SPR) continue to be developed. New biosensing technologies and modalities have also been developed, including the use of nanomaterials for biosensors, fiber-optic-based biosensors, genetic code based sensors, field effect transistors, and the use of mobile communication device-based biosensors. Although it is impossible to describe the fast-moving field of biosensing in a single publication, this book presents descriptions of methods and uses for some of the basic types of biosensors while also providing the reader a sense of the enormous importance and potential for these devices. In order to present a more comprehensive overview, the book also describes other biodetection technologies.

Dr. Leland C. Clark, who worked on biosensors in the early 1960s, provided an early reference to the concept of a biosensor by developing an "enzyme electrode" for glucose concentration measurement using the enzyme glucose oxidase (GOD) [2]. Glucose monitoring is essential for diabetes patients, and even today the most common clinical biosensor technology for glucose analysis is the electrochemical detection method envisioned by Clark more than 50 years ago. Today, glucose monitoring is performed using rapid point of care biosensors made possible through advances in electronics that have enabled sensor miniaturization. The newest generation of biosensors includes phone-based optical detectors with high-throughput capabilities.

The Use of Biosensors

Biosensors have several potential advantages over other methods of biodetection, including increased assay speed and flexibility. Rapid, real-time analysis can provide immediate interactive information to health-care providers that can be incorporated into the planning of patient care. In addition, biosensors allow multi-target analyses, automation, and reduced testing costs. Biosensor-based diagnostics may also facilitate screening for cancer and other diseases by improving early detection and therefore improving prognosis. Such technology may be extremely useful for enhancing health-care delivery to underserved populations and in community settings.

The main advantages of biosensors include:

Rapid or real-time analysis: Direct biosensors such as those employing surface plasmon resonance (SPR) enable rapid or real-time label-free detection and provide almost immediate interactive sample information. This enables facilities to take corrective measures before a product is further processed or released for consumption.

Point of care detection capabilities: Biosensors can be used for point of care testing. This enables state-of-the-art molecular analysis without requiring a laboratory.

Continuous flow analysis: Many biosensors are designed to allow analysis of bulk liquids. In such biosensors the target analyte is injected onto the sensor using a continuous flow system immobilized in a flow cell or column, thereby enhancing the efficiency of analyte binding to the sensor and enabling continuous monitoring.

Miniaturization: Increasingly, biosensors are being miniaturized for incorporation into equipment for a wide variety of applications including clinical care, food and dairy analyses, agricultural and environmental monitoring, and in vivo detection of a variety of diseases and conditions.

Control and automation: Biosensors can be integrated into online process monitoring schemes to provide real-time information about multiple parameters at each production step or at multiple time points during a process, enabling better control and automation of biochemical facilities.

Biosensor Classification

In general biosensors can be divided into two groups: direct recognition sensors in which the biological interaction is directly measured and indirect detection sensors which rely on secondary elements (often catalytic) such as enzymes or fluorescent tags for measurements. Figure 1 illustrates the two types of biosensors. In each group there are several types of

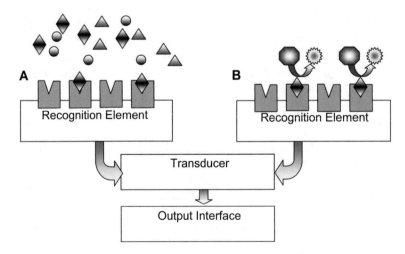

Fig. 1 General schematic of biosensors. (**A**) Direct detection biosensors where the recognition element is label-free; (**B**) indirect detection biosensors using "sandwich" assay where the analyte is detected by labeled molecule. Direct detection biosensors are simpler and faster but typically yield a higher limit of detection compared to indirect detection systems

optical, electrochemical, or mechanical transducers. Although the most commonly used ligands are antibodies, other ligands are being developed including aptamers (protein-binding nucleic acids) and peptides.

There are numerous types of direct and indirect recognition biosensors, and choice of a suitable detector is complex and based on many factors. These include the nature of the application, type of labeled molecule (if used), sensitivity required, number of channels (or area) measured, cost, technical expertise, and speed of detection. In this book we describe many of these detectors, their application to biosensing, and their fabrication.

The transducer element of biosensors converts the biochemical interactions of the ligand into a measurable electronic signal. The most important types of transducer used today are optical, electrochemical, and mechanical.

Direct Label-Free Detection Biosensors

Direct recognition sensors, in which the biological interaction is directly measured in real time, typically use non-catalytic ligands such as cell receptors or antibodies. Such detectors typically measure directly physical changes (e.g., changes in optical, mechanical, or electrical properties) induced by the biological interaction and do not require additional labeled molecules (i.e., are label-free) for detection. The most common direct detection biosensors are optical biosensors including biosensors which employ evanescent waves generated when a beam of light is incident on a surface at an angle yielding total reflection. Common evanescent wave biosensors are surface plasmon resonance (SPR) or resonant mirror sensors. Other direct optical detectors include interferometric sensors or grating coupler. Nonoptical direct detection sensors are quartz resonator transducers that measure change in resonant frequency of an oscillating piezoelectric crystal as a function of mass (e.g., analyte binding) on the crystal surface, microcantilevers used in microelectromechanical systems (MEMS) measuring bending induced by the biomolecular interactions, or field effect transistor (FET) biosensors, a transistor gated by biological molecules. When biological molecules bind to the FET gate, they can change the gate charge distribution resulting in a change in conductance of the FET.

Indirect Label-Based Detection Biosensors

Indirect detection sensors rely on secondary elements for detection and utilize labeling or catalytic elements such as enzymes. Examples of such secondary elements are the enzyme alkaline phosphatase and fluorescently tagged antibodies that enhance detection of a sandwich complex. Unlike direct sensors, which directly measure changes induced by biological interaction and are "label-free," indirect sensors require a labeled molecule bound to the target. Most optical indirect sensors are designed to measure fluorescence; however, such sensors can also measure densitometric and colorimetric changes as well as chemiluminescence, depending on the type of label used.

Electrochemical transducers measure the oxidation or reduction of an electroactive compound on the secondary ligand and are one common type of indirect detection sensor. Several types of electrochemical biosensors have been developed including amperometric devices, which detect ions in a solution based on electric current or changes in electric current when an analyte is oxidized or reduced. Another common indirect detection biosensor employs optical fluorescence, detecting fluorescence of the secondary ligand via CCD, PMT, photodiode, and spectrofluorometric analysis. In addition, visual measurement such as change of color or appearance of bands (e.g., lateral flow detection) can be used for indirect detection.

Indirect detection can be combined with direct detection to increase sensitivity or to validate results; for example, the use of secondary antibody in combination with an SPR immunosensor. Using a sandwich assay, the analyte captured by the primary antibody is immobilized on the SPR sensor and generates a signal which can be amplified by the binding of a secondary antibody to the captured analyte.

Ligands for Biosensors

Ligands are molecules that bind specifically with the target molecule to be detected. The most important properties of ligands are affinity and specificity. Of the various types of ligands used in biosensors, immunosensors—particularly antibodies—are the most common biosensor recognition element. Antibodies (Abs) are highly specific and versatile and bind strongly and stably to specific antigens. However, Ab ligands have limited long-term stability and are difficult to produce in large quantities for multi-target biosensor applications where many ligands are needed.

Other types of ligands such as aptamers and peptides are more suited to high-throughput screening and chemical synthesis. Aptamers are protein-binding nucleic acids (DNA or RNA molecules) selected from random pools based on their ability to bind other molecules with high affinity. Peptides are another potentially important class of ligand suitable for high-throughput screening due to their ease of selection. However, the affinity of peptides is often lower than that of antibodies or aptamers, and peptides vary widely in structural stability and thermal sensitivity.

New Trends in Biosensing

While the fundamental principles and the basic configuration of biosensors have not changed in the last decade, this book expands the application of these principles using new technologies such as nanotechnology, integrated optics (IO) bioelectronics, portable imaging, new fluidics and fabrication methodologies, and new cellular and molecular approaches.

Integration of nanotechnology. There has been great progress in nanotechnology and nano-material in recent years. New nanoparticles have been developed having unique electric conductivity, optical, and surface properties. For example, in several chapters new optical biosensors are described that integrate nanomaterials in SPR biosensor configurations such as localized surface plasmon resonance (LSPR), 3D SPR plasmonic nanogap arrays, or gold nanoparticle SPR plasmonic peak shift. In addition to SPR biosensors, nanomaterials are also applied to fluorescence detection utilizing fluorescence quantum dot or silica nanoparticles to increase uniform distribution of enzyme and color intensity in colorimetric biosensors or to improve lateral flow detection. In addition to optical sensors, gold nanoparticles (AuNPs) have been integrated into electrochemical biosensors to improve electrochemical performance, and magnetic nanoparticles (mNPs) have been used to improve sample preparation. Nanoparticle-modified gate electrodes have been used in the fabrication of organic electrochemical transistors.

Bioelectronics. Several chapters described the integration of biological elements in electronic technology including the use of semiconductors in several configurations of field effect transistors and light-addressable potentiometric sensors.

Application of imaging technologies: The proliferation of high-resolution imaging technologies has enabled better 2D image analysis and increases in the number of analytical channels available for various modalities of optical detection. These include two-dimensional surface plasmon resonance imaging (2D-SPRi) utilizing CCD cameras or 2D photodiode arrays. The use of smartphones for both fluorescence and colorimetric detectors is described in several manuscripts.

Integrated optics (IO): Devices with photonic integrated circuits are presented which integrate several optical and often electronic components. Examples include an integrated optical (IO) nano-immunosensor based on a bimodal waveguide (BiMW) interferometric transducer integrated into a complete lab-on-a-chip (LOC) platform.

New fluidics and fabrication methodologies: Fluidics and fluid delivery are important components of many biosensors. In addition to traditional polymer fabricated microfluidics systems, inkjet-printed paper fluidics are described that may play an important role in LOCs and medical diagnostics. Such technologies enable low-cost mass production of LOCs. In addition, several chapters describe the use of screen printing for device fabrication.

Cellular and molecular approaches: Molecular approaches are described for aptamer-based biosensors (aptasensors), synthetic cell-based sensors, loop-mediated DNA amplification, and circular strand displacement for point mutation analysis.

While "classic" transducer modalities such as SPR, electrochemical, or piezoelectric remain the predominant biosensor platforms, new technologies such as nanotechnology, integrated optics, or advanced fluidics are providing new capabilities and improved sensitivity.

Aims and Approaches

This book attempts to describe the basic types, designs, and applications of biosensors and other biodetectors from an experimental point of view. We have assembled manuscripts representing the major technologies in the field and have included enough technical information so that the reader can both understand the technology and carry out the experiments described.

The target audience for this book includes engineering, chemistry, biomedical, and physics researchers who are developing biosensing technologies. Other target groups are biologists and clinicians who ultimately benefit from development and application of the technologies.

In addition to research scientists, the book may also be useful as a teaching tool for bioengineering, biomedical engineering, and biology faculty and students. To better represent the field, most topics are described in more than one chapter. The purpose of this redundancy is to bring several experimental approaches to each topic, to enable the reader to choose an appropriate design, to combine elements from different designs in order to better standardize methodologies, and to provide readers more detailed protocols.

Organization

The publication is divided into two volumes. Volume I (Springer Vol. 1571) focuses on optical-based detectors, while Vol. II (Springer Vol. 1572) focuses on electrochemical, bioelectronic, piezoelectric, cellular, and molecular biosensors.

Volume I (Springer Vol. 1571)

Optical-based detection encompasses a broad array of technologies including direct and indirect methods as discussed above. Part I of Vol. I describes various optical-based direct detectors, while Part II focuses on indirect optical detection. Three types of direct optical detection biosensors are described: evanescent wave (SPR and resonant waveguide grating), interferometers, and Raman spectroscopy sensors.

The second part of Vol. I describes various indirect optical detectors as discussed above. Indirect directors require a labeled molecule to be bound to the signal-generating target. For optical sensors such molecules emit or modify light signals. Most indirect optical detectors are designed to measure fluorescence; however, such detectors can also measure densitometric and colorimetric changes as well as chemiluminescence, and detection depends on the type of label used. Such optical signals can be measured in various ways as described in Part II. These include various CCD-based detectors which are very versatile, inexpensive, and relatively simple to construct and use. Other optical detectors discussed in Part II are photodiode-based detection systems and mobile phone detectors. Lateral flow systems that rely on visual detection are included in this section. Although lateral flow devices are not "classical" biosensors with ligands and transducers, they are included in this book because of their importance for biosensing. Lateral flow assays use simple immunodetection (or DNA hybridization) devices, such as competitive or sandwich assays, and are used mainly for medical diagnostics such as laboratory and home testing or any other point of care (POC) detection. A common format is a "dipstick" in which the test sample flows on an absorbant matrix via capillary action; detection is accomplished by mixing a colorimetric reagent with the sample and binding to a secondary antibody to produce lines or zones at specific locations on the absorbing matrix.

Volume II (Springer Vol. 1572)

Volume II describes various electrochemical, bioelectronic, piezoelectric, and cellular- and molecular-based biosensors.

In Part I of Vol. II, we describe several types of electrochemical and bioelectronic detectors. Electrochemical biosensors were the first biosensors developed and are the most commonly used biosensors in clinical settings (e.g., glucose monitoring). Also included are several electronic/semiconductor sensors based on the field effect. Unlike electrochemical sensors, which are indirect detectors and require labeling, electronic/semiconductor biosensors are label-free.

In Part II we describe "mechanical detectors" which modify their mechanical properties as a result of biological interactions. Such mechanical direct biosensors include piezoelectric biosensors which change their acoustical resonance and cantilevers which modify their movement.

Part III describes a variety of biological sensors including aptamer-based sensors and cellular and phage display technologies.

Part IV describes several microfluidics technologies for cell isolation. In addition, a number of related technologies including Raman spectroscopy and high-resolution micro-ultrasound are described.

The two volumes provide comprehensive and detailed technical protocols on current biosensor and biodetection technologies and examples of their applications and capabilities.

Rockville, MD   *Avraham Rasooly*
Rockville, MD   *Ben Prickril*

References

1. International Union of Pure and Applied Chemistry (1992) IUPAC compendium of chemical terminology, 2nd edn (1997). International Union of Pure and Applied Chemistry, Research Triangle Park, NC
2. Clark LC Jr, Lyons C (1962) Electrode systems for continuous monitoring in cardiovascular surgery. Ann N Y Acad Sci 102:29–45

Contents

Contributors

MUSTAFA AL-ADHAMI • *University of Maryland Baltimore County, Baltimore, MD, USA*

ANTJE J. BAEUMNER • *Department of Biological and Environmental Engineering, Cornell University, Ithaca, NY, USA; Institute for Analytical Chemistry, Chemo- and Biosensors, University of Regensburg, Regensburg, Germany*

JOSHUA BALSAM • *Center for Devices and Radiological Health, FDA, Silver Spring, MD, USA*

DUMITRU BRATU • *International Centre of Biodynamics, Bucharest, Romania*

HUGH A. BRUCK • *University of Maryland College Park (UMCP), College Park, MD, USA*

THIAGO M.G. CARDOSO • *Instituto de Química, Universidade Federal de Goiás, Goiânia, GO, Brazil*

CHRISTOPHE CAUCHETEUR • *Université de Mons, Mons, Belgium*

SIDDARTH CHANDRASEKARAN • *Biochemical Technologies, Corning Research and Development Corporation, Corning Incorporated, Corning, NY, USA; Department of Biomedical Engineering, Cornell University, Ithaca, NY, USA*

GANGYI CHEN • *Natural Products Research Center, Chengdu Institution of Biology, Chinese Academy of Science, Chengdu, China*

WENDELL K.T. COLTRO • *Instituto de Química, Universidade Federal de Goiás, Goiânia, GO, Brazil; Instituto Nacional de Ciência e Tecnologia de Bioanalítica (INCTBio), Campinas, SP, Brazil*

JOHN H. CONNOR • *Microbiology Department, Boston University School of Medicine, Boston, MA, USA*

SORIN DAVID • *International Centre of Biodynamics, Bucharest, Romania*

JOHN R. DAY • *Illumina, Inc., San Diego, CA, USA*

SAMER DOUGHAN • *Chemical Sensors Group, Department of Chemical and Physical Sciences, University of Toronto Mississauga, Mississauga, ON, Canada*

FENG DU • *Natural Products Research Center, Chengdu Institution of Biology, Chinese Academy of Science, Chengdu, China*

KATIE A. EDWARDS • *Department of Biological and Environmental Engineering, Cornell University, Ithaca, NY, USA*

ELIZABETH EVANS • *Department of Chemistry, , Clemson University, Clemson, SC, USA*

YE FANG • *Biochemical Technologies, Corning Research and Development Corporation, Corning Incorporated, Corning, NY, USA*

NICOLE K. FEBLES • *Biochemical Technologies, Corning Research and Development Corporation, Corning Incorporated, Corning, NY, USA; NanoScience Technology Center, Department of Mechanical, Materials and Aerospace Engineering, University of Central Florida, Orlando, FL, USA*

PETER FECHNER • *Biametrics GmbH, Tuebingen, Germany*

ELLEN FLÁVIA MOREIRA GABRIEL • *Instituto de Química, Universidade Federal de Goiás, Goiânia, GO, Brazil*

PAULO T. GARCIA • *Instituto de Química, Universidade Federal de Goiás, Goiânia, GO, Brazil*

CARLOS D. GARCIA • *Department of Chemistry, Clemson University, Clemson, SC, USA*

MARSHA A. GASTON • *Department of Molecular Genetics, Biochemistry and Microbiology, University of Cincinnati, Cincinnati, OH, USA*

GUENTER GAUGLITZ • *Institute of Physical and Theoretical Chemistry, University of Tuebingen, Tuebingen, Germany*

ADRIÁN FERNÁNDEZ GAVELA • *Nanobiosensors and Bioanalytical Applications Group, Catalan Institute of Nanoscience and Nanotechnology (ICN2), CSIC, The Barcelona Institute of Science and Technology, Bellaterra, Barcelona, Spain*

EUGEN GHEORGHIU • *International Centre of Biodynamics, Bucharest, Romania; University of Bucharest, Bucharest, Romania*

MIHAELA GHEORGHIU • *International Centre of Biodynamics, Bucharest, Romania*

DAYONG GU • *Department of Electronic Engineering, Center for Advanced Research in Photonics, The Chinese University of Hong Kong, N.T. Hong Kong SAR, China*

CHANDRASEKHAR GURRAMKONDA • *University of Maryland Baltimore County, Baltimore, MD, USA*

YI HAN • *Chemical Sensors Group, Department of Chemical and Physical Sciences, University of Toronto Mississauga, Mississauga, ON, Canada*

SONIA HERRANZ • *Nanobiosensors and Bioanalytical Applications Group, Catalan Institute of Nanoscience and Nanotechnology (ICN2), CSIC, The Barcelona Institute of Science and Technology, Bellaterra, Barcelona, Spain*

HO PUI HO • *Department of Electronic Engineering, Center for Advanced Research in Photonics, The Chinese University of Hong Kong, N.T. Hong Kong SAR, China*

FULYA EKIZ KANIK • *Electrical and Computer Engineering Department, Boston University, Boston, MA, USA*

DONGHYUN KIM • *School of Electrical and Electronic Engineering, Yonsei University, Seoul, Republic of Korea*

SIU KAI KONG • *Department of Electronic Engineering, Center for Advanced Research in Photonics, The Chinese University of Hong Kong, N.T. Hong Kong SAR, China*

RICKI KORFF • *Department of Biological and Environmental Engineering, Cornell University, Ithaca, NY, USA*

YORDAN KOSTOV • *University of Maryland Baltimore County, Baltimore, MD, USA*

ULRICH J. KRULL • *Chemical Sensors Group, Department of Chemical and Physical Sciences, University of Toronto Mississauga, Mississauga, ON, Canada*

TIAN LAN • *Glucosentient Inc., Champaign, IL, USA*

LAURA M. LECHUGA • *Nanobiosensors and Bioanalytical Applications Group, Catalan Institute of Nanoscience and Nanotechnology (ICN2), CSIC, The Barcelona Institute of Science and Technology, Bellaterra, Barcelona, Spain; CIBER-BBN, Campus UAB, Ed-ICN2, Bellaterra, Barcelona, Spain*

CHANGWON LEE • *Department of Chemistry, University of Cincinnati, Cincinnati, OH, USA*

WONJU LEE • *School of Electrical and Electronic Engineering, Yonsei University, Seoul, Republic of Korea*

CHANGHUN LEE • *School of Electrical and Electronic Engineering, Yonsei University, Seoul, Republic of Korea*

SHUANG LI • *Biosensor National Special Laboratory, Key Laboratory for Biomedical Engineering of Education Ministry, Department of Biomedical Engineering, Zhejiang University, Hangzhou, China*

MEI LI • *Natural Products Research Center, Chengdu Institution of Biology, Chinese Academy of Science, Chengdu, China*

YONGBIN LIN • *Center for Applied Optics, University of Alabama at Huntsville, Huntsville, AL, USA*

QINGJUN LIU • *Biosensor National Special Laboratory, Key Laboratory for Biomedical Engineering of Education Ministry, Department of Biomedical Engineering, Zhejiang University, Hangzhou, China*

FONG CHUEN LOO • *Department of Electronic Engineering, Center for Advanced Research in Photonics, The Chinese University of Hong Kong, N.T. Hong Kong SAR, China*

YANLI LU • *Biosensor National Special Laboratory, Key Laboratory for Biomedical Engineering of Education Ministry, Department of Biomedical Engineering, Zhejiang University, Hangzhou, China*

YI LU • *Department of Chemistry, University of Illinois at Urbana-Champaign, Urbana, IL, USA*

JOHN H.T. LUONG • *Innovative Chromatography Group, Irish Separation Science Cluster (ISSC), Department of Chemistry and Analytical, Biological Chemistry Research Facility (ABCRF), University College Cork, Cork, Ireland*

VIERA MALACHOVSKA • *B-SENS, Mons, Belgium*

E. MARION SCHNEIDER • *Sektion Experimentelle Anaesthesiologie, University Hospital Ulm, Ulm, Germany*

GORAN MARKOVIC • *Biametrics GmbH, Tuebingen, Germany*

DEBAPRIYA MAZUMDAR • *ANDalyze Inc., Champaign, IL, USA*

ZHONG MEI • *Department of Biomedical Engineering, University of Texas at San Antonio, San Antonio, TX, USA*

TANVEER AHMAD MIR • *Graduate School of Science and Engineering for Research, University of Toyama, Toyama, Japan; Graduate School of Innovative Life Sciences for Education, University of Toyama, Toyama, Japan; Institutes for Analytical Chemistry, Chemo- and Biosensor, University of Regensburg, Regensburg, Germany; Institute of BioPhysio Sensor Technology (IBST), Pusan National University, Busan, South Korea*

M. OMAIR NOOR • *Chemical Sensors Group, Department of Chemical and Physical Sciences, University of Toronto Mississauga, Mississauga, ON, Canada*

YONGJIN OH • *School of Electrical and Electronic Engineering, Yonsei University, Seoul, Republic of Korea*

THOMAS VAN OORDT • *Hahn-Schickard, Freiburg, Germany*

MIGUEL OSSANDON • *National Cancer Institute, NIH/NCI, Rockville, MD, USA*

CRISTINA POLONSCHII • *International Centre of Biodynamics, Bucharest, Romania*

BEN PRICKRIL • *National Cancer Institute, National Institutes of Health, Rockville, MD, USA*

AASHISH PRIYE • *Artie McFerrin Department of Chemical Engineering, Texas A&M University, College Station, TX, USA*

FLORIAN PROELL • *Institute of Physical and Theoretical Chemistry, University of Tuebingen, Tuebingen, Germany; Biametrics GmbH, Tuebingen, Germany*

GUENTHER PROLL • *Institute of Physical and Theoretical Chemistry, University of Tuebingen, Tuebingen, Germany; Biametrics GmbH, Tuebingen, Germany*

GOVIND RAO • *University of Maryland Baltimore County, Baltimore, MD, USA*

REUVEN RASOOLY • *Western Regional Research Center, Agricultural Research Service, U.S. Department of Agriculture, Albany, CA, USA*

AVRAHAM RASOOLY • *National Cancer Institute, National Institutes of Health Rockville, MD, USA*

CLOTILDE RIBAUT • *B-SENS, Mons, Belgium*

ELIF SEYMOUR • *Biotechnology Research Program Department, ASELSAN Research Center, Ankara, Turkey*

HIROAKI SHINOHARA • *Graduate School of Science and Engineering for Research, University of Toyama, Toyama, Japan; Graduate School of Innovative Life Sciences for Education, University of Toyama, Toyama, Japan*

TAEHWANG SON • *School of Electrical and Electronic Engineering, Yonsei University, Seoul, Republic of Korea*

JOON MYONG SONG • *College of Pharmacy, Seoul National University, Seoul, South Korea*

FELIX VON STETTEN • *Hahn-Schickard, Freiburg, Germany; Laboratory for MEMS Applications, Department of Microsystems Engineering—IMTEK, University of Freiburg, Freiburg, Germany*

ANNIE AGNES SUGANYA SAMSON • *College of Pharmacy, Seoul National University, Seoul, South Korea*

JIAN-HE SUN • *Key Laboratory of Veterinary Biotechnology, School of Agriculture and Biology, Shanghai Jiao Tong University, Shanghai, China*

LIANG TANG • *Department of Biomedical Engineering, University of Texas at San Antonio, San Antonio, TX, USA*

ZHUO TANG • *Natural Products Research Center, Chengdu Institution of Biology, Chinese Academy of Science, Chengdu, China*

DAGMAWI TILAHUN • *University of Maryland Baltimore County, Baltimore, MD, USA*

VICTOR M. UGAZ • *Artie McFerrin Department of Chemical Engineering, Texas A&M University, College Station, TX, USA; Department of Biomedical Engineering, Texas A&M University, College Station, TX, USA*

NESE LORTLAR ÜNLÜ • *Biomedical Engineering Department, Boston University, Boston, MA, USA; Faculty of Medicine, Bahcesehir University, Istanbul, Turkey*

M. SELIM ÜNLÜ • *Biomedical Engineering Department, Boston University, Boston, MA, USA; Electrical and Computer Engineering Department, Boston University, Boston, MA, USA; Microbiology Department, Boston University School of Medicine, Boston University, Boston, MA, USA*

SANDEEP KUMAR VASHIST • *Hahn-Schickard, Freiburg, Germany; Laboratory for MEMS Applications, Department of Microsystems Engineering—IMTEK, University of Freiburg, Freiburg, Germany; Immunodiagnostics Systems, Liege, Belgium*

A.G. VENKATESH • *Department of Electrical and Computer Engineering, Jacobs School of Engineering, University of California San Diego, San Diego, CA, USA*

PENG WANG • *Department of Chemistry, University of Cincinnati, Cincinnati, OH, USA*

YANYAN WANG • *Department of Biomedical Engineering, University of Texas at San Antonio, San Antonio, TX, USA*

YUAN-KAI WANG • *Key Laboratory of Veterinary Biotechnology, School of Agriculture and Biology, Shanghai Jiao Tong University, Shanghai, China*

RUDDY WATTIEZ • *B-SENS, Mons, Belgium*

JIANJUN WEI • *Department of Nanoscience, Joint School of Nanoscience and Nanoengineering (JSNN), University of North Carolina at Greensboro, Greensboro, NC, USA*

ALISON A. WEISS • *Department of Molecular Genetics, Biochemistry and Microbiology, University of Cincinnati, Cincinnati, OH, USA*

SHU YUEN WU • *Department of Electronic Engineering, Center for Advanced Research in Photonics, The Chinese University of Hong Kong, N.T. Hong Kong SAR, China*

YA-XIAN YAN • *Key Laboratory of Veterinary Biotechnology, School of Agriculture and Biology, Shanghai Jiao Tong University, Shanghai, China*

LI YANG • *Natural Products Research Center, Chengdu Institution of Biology, Chinese Academy of Science, Chengdu, China*

YAO YAO • *Biosensor National Special Laboratory, Key Laboratory for Biomedical Engineering of Education Ministry, Department of Biomedical Engineering, Zhejiang University, Hangzhou, China*

AFSHAN YASMEEN • *Natural Products Research Center, Chengdu Institution of Biology, Chinese Academy of Science, Chengdu, China*

KEN-TYE YONG • *Department of Electronic Engineering, Center for Advanced Research in Photonics, The Chinese University of Hong Kong, N.T. Hong Kong SAR, China*

ZHENG ZENG • *Department of Nanoscience, Joint School of Nanoscience and Nanoengineering (JSNN), University of North Carolina at Greensboro, Greensboro, NC, USA*

ROLAND ZENGERLE • *Hahn-Schickard, Freiburg, Germany; Laboratory for MEMS Applications, Department of Microsystems Engineering—IMTEK, University of Freiburg, Freiburg, Germany*

QIAN ZHANG • *Biosensor National Special Laboratory, Key Laboratory for Biomedical Engineering of Education Ministry, Department of Biomedical Engineering, Zhejiang University, Hangzhou, China*

PENG ZHANG • *Department of Chemistry, University of Cincinnati, Cincinnati, OH, USA*

DIMING ZHANG • *Biosensor National Special Laboratory, Key Laboratory for Biomedical Engineering of Education Ministry, Department of Biomedical Engineering, Zhejiang University, Hangzhou, China*

Chapter 1

Localized Surface Plasmon Resonance (LSPR)-Coupled Fiber-Optic Nanoprobe for the Detection of Protein Biomarkers

Jianjun Wei, Zheng Zeng, and Yongbin Lin

Abstract

Here is presented a miniaturized, fiber-optic (FO) nanoprobe biosensor based on the localized surface plasmon resonance (LSPR) at the reusable dielectric-metallic hybrid interface with a robust, gold nano-disk array at the fiber end facet. The nanodisk array is directly fabricated using electron beam lithography (EBL) and metal lift-off process. The free prostate-specific antigen (f-PSA) has been detected with a mouse anti-human prostate-specific antigen (PSA) monoclonal antibody (mAb) as a specific receptor linked with a self-assembled monolayer (SAM) at the LSPR-FO facet surfaces. Experimental investigation and data analysis found near field refractive index (RI) sensitivity at ~226 nm/RIU with the LSPR-FO nanoprobe, and demonstrated the lowest limit of detection (LOD) at 100 fg/mL (~3 fM) of f-PSA in PBS solutions. The SAM shows insignificant nonspecific binding to the target biomarkers in the solution. The control experimentation using 5 mg/mL bovine serum albumin in PBS and nonspecific surface test shows the excellent specificity and selectivity in the detection of f-PSA in PBS. These results indicate important progress toward a miniaturized, multifunctional fiber-optic technology that integrates informational communication and sensing function for developing a high-performance, label-free, point-of-care (POC) device.

Key words Fiber optics, Protein biomarker biosensors, Nanofabrication, Au nanodisk array, Localized surface plasmon resonance (LSPR), Signal transduction

1 Introduction

Recent advances of biomarker detection have been made in optical fluorescence [1], light scattering [2], surface enhanced Raman spectroscopy (SERS) [3], electrochemical [4], functional quartz crystal microbalance [5], microcantilevers [6], and surface plasmon resonance (SPR) imaging [7] sensors. Harnessing the advances in biological ligand interactions, the fiber-optic (FO) technology incorporating localized surface plasmon resonance (LSPR) nanoprobe sensing may provide an alternative tool for effective biomarker diagnosis via a compact, label-free format that does not require a

Avraham Rasooly and Ben Prickril (eds.), *Biosensors and Biodetection: Methods and Protocols Volume 1: Optical-Based Detectors*, Methods in Molecular Biology, vol. 1571, DOI 10.1007/978-1-4939-6848-0_1,
© Springer Science+Business Media LLC 2017

dedicated laboratory facility or highly trained personnel. Furthermore, there is a need to develop advanced LSPR-FO biosensors that may avoid the use of bulky optics and high-precision mechanics for angular or wavelength interrogation of metal films in contact with analytes, and provide high-performance, e.g., enhanced stability and high RI sensitivity, and overcome unwanted doping or weak adhesion.

Surface plasmon resonance (SPR) is the resonance oscillation of conduction electrons at the interface between a negative and a positive permittivity material excited by an electromagnetic radiation, e.g., light. The surface plasmon polaritons (SPPs) launched upon the radiation can be propagating along the metal-dielectric interface and decay evanescently at the normal direction for a flat surface. Surface plasmons [8] (SPs) are very sensitive to the near surface refractive index (RI) changes and well suited to the detection of surface-binding events. The basic methodology of SPR sensing is based on the Kretschmann configuration (Fig. 1) where a prism is used for the light-SP coupling at the surface of a thin metal film. The probe light undergoes total internal reflection on the inner surface of the prism. At a defined SPR angle, an evanescent light field travels through the thin gold film and excites SPs at the metal-dielectric interface, reducing the intensity of the reflected light at the resonance wavelength or changing the phase of the incident light. The intensities of scattered and transmitted light fields are used to determine the thickness and/or dielectric constant of the coating [9]. The control variables for SPR sensor applications, i.e., the wavelength of incident light, the thickness of the metal film, the physical and optical properties of the prism, and the RI of the medium near the metallic interface have been well studied [10]. However, the advantages of averaging over a large surface area and the challenges of miniaturizing the optics limit the integration of SPR-based sensing.

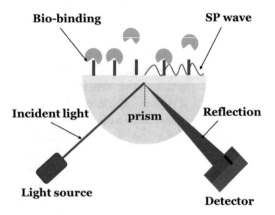

Fig. 1 A conventional surface plasmon resonance (SPR) configuration and setup for biological sensing

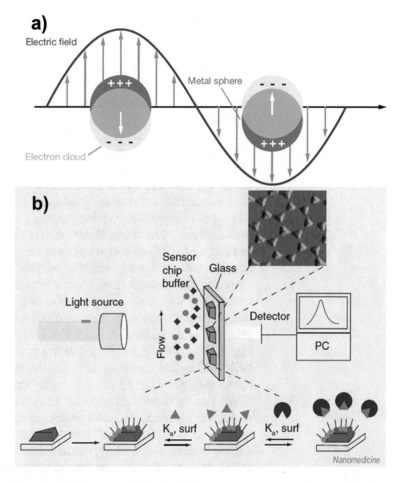

Fig. 2 Schematic diagrams illustrating (**a**) a localized surface plasmon [18], (**b**) a configuration of representative experimental setup and procedure for LSPR sensing [19]

LSPR is caused by resonant surface plasmons localized in nano-scale systems when light interacts with particles much smaller than the incident wavelength (Fig. 2a). Similar to the SPR, the LSPR is sensitive to changes in the local dielectric environment near the nanoparticle surfaces. Usually, the sense changes are measured in the local environment through an LSPR wavelength shift of the scattering light via reflection or transmission (Fig. 2b). Nanoscale LSPR makes possible for the development of a portable device for point-of-care (POC) detection regarding its requirements, such as robust to handle, small volume sample, and little to no sample pretreatment, label-free and rapid response time, and compact size.

Incorporation of the LSPR to fiber optics (FO) offers a few advantages in terms of avoiding the use of bulky optics and high-precision mechanics for angular or wavelength interrogation of metal films in contact with analytes, including immobilization of Au or Ag nanoparticles (NPs) to optical fiber probes for LSPR detection [11, 12]. It may allow realizing an optical communication and analytical tool for a wide spectrum of applications. However, more controllable and stable LSPR nanoscale systems for fiber optic integration are desirable to develop robust, reliable, portable LSPR biosensors.

In this work, the progress of developing a miniaturized LSPR-FO probe and demonstration of sensitive, label-free detection of a protein cancer biomarker, free prostate-specific antigen (f-PSA) are reported, which involves three major steps: (1) a low-cost lift-off process adapted to fabricate gold nanodisk arrays at the end of tip-facet, providing a very stable, robust, and clean LSPR fiber tip probe, (2) the probe was functionalized via a facile self-assembled monolayer (SAM) of alkanethiolates on the gold nanodisk array to attach a capture ligand, anti-PSA antibody, as a selective immuno-assay for the detection of the f-PSA, (3) the FO-based sensing technology was used as a powerful analytical tool by integrating the LSPR nanoprobe to communicative fiber optics. The sensing principle and configuration of the LSPR nanoprobe is shown in Fig. 3. The white light guided in the optical fiber using as incident light to the gold nanodisk arrays at the end of the fiber tip surface for excitation of the LSPR. The reflectance of the light scattering from the nanodisk arrays was recorded before and after the biological binding. The correlation of the changes of reflectance spectra to the binding of the analytes to the nanodisk arrays, corresponding to the analyte concentrations in samples, was established for detection. The reported label-free LSPR fiber biosensor may allow an alternative approach for direct discrimination of the

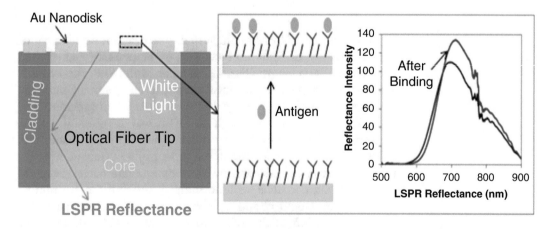

Fig. 3 A diagram illustrating the principle of LSPR coupling on fiber optic probe for biosensing. The nanodisk arrays are fabricated on optical fiber tip end. The LSPR reflectance is recorded with a white incident light guided in the fiber

cancer biomarkers, and potentially developing a miniaturized, point-of-care (POC) device for early disease diagnosis.

This work suggests that: (1) as an emerging technology, the LSPR-FO sensor has ultra-high sensitivity in molecular adsorption detection; (2) the LSPR Au nano-array fabricated at the end facet of the fiber tip for sensing is very robust and reusable; (3) the primary resonant wavelength can be tuned to a desired range (e.g., NIR) by tailoring the nano-array structure to enhance the sensitivity; (4) tailored surface functionalization harnessing the advances in biological ligand interactions (e.g., immunoassays) enables signal amplification and a label-free, selective detection; and (5) the target molecules are immobilized by dipping the fiber-optic probe in sample/reagent solution, contrary to pouring of sample solution in conventional methods (ELISA, Electrochemiluminscence, Radioimmunoassay-RIA), resulting in drastic reduction of the amount of sample/reagents needed and decreases the washing time of probes.

2 Materials and Equipment

1. Single-mode optical fiber for 633 nm wavelength was purchased from Newport Corporation.

2. Electron beam resist (ZEP-520A), thinner (ZEP-A), developer (ZED-N50), resist remover (ZEDMAC) were purchased from ZEON Corporation, Japan, and used without further purification.

3. 11-Mecaptoundecanoic acid (HSC10COOH, 99%), 8-Mercapto-1-Octanol (HSC8OH, 98%), N-(3-Dimethylaminopropyl)-N'-ethylcarbodiimide hydrochloride (EDC), N-Hydroxysuccinimide (NHS), and glycine were purchased from Sigma-Aldrich (Milwaukee, WI) and used without further purification.

4. Mouse anti-human PSA monoclonal antibody (capture mAb), human free-PSA, and ELISA kits for CA125 were obtained from Anogen-YES Biotech Laboratories Ltd. (Mississauga, Canada).

5. The 5 ng/mL free-PSA standard solution was used for preparation of free-PSA solutions with lower concentrations obtained using sample dilutant provided in the ELISA kit. The 5 ng/mL free-PSA standard solution was prepared in a protein matrix solution according to the World Health Organization (WHO) standard [13] by the vendor.

6. Chrome etchant was obtained from Microchem GmbH, Germany.

7. The fiber clamp for hoisting fiber for vibration was obtained from Newport Corporation, CA, USA.

8. Ultra clean convection oven (Ultra-clean 100) was obtained from Thermo Fisher Scientific, OH, USA.

9. Nano pattern generation system (NPGS) was obtained from JC Nabity Lithography Systems, MT, USA.

10. Field emission scanning electron microscope (FESEM) was a LEO 1550 model.

11. Three cathode vacuum sputter system (Denton Discovery-18 Sputter System) was obtained from Denton Vacuum LLC, NJ, USA.

12. Vacuum thermal evaporate system was donated to University of Alabama in Huntsville by US Army.

13. Reactive Ion Etching (RIE) system (Plasma-Therm 790) was obtained from Plasma-Therm, FL, USA.

14. Optical fiber coupled Tungsten Halogen light source (LS-1) was obtained from Ocean Optics, FL, USA.

15. Mini-spectrometer (USB2000-VIS-NIR) was obtained from Ocean Optics, FL, USA.

16. 2 × 2 Single-mode fiber optic couplers for 633 nm wavelength were obtained from Newport Corporation, CA, USA.

3   Methods

3.1   Fabrication of LSPR-FO Nanoprobe

1. Typical semiconductor fabrication technologies [14, 15] with lift-off method are utilized in the fabrication development for Au nanostructures on fiber end face, as schematically illustrated in Fig. 4. There are four main technological steps: (1) deposition of positive electron beam resist (ZEP520A, Zeon Chemicals, Japan) on the fiber end face with uniform thickness (Fig. 4a–c), (2) nano-patterning on the E-beam resist using EBL method (Fig. 4d), (3) vacuum deposition of functional metallic materials over the e-beam resist using cleanroom

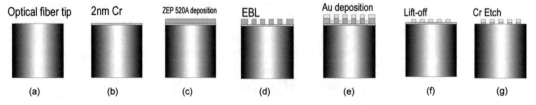

Fig. 4 A schematic flow-chart displays the fabrication procedure for nanodisk arrays at the optical fiber end facet. (**a**) optical fiber tip, (**b**) 2 nm Cr deposition, (**c**) resist deposition, (**d**) EBL process, (**e**) gold film deposition, (**f**) lift-off process, (**g**) Cr etch to get the nanodisk arrays

thermal evaporation (Fig. 4e), and (4) nano-pattern transfer using standard liftoff method (Fig. 4f, g).

2. An optical fiber for single-mode wavelength of 633 nm is employed in this work, which has a core diameter of 4 μm, a cladding diameter of 125 μm, and a polymer buffer coating diameter of 250 μm (Fig. 4a). Compared to standard optical fiber operated at single-mode wavelength of 1310 nm, the advantage of using this small core fiber is that it provides more spectral stability in the wavelength ranges of 600–750 nm, in which locate the resonance peak of our fabricated fiber tip LSPR sensors. Small core size also means higher fabrication yield and less nanoparticles involved in the sensing, thus requires smaller amount of target samples.

3. The preparation of optical fiber tip includes stripping off the polymer buffer layer 4 cm from the end and cleaving the end face with a fiber cleaver, followed by cleaning with acetone and isopropyl alcohol (IPA) rinse for 5 min.

4. Two nanometers of Cr film is firstly deposited on the fiber end face using the vacuum sputtering method [16] to provide a conductive layer for the e-beam resist in the EBL process (Fig. 4b).

5. A simple and unique wet resist coating method called "dip and vibration" technique has been developed in a nanofabrication lab to deposit e-beam resist on the optical fiber tip (Fig. 4c), and its procedure is schematically represented in Fig. 5. The optical fiber is dipped into the diluted e-beam resist (ZEP520A diluted with ZEP thinner at a ratio of 1:3) for 10 s (Fig. 5a). Then, it is removed from the resist and hoisted into a vertically upward position using a Newport fiber clamp, with ~15 mm of fiber tip outside the clamp at the top (Fig. 5b, c).

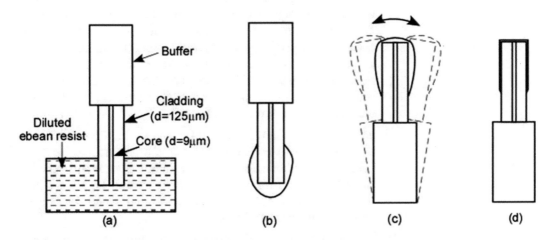

Fig. 5 Illustrations of the procedure called "dip and vibration" technique, (**a**) dip in the diluted e-beam resist, (**b**) after dip, (**c**) vibration, and (**d**) after bake

The fiber tip is then vibrated manually by pulling the fiber tip to one side and then releasing to get rid of excessive resist by means of cantilever beam free vibration. The vibration frequency and strength is controlled by the length of fiber tip outside the fiber clamp and the initial displacement of the fiber tip. The thickness of resist on the optical fiber tip is dependent on the ZEP dilution ratio and vibration strength. The iterative method is used to optimize the vibration process. The initial fiber tip displacement for the vibration is 2–5 mm, and the fiber tip length is 15–25 mm. The ZEP dilution ratio used is from 20% to 40%, diluted in ZEP thinner. The resulted e-beam resist layer thickness on the fiber tip is 100–200 nm, measured by SEM observation. The fiber tip is held vertically upward and baked in a 120° C oven for 60 min (Fig. 5d).

6. The EBL process based on the Nano Pattern Generation System (NPGS) and a field emission scanning electron microscope (FESEM) is used to create nanodot array pattern on the e-beam resist on the fiber end face (Fig. 4d). A Newport fiber clamp is used to hold the fiber vertically on the translation stage in the SEM chamber. The EBL voltage is 30 kV and the exposure dose is 70 $\mu C/cm^2$. The fiber tip is developed by dipping in resist developer (ZEP N50) for 1 min. The fiber tip is then rinsed in DI water and baked in the 120° C oven for 10 min for dehydration.

7. An oxygen plasma descum procedure in a reactive ion etcher (RIE, 25 watts power for 3 min) is used to remove the thin residual layers of photoresist following the photoresist development step (*see* **Note 1**).

8. The deposition of 40 nm Au overlay over the patterned area is carried out by using the standard thermal evaporation coating technique (Fig. 4e). The 2 nm Cr layer previously deposited as a conductive layer for the EBL process is now served as an adhesive layer for Au overlay. To lift off the e-beam resist, the fiber tip is dipped in the ZEP remover for 10 min, followed by a 1-min ultrasonic bath to assist the lift-off process (Fig. 4f). The fiber tip is rinsed in DI water and checked under an optical microscope to make sure that the resist layer on the fiber end face has been removed (*see* **Note 2**).

9. The fiber tip is dipped into the Cr remover solution for 30 s to remove the Cr layer that is not covered by the Au nanoparticles (Fig. 4g). The fiber tip is rinsed again in DI water and baked in the 120 °C oven for 10 min for dehydration. Figure 6 shows the scanning electron micrographs of a gold nanodisks array on an optical fiber tip end facet.

10. Finally, the Au nanodisk array on fiber end face is annealed at 530 °C for 5 min and ready for next usage (*see* **Note 3**).

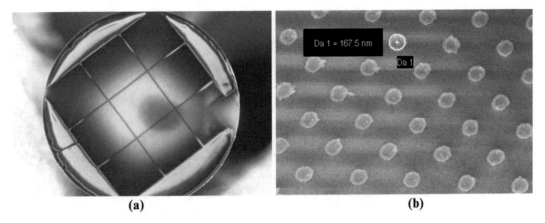

Fig. 6 Images of (**a**) overview of the optical fiber tip end, (**b**) SEM images of gold nanodisk arrays on the optical fiber facet after the lift-off process

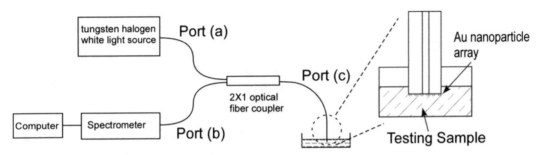

Fig. 7 A schematic diagram illustrating optical setup for the fiber tip LSPR sensor based on reflection spectrum measurement

3.2 Optical Measurement

1. In this optical setup (Fig. 7), a 2×1 single-mode optical fiber coupler for 633 nm wavelength is used, which has three light connection ports. In Fig. 7, port (a) and port (b) are two connections on one side of optical fiber coupler, and port (c) is connection on the opposite side. Port (a) is connected to the tungsten halogen white light source; port (b) is connected to a mini-spectrometer; and port (c) is connected to the LSPR-FO probe using a fusion splicer. The white light (450–950 nm) is launched from port (a), propagated to port (c), and reflected from the end facet with the Au nanodisks arrays. The reflection light propagated to port (b), where the spectrum of the reflection light is measured by the mini-spectrometer, which is connected to a computer for data acquisition and processing. Figure 8 shows the photograph of the Optical setup for the fiber tip LSPR sensor based on reflection spectrum measurement.

2. During fiber optic LSPR detection, the light is launched and guided to the LSPR fiber tip facet. The light wave propagates along the core in the center of the optical fiber tip where the

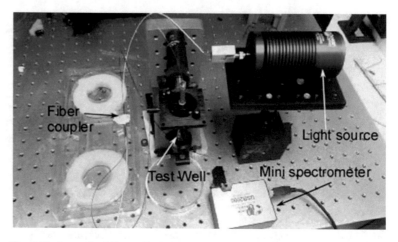

Fig. 8 Photograph of the optical components (optical fiber, fiber coupler, mini-spectrometer, light source) and setup for the fiber tip LSPR sensor. Computer connection to the mini spectrometer via USB cable is not shown

nanodisk array is located. The guided white light interacts with the Au nanodisk array and excites the localized surface plasmon giving raise to enhanced scattering around the resonance wavelength. The LSPR wavelength strongly depends on interface refractive properties [16]. The light reflected from the LSPR interface is guided in the fiber and propagated to the minispectrometer, and the spectrum of the reflected light is recorded as a function of wavelength. The detection of RI change at the interface is accomplished by observing the primary resonance peak shift in the spectrum.

3. Before starting the experiment, the sensor tip must be cleaned of any impurities or other contaminants. The sensor tip is rinsed with ethanol to clean the tip. A baseline wavelength is achieved if the sensor tip returns to this wavelength three times after being washed in deionized water. If the tip spectra do not return to this line, acetone and isopropyl alcohol (IPA) dip for 5 min may be used to further clean the tip of lingering impurities.

4. In order to get the LSPR spectrum response due to Au nanostructures on the fiber end facet, a reference spectrum and a dark spectrum need to be recorded. The reference spectrum is acquired without fiber LSPR probe on port (c), and the fiber end face of port (c) is perpendicularly cleaved. The dark spectrum is obtained by turning off both the tungsten halogen light source and room light. The measured reflection spectra (M_λ) of the fiber tip sensor probe is obtained by the equation:

$$M_\lambda = (S_\lambda - D_\lambda)/(R_\lambda - D_\lambda) \times 100\ \%\ ,$$

where S_λ is the sample intensity, D_λ is the dark intensity, and R_λ is the reference intensity at wavelength λ (the intensity defines as photon counts).

3.3 LSPR-FO Nanoprobe Sensitivity Characterization

1. The nanoprobe sensitivity is used to determine if the sensor tip in experimentation would be sensitive enough to determine small wavelength shifts accurately.

2. The solvents used in determining the bulk RI sensitivity of the LSPR tip are acetone, methanol, ethanol, isopropyl alcohol, and water of different RIs.

3. The LSPR-FO nanoprobe is dipped in the various solvents, and the spectra of the reflected light were recorded, respectively, as shown in Fig. 9a (*see* **Note 4**).

4. To determine the peak wavelength in the LSPR reflection spectrum, a Matlab program was created to fit the data to an eighth-order polynomial function over a range of 80 nm [17], centered at the wavelength of maximum reflection in the raw data.

5. The calculated LSPR wavelength red shifted as the RI of the solvent increased. A linear relationship between the LSPR peak wavelength and the bulk RI of the medium is obtained (Fig. 9b black dots), with the sensitivity of 226 nm/RIU (RIU, refractive index unit) used in this study.

6. The light intensity-based RI sensitivity is investigated as well (Fig. 9b red diamonds). The return light intensities at the LSPR peak wavelength are plotted against the bulk RI of the media. The LSPR peak intensity is seen to be linearly proportional to the RI, and the gradient of the line is 123 per RIU. The return light intensity of a single-wavelength laser can be used as a probing light for the refractive index change due to

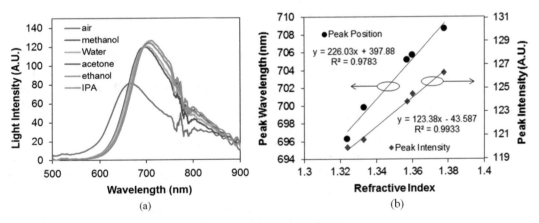

(a) (b)

Fig. 9 Sensitivity measurements of the LSPR-FO nanoprobe. (**a**) Measured reflectance spectra for the LSPR-FO probe in various solvents of different refractive index. (**b**) Correlation of the LSPR peak wavelengths (*left axis*) and LSPR peak intensity (*right axis*) with the refractive index (RI) of the solvents, the linear fit of the changes vs. the RI gives the sensing sensitivity. In the inset equation, *x* and *y* represent the values of *x* and *y* axis. R^2 is the coefficient of determination in the linear fitting

biochemistry binding events, and thus eliminate the need for a broadband light source and optical spectrometer for an LSPR biosensing.

3.4 Functionalization of the LSPR-FO Nanoprobe

1. To achieve the immobilization of mAb at the LSPR-FO sensor surface, the first step is to incubate a cleaned gold nanodisck fiber tip in a mixture of 1 mM $HS(CH_2)_{10}COOH$ and $HS(CH_2)_8OH$ with mole ratio 1:5 in absolute ethanol solution for 1 h. A mixed self-assembled monolayer is formed at the gold nanodisck surfaces.

2. The self-assembled monolayer (SAM) tethered with carboxylic acid groups then is activated by incubation in a pH 7.0, 10 mM phosphate buffer solution containing 0.5 mM of EDC/NHS, respectively, for 1 h.

3. The activated SAM is rinsed with the distilled water and immediately moved to a freshly prepared 10 mM PBS containing 10 µg/mL of the detector mAb for 1 h incubation.

4. The fiber probe is finally rinsed with the PBS and followed by dipping in a 0.2 M glycine PBS solution for 10 min to deactivate the remaining active sites at the SAM.

3.5 Detection of f-PSA Biomarker

1. During the experiments, PBS (containing 0.05% tween-20) is used as a running buffer to help minimize the nonspecific adsorption of f-PSA at the fiber tips.

2. In order to test binding of PSA to the anti-PSA mAb modified surfaces, five tenfold-dilutions of a 5 ng/mL concentration of f-PSA are prepared in PBS solution for the desired concentrations (i.e., 5 ng/mL, 0.5 ng/mL, 0.05 ng/mL, 5 pg/mL, 0.5 pg/mL, 0.05 pg/mL). The sensor tip is placed in the f-PSA PBS for 10 min in the sequence at the lowest concentration of 50 fg/mL to measure the reflectance spectrum.

3. After each measurement, the fiber tip is rinsed thoroughly with DI water and dried in air. All spectra are obtained after the fiber tip was cleaned and dried in air. Figure 10a shows the representative reflectance spectra of f-PSA detection sensing various f-PSA concentrations from 0 to 0.5 ng/mL.

4. Figure 10b shows the dependence of the wavelength peak shift on the f-PSA concentration. The peak shift for each point is obtained by averaging three measurements. A wavelength shift of twice the largest standard deviation (1.2 nm) is used to determine LOD of the fiber LSPR sensors, which corresponds to 100 fg/mL of f-PSA in PBS solution.

5. Control experiments have been designed and carried out to evaluate the specificity/selectivity of the immunoassay using the LSPR-FO devices. Specifically, the binding between the detector mAb and bovine serum albumin (BSA), similar to

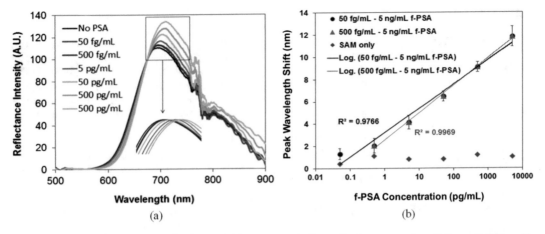

Fig. 10 Detection of protein biomarker. (**a**) The representative reflectance spectra of the anti-PSA mAb modified LSPR-FO nanoprobe in various concentrations of f-PSA (from 0 up to 0.5 ng/mL). Insert plot shows spectra normalized to peak intensity to show the LSPR reflectance peak shifts. (**b**) Correlation of f-PSA concentration to the reflectance light of resonance peak wavelength shift

human serum albumin, at concentrations of 5 mg/mL has been evaluated. Additionally, to evaluate the nonspecific binding to the SAM, the LSPR-FO devices modified only with the mix SAM of HSC10COOH and HSC8OH SAM without antibody attached are tested by following the f-PSA detection process, as shown in Fig. 10b (blue diamond markers).

4 Notes

1. The optional step after the oxygen descum is to dip the fiber tip in the chrome remover solution for 60 s to remove the thin Cr layer on the nanodisk area. This step will eliminate the Cr adhesion layer under the Au nanodisk, avoiding the undesired side effects of Cr layer under Au nanodisk in the LSPR sensing process, such as spectrum broadening and dephasing. However, this step will weaken the adhesion strength between the Au nanodisks and the fiber end glass substrate, and will reduce the reusability of the fiber LSPR probe.

2. If the lift-off process is not complete, ZEP remover and ultrasonic bath may be used to further lift off the e-beam resist.

3. The fiber probe is inserted horizontally into the oven through an access hole on the sidewall of a high-temperature oven. In order to protect the polymer buffer on the fiber, only 10–15 mm of fiber probe needs to be inside the oven.

4. Note that there is a noisy region in all spectra from ~760 to 800 nm, which is caused by the transmission minimum for the optical fiber coupler (single mode at 633 nm) in the same wavelength range.

Reference

1. Mukundan H et al (2009) Optimizing a waveguide-based sandwich immunoassay for tumor biomarkers: evaluating fluorescent labels and functional surfaces. Bioconjug Chem 20 (2):222–230

2. Liu X et al (2008) A one-step homogeneous immunoassay for cancer biomarker detection using gold nanoparticle probes coupled with dynamic light scattering. J Am Chem Soc 130 (9):2780–2782

3. Sekhar PK, Ramgir NS, Bhansali S (2008) Metal-decorated silica nanowires: an active surface-enhanced raman substrate for cancer biomarker detection. J Phys Chem C 112 (6):1729–1734

4. Mani V et al (2009) Ultrasensitive immunosensor for cancer biomarker proteins using gold nanoparticle film electrodes and multienzyme-particle amplification. ACS Nano 3 (3):585–594

5. Henne WA et al (2006) Detection of folate binding protein with enhanced sensitivity using a functionalized quartz crystal microbalance sensor. Anal Chem 78(14):4880–4884

6. Baker GA, Desikan R, Thundat T (2008) Label-free sugar detection using phenylboronic acid-functionalized piezoresistive microcantilevers. Anal Chem 80(13):4860–4865

7. Lee HJ, Nedelkov D, Corn RM (2006) Surface plasmon resonance imaging measurements of antibody arrays for the multiplexed detection of low molecular weight protein biomarkers. Anal Chem 78(18):6504–6510

8. Raether H (1988) Surface plasmons on smooth and rough surfaces and on gratings. Springer tracts in modern physics. Springer, Berlin

9. Leskova TA, Maradudin AA, Zierau W (2005) Surface plasmon polariton propagation near an index step. Optics Commun 249(1–3):23–35

10. Jung LS et al (1998) Quantitative interpretation of the response of surface plasmon resonance sensors to adsorbed films. Langmuir 14 (19):5636–5648

11. Mitsui K, Handa Y, Kajikawa K (2004) Optical fiber affinity biosensor based on localized surface plasmon resonance. Appl Phys Lett 85 (18):4231–4233

12. Chang T-C et al (2012) Using a fiber optic particle plasmon resonance biosensor to determine kinetic constants of antigen–antibody binding reaction. Anal Chem 85(1):245–250

13. 1985. http://apps.who.int/iris/bitstream/10665/38405/1/WHO_TRS_725.pdf.

14. Nishi Y, Doering R (2000) Handbook of semiconductor manufacturing technology. CRC Press, Boca Raton, FL

15. May GS, Spanos CJ (2006) Fundamentals of semiconductor manufacturing and process control. John Wiley & Sons, New York, NY

16. Obando LA, Booksh KS (1999) Tuning dynamic range and sensitivity of white-light, multimode, fiber-optic surface plasmon resonance sensors. Anal Chem 71(22):5116–5122

17. Wu H-J et al (2012) Membrane-protein binding measured with solution-phase plasmonic nanocube sensors. Nat Methods 9(12):1189–1191

18. Willets KA, Van Duyne RP (2007) Localized surface plasmon resonance spectroscopy and sensing. Annu Rev Phys Chem 58(1):267–297

19. Zhao J et al (2006) Localized surface plasmon resonance biosensors. Nanomedicine 1 (2):219–228

Chapter 2

Ultra-Sensitive Surface Plasmon Resonance Detection by Colocalized 3D Plasmonic Nanogap Arrays

Wonju Lee, Taehwang Son, Changhun Lee, Yongjin Oh, and Donghyun Kim

Abstract

Ultra-sensitive detection based on surface plasmon resonance (SPR) was investigated using 3D nanogap arrays for colocalization of target molecular distribution and localized plasmon wave in the near-field. Colocalization was performed by oblique deposition of a dielectric mask layer to create nanogap at the side of circular and triangular nanoaperture, where fields localized by surface plasmon localization coincide with the spatial distribution of target molecular interactions. The feasibility of ultra-sensitivity was experimentally verified by measuring DNA hybridization. Triangular nanopattern provided an optimum to achieve highly amplified angular shifts and led to enhanced detection sensitivity on the order of 1 fg/mm^2 in terms of molecular binding capacity. We confirmed improvement of SPR sensitivity by three orders of magnitude, compared with conventional SPR sensors, using 3D plasmonic nanogap arrays.

Key words Surface plasmon resonance, Localized surface plasmon resonance, Surface plasmon resonance detection, Nanogap arrays, Nanoapertures, Colocalization, DNA hybridization

1 Introduction

In recent years, diverse optical techniques have attracted tremendous interests for ultra-sensitive real-time detection of various phenomena involving biomolecular interactions. Most of these techniques have been based on fluorescence. Fluorescence-based sensing, however, suffers from fluorescence interference or chemotoxicity issues. In contrast, surface plasmon resonance (SPR) biosensing has been widely investigated as a representative label-free technique, by which specific molecular interactions can be monitored in real time and kinetic characteristics related to molecular binding are conveniently measured on a quantitative basis.

Surface plasmon (SP) is a collective logitudinal oscillation of electrons existing at the metal/dielectric interface. The oscillation of electron can be coupled with a TM-polarized incident light.

Avraham Rasooly and Ben Prickril (eds.), *Biosensors and Biodetection: Methods and Protocols Volume 1: Optical-Based Detectors*, Methods in Molecular Biology, vol. 1571, DOI 10.1007/978-1-4939-6848-0_2,
© Springer Science+Business Media LLC 2017

By tuning angle of incidence or incident wavelength, the momentum matching condition between incident light and SP can be satisfied. At resonance, due to energy transfer from incident light to a longitudinal surface wave, a narrow dip in the reflection characteristics is observed with respect to wavelength or angle of incidence. A small change of dielectric medium refractive index induced from a biochemical interaction affects momentum matching condition, followed by the shift of resonance dip in the reflection characteristics [1].

Despite broad uses of SPR sensors, conventional SPR sensors have relatively poor detection limit on the order of 1 pg/mm^2 in binding capacity [2]. As an optical sensing technique other than SPR, metal nanoparticle-based surface-enhanced Raman scattering and microresonator-based whispering gallery modes have emerged for ultra-sensitive label-free detection [3, 4].

In this chapter, we describe self-aligned colocalization using three-dimensional plasmonic nanogap arrays for ultra-sensitive SPR biosensors [5]. Note that colocalization indicates the spatial overlap between the area of localized electromagnetic field and molecular interaction. Colocalization is preferred in terms of sensitivity enhancement per molecule, because much smaller number of molecules are involved at the interaction. In previous studies, use of plasmonic nanostructures was investigated to localize SP and evanescent near-field to enhance detection sensitivity [6, 7]. Random and periodic nanostructure was also employed for specific and nonspecific detection based on enhanced surface plasmon resonance [8, 9]. It was also shown that when biomolecules are spatially aligned for colocalization with the localized field that is defined by 2D linear nanopattern arrays, SPR signals can be effectively amplified, which leads to efficient improvement of detection sensitivity [10–12]. Moreover, theoretical investigation of nonspecific, noncolocalized, and colocalized detection models was performed based on silver nanoislands, which confirmed possibility of further amplification of optical signature [13]. Meanwhile, SPR sensor characteristics using grapheme-related materials were also studied [14, 15].

Here, we describe use of 3D plasmonic nanogap arrays, illustrated in a schematic diagram of Fig. 1, for colocalized detection of bio-interactions of target molecules to achieve even more dramatic enhancement of sensitivity in SPR biosensors. In particular, circular and triangular nanoholes were developed lithographically and 3D nanogaps were then formed for colocalization by the shadowing of obliquely evaporated dielectric mask layer on the metallic nanostructures. This enables target molecules to directly access the underlying metal film. Oblique evaporation was previously used to fabricate molecular electronic devices based on nanoscale gap structures [16]. Since localized near-field is formed at the ridge of the nanostructure, localized fields and nanogap where target

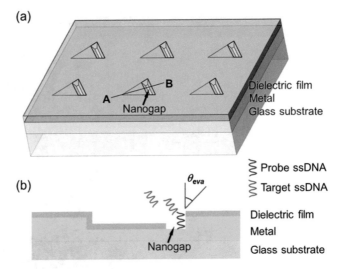

Fig. 1 (**a**) Schematic illustration of 3D triangular nanogap aperture arrays that may be produced by oblique evaporation for colocalized biomolecular interaction. (**b**) Lateral profile across the *solid line* (*A–B*) in (**a**)

molecules can bind are self-aligned for colocalization. 3D plasmonic nanostructures induce colocalization area to be much smaller than what 2D linear grating patterns may produce. Therefore, detection of much weaker interactions and/or those involve a smaller number of molecules would be feasible, which allows extreme sensitivity enhancement in SPR sensing.

2 Materials

2.1 Nanogap Fabrication

1. A glass substrate (SF10) with a 2-nm-thick chromium adhesion layer and a 40-nm-thick gold film.

2. Electron beam (VEGA II LSH; TESCAN, Brno, Czech).

3. Scanning electron microscope (Elphy Quantum; Raith, Dortmund, Germany).

4. A polymethylmethacrylate (PMMA) photoresist (AR-N 7520.18; ALLRESIST, Strausberg, Germany).

5. Spin-coater (ACE-200; DONG AH Trade Corp, Seoul, Korea).

6. A dielectric mask layer using ITO or SiO_2.

7. Developer (AR 300-47; ALLRESIST, Strausberg, Germany).

8. Remover (AR 300-70; ALLRESIST, Strausberg, Germany).

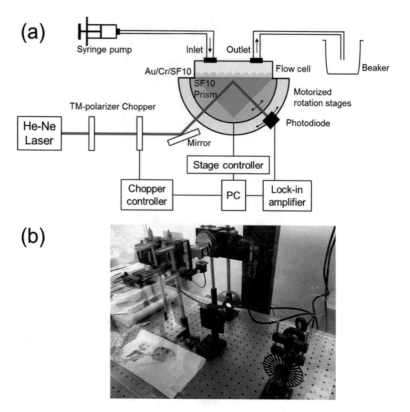

Fig. 2 (**a**) Schematics of optical setup for a SPR sensor based on colocalization using plasmonic nanogap arrays and (**b**) photograph of the experimental setup

2.2 Optical Setup

A schematic illustration of the optical setup for colocalized SPR detection is presented in Fig. 2 (*see* **Note 1**).

1. He-Ne laser (36 mW, $\lambda = 632.8$ nm, Nominal beam diameter: 650 μm; Melles-Griot, Carlsbad, CA).

2. Glan-Thompson Linear Polarizer (Thorlabs, Inc., Newton, NJ).

3. Chopper and controller (SR540; Stanford Research Systems, Sunnyvale, CA).

4. Two concentric motorized rotation stages (URS75PP and ESP330; Newport, Irvine, CA) (*see* **Note 2**).

5. Low-noise lock-in amplifier (Model SR830; Stanford Research Systems, Sunnyvale, CA).

6. A p-i-n photodiode (818-UV; Newport, Irvine, CA).

7. Lab ViewLabVIEW (National Instruments, Austin, TX).

8. Index-matching gel ($n = 1.725$; Cargille Laboratories, Cedar Grove, NJ).

2.3 DNA Preparation

1. HPLC purified 24-mer sequence length capture probe and target oligonucleotides (IDT, Coralville, IA).

2. The sequence of single-stranded probe DNA (p-DNA) was 5′-TTT TTT CGG TAT GCA TGC CAT GGC-3 modified with thiol at 5′.

3. The sequence of single-stranded target DNA (t-DNA) was 5′-GCC ATG GCA TGC ATA CCG AAA AAA-3′.

4. Plasma cleaner (Harrick Scientific Products, Pleasantville, NY).

5. Micropump (KD Scientific, Holliston, MA).

6. Acetone (1009-4404; DAEJUNG CHEMICAL & METALS CO., Siheung, Korea).

7. Phosphate Buffered Saline (PBS) buffer (pH = 7.4, BP3991; Fisher Scientific, Pittsburgh, PA).

3 Methods

3.1 Nanogap Fabrication

To make linear nanograting-based 2D nanogap arrays, the overall fabrication steps are explained below.

1. A 40-nm-thick gold film was evaporated on an SF10 glass substrate with a 2-nm-thick chromium adhesion layer.

2. Nanograting array patterns were defined by electron beam (e-beam) lithography using PMMA photoresist that was spin-coated on the gold film.

3. Metallic nanograting arrays were transferred by a lift-off process after gold evaporation (*see* Fig. 3a).

4. 2D nanogap arrays were created on nanograting arrays by oblique evaporation of a dielectric mask layer using ITO or SiO$_2$ (*see* Fig. 3b).

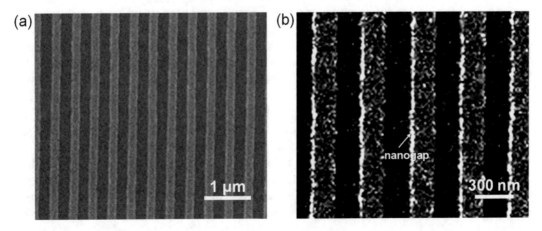

Fig. 3 (**a**) SEM image of fabricated linear nanograting arrays and (**b**) magnified image of 2D nanogap arrays

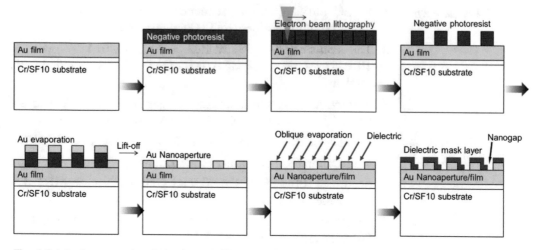

Fig. 4 Fabrication procedure to implement 3D plasmonic nanogap arrays

To create 3D nanogap arrays, the overall fabrication process is presented in Fig. 4.

1. A 40-nm gold film was evaporated on an SF10 glass with a 2-nm chromium adhesion layer.

2. Circular and triangular nanostructures with a 2-µm array period and an aperture size of 600 nm were defined by electron beam lithography.

3. 20-nm-thick nanohole arrays were formed after a lift-off process that involves e-beam lithography, evaporation of gold, and removal of polymer resist.

4. 5-nm-thick ITO layer was obliquely deposited with the deposition angle varied at 30°, 45°, and 60° to adjust the gap area differently (*see* **Note 3**).

3.2 Optical Setup

A He-Ne laser beam (36 mW, $\lambda = 632.8$ nm, nominal beam diameter: 650 µm, Melles-Griot, Carlsbad, CA) is TM-polarized by linear polarizer and temporally modulated by optical chopper (SR540; Stanford Research Systems, Sunnyvale, CA), which alters the state of light on and off at a specific frequency. The chopping frequency is used as the reference input of a lock-in amplifier composed with usually set to 0.6 kHz lock-in amplifier and feedback system. Among the signals captured at a photodiode, noise signal at frequency other than the chopping can be removed. Two motorized rotation stages (URS75PP; Newport, Irvine, CA) are employed; one is for rotating a prism on which a plasmonic nanostructured sample is located, and the other is for rotating the photodiode to keep laser alignment to photodiode. The angle of the stage on which the nanostructure sample located can be corrected by using the calibration constant and the other stage with the photodiode follows the former stage. The nanostructured sample is located on the SF10 prism with index matching gel ($n = 1.725$; Cargille Laboratories, Cedar Grove, NJ), and covered with flow cell system head. Using one motor controller (ESP330, Newport), two stages are controlled at the same time with a minimum angle increment of $0.0002°$, uni-directional repeatability of $0.001°$, absolute accuracy of $\pm 0.008°$. A p-i-n photodiode (818-UV, Newport) is used to detect refractance change by the angle scanning. The value measured by the diode is used as input for the feedback system of lock-in amplifier. Therefore, the signal from photodiode is noise removed due to the phase-sensitive detection. The stage and chopper controller are connected to PC and controlled by LabVIEW, so whole system is fully automated.

3.3 Estimation of Colocalization Areas

3.3.1 2D Nanogap Production

A 2D nanogap area $G_{\Lambda,2D}$ produced by linear nanograting per unit grating length with the thickness d_g can be calculated as

$$G_{\Lambda,2D} = d_g(1 + \tan\theta_{eva}) \tag{1}$$

where θ_{eva} is the oblique evaporation angle (*see* Fig. 1b) in Eq. 1, the first term d_g and the second term $d_g \tan\theta_{eva}$ come from vertical nanograting edge and horizontal surface which are not evaporated with dielectric, respectively. For example, when the nanograting with $d_g = 20$ nm and $\theta_{eva} = 30°$ produces a nanogap with an area of 31.56 nm times the grating length.

3.3.2 Nanogap Reduction of the Colocalization Area

3D nanoaperture arrays can reduce the colocalization area, in which target molecular binding overlaps with localized fields to amplify optical signatures of the target interaction. A 3D nanogap area defined by an equi-triangular aperture with a side length (L) is calculated as

$$G_{\Lambda,3D} = Ld_g \times \left[(1 + \tan\theta_{eva}) + \frac{d_g \tan\theta_{eva}}{\sqrt{3}L}(2 - \tan\theta_{eva}) \right] \tag{2}$$

where the direction of oblique evaporation is parallel to a side of a triangular pattern. The first term Ld_g represents the area open to the side, while the second term $Ld_g \tan\theta_{eva}$ is the area formed at nanogap surface. The last term in Eq. 2 is the correction due to the triangular shape. If L is much longer than d_g, the above equation can be simplified as

$$G_{\Lambda,3D} \approx Ld_g \times (1 + \tan\theta_{eva}). \tag{3}$$

In the case of using circular nanoholes, the nanogap area with a nanohole diameter (ϕ) is changed to

$$G_{\Lambda,3D} \approx \frac{\pi\phi}{2} d_g \times \left(1 + \frac{1}{2}\tan\theta_{eva}\right). \qquad (4)$$

Using 3D nanogap arrays, the number of molecular interactions can be adjusted by changing the size to period ratio L/Λ.

3.4   Numerical Studies

In this section, the effectiveness of colocalization is evaluated from numerically calculated near-field distribution and the experimentally determined nanogap area. Electromagnetic field distribution at the nanogap was theoretically calculated based on rigorous coupled wave analysis (RSoft DiffractMOD™). Numerical calculation with DNA molecules can be done as below.

1. DNA molecules are assumed to be uniformly distributed on the nanogap.
2. The thickenss of hybridized DNA layers is modeled to 8.16 nm, which corresponds to the length of oligonucleotides.
3. The refractive indices of ssDNA and dsDNA are set to 1.449 and 1.517 [17].

As shown in Fig. 5, it is clear that fields are highly localized at nanogap. The presence of side lobe peaks was also observed on the opposite side of nanogaps, which cannot be neglected compared with the main peaks of localized fields. The suppression ratio, defined as the intensity ratio of the main peak to that of side lobe, in circular and triangular nanopatterns was estimated to be 2.3–2.5 dB. In the case of colocalization, optical signature detected at the main peak can be dominant because of spatial selectivity of

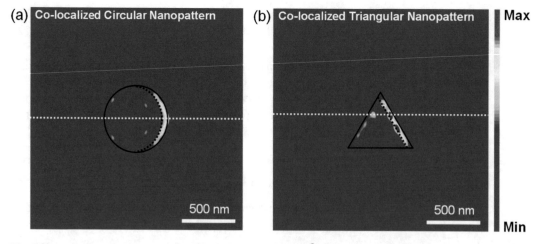

Fig. 5 Numerically calculated near-field intensity distribution $|E|^2$ of (**a**) circular and (**b**) triangular nanopatterns with ϕ, $L = 600$ nm at $\Lambda = 2$ μm. A 5-nm-thick ITO layer was assumed to be deposited on the nanopattern with $\theta_{eva} = 30°$

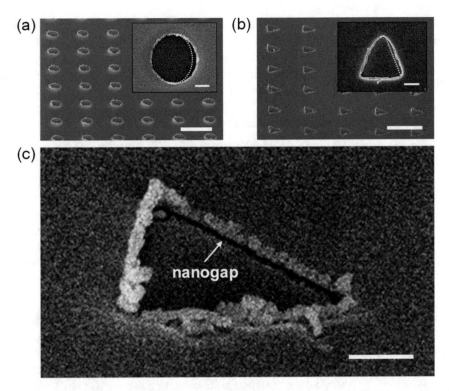

Fig. 6 SEM images: (**a**) circular and (**b**) triangular nanopattern arrays. (**c**) A magnified image of a triangular nanogap pattern. A very narrow nanogap is clearly visible at the side ridge of a triangular nanoaperture. Each inset of (**a**) and (**b**) shows the nanogap created on a circular and triangular aperture (scale bars: 2 μm for (**a**) and (**b**), 200 nm for (**c**), insets of (**a**) and (**b**))

target binding in the nanogap. Nonspecific binding that may be present is estimated to be insignificant.

SEM images of fabricated 3D nanogap arrays formed on circular and triangular patterns were presented in Fig. 6, in which the existence of nanogap at the ridge of nanoaperture can be corroborated. When d_g is fixed, the angle of oblique evaporation affects the underlying nanogap size, which is directly related to the amount of target molecules, i.e., DNA immobilization and eventually hybridization. Each colocalization area of experimentally fabricated circular and triangular nanopatterns was estimated from the SEM images in Fig. 6a, b to be $24,600 \pm 7000$ and $19,300 \pm 4600$ nm^2 without considering the sidewall area of the apertures. This is in good agreement with the numerically estimated gap area (*see* **Note 4**).

(a) Before immobilization of p-DNAs, first, sample chips were cleaned in acetone for 5 min and in ethanol.

(b) The sample chip surface was treated in plasma cleaner (Harrick Scientific Products, Pleasantville, NY).

3.5 Experimental Measurements of DNA Hybridization

3.5.1 DNA Preparation

(c) 1 ml Phosphate buffered saline (PBS) buffer (pH = 7.4) including 4 µM thiol hexane labeled probe ssDNA (HSssDNA) was injected to flow channel by a sylinge pump. The sample chip was immersed in the buffer at 4 °C for 4 h to immobilize the p-DNAs at surface (*see* **Note 5**). In order to obtain SPR angle shift, the reference SPR angle was measured and smoothed by 50th order Fourier filter.

(d) For DNA hybridization, PBS buffer including 4 µM t-DNAs was injected on the p-DNA-immobilized sample chip surface for 120 min by a fluidic channel of a micropump (KD Scientific, Holliston, MA) at a volume flow rate of 150 µL/hr. The sample chip was incubated to improve the efficiency of hybridization through mild heating. For 2 h, the solution of t-DNAs is warmed to 65 °C and slowly cooled down to room temperature. Finally, the SPR angle shift was obtained by comparing SPR angles before and after the DNA hybridization.

SPR shifts were measured using linear nanogratings and 3D nanopattern arrays to analyze the sensor performance, as shown in Figs. 7 and 8 for DNA hybridization. Two metrics were set to evaluate the sensor performance: (1) sensitivity enhancement factor, i.e., resonance shift with respect to the shift obtained from conventional SPR detection, and (2) molecular binding capacity measured by colocalization in the nanogap.

3.5.2 Sensitivity Enhancement

First, resonance angle shifts were measured based on colocalization using plasmonic nanogap arrays, such as nanogratings, circular and triangular nanopatterns, compared to non-colocalized SPR detection without nanogap structures. In the non-colocalized detection, DNA hybridization occurs on the overall metal surface of nanostructures. This leads to inefficient detection of target molecules, majority of which may be distributed in the region with weak near-fields.

Figure 7a shows raw data of measured SPR angle shifts θ_{SPR} in various platforms. In case of non-colocalized detection, resonance angle shifts made on nanostructures were detected to be slightly larger than that of the thin film control. The larger resonance shift is associated with the increase of the total area available for DNA hybridization due to the presence of nanopatterns. For colocalized detection, SPR angle shift tends to be larger with an increase of the ITO-evaporation angle (θ_{eva}). A larger θ_{eva} enlarges the nanogap surface area, which allows more probe-DNAs (p-DNAs) to immobilize at the nanogap surface. Increased resonance shifts were observed for colocalization by nanogratings because the nanogap area by grating patterns is much larger than that of 3D nanogap patterns and therefore induces more p-DNA targets to participate in hybridization.

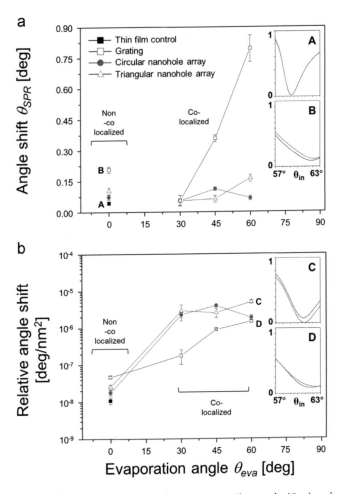

Fig. 7 (a) Measured SPR angle shifts as the evaporation angle (θ_{eva}) varies. **(b)** Relative angle shifts (*RAS*) after normalization by the nanogap area. Reproduced from Y. Oh et al. 2014 with permission from Elsevier

The angular shift made of 3D nanogap patterns may not be clear in Fig. 7a. To consider the total area of nanogap to which target molecules can bind for biomolecular interactions, a new parameter was introduced as the resonance shift normalized by the amount of molecular interactions, i.e., *relative angle shift (RAS)* = $\theta_{SPR}/nanogap\ area$, where the amount of molecular interactions is presumed to be proportionate to the nanogap area. Mearsured *RAS*, corresponding to each case in Fig. 7a, is presented in Fig. 7b. First, RAS_{3D} increases by more than two orders based on colocalization with 3D nanogap structures such as circular and triangular patterns, compared to *RAS* of the non-colocalized control. In the case of using 2D grating-based nanogap arrays, RAS_{2D} is increased by one order of magnitude. Colocalization by 3D triangular patterns produces the maximum RAS_{3D} at $\theta_{eva} = 60°$,

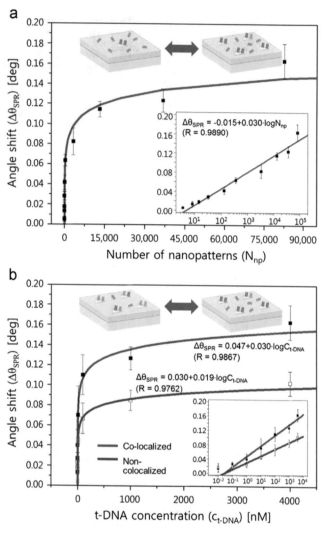

Fig. 8 (**a**) Angular SPR shifts ($\Delta\theta_{SPR}$) measured as the array period (Λ) is varied. 3D nanogap arrays were created by triangular nanopatterns at $\theta_{eva} = 60°$ and the concentration of t-DNA was fixed at 4 μM. Inset shows a linear relation between $\Delta\theta_{SPR}$ and $\log(N_{np})$. (**b**) Angular shifts measured as the concentration of t-DNAs (C_{t-DNA}) varied with the fixed p-DNA concentration at 4 μM. *Inset* also shows a linear relation between $\Delta\theta_{SPR}$ and $\log(C_{t-DNA})$. Reproduced from Y. Oh et al. 2014 with permission from Elsevier

which is 460 times larger than non-colocalized RAS measured at thin film control. The result is associated with efficient localization of SP at triangular nanopatterns as well as reduced nanogap area that is optimized to the size of localized fields [18, 19]. Note that the optimum evaporation angle for 3D nanogap arrays is not directly proportional to the nanogap area. The evaporation angle θ_{eva} may also affect the peak near-field intensity. The insets of Fig. 7.

show the resonance characteristics: A and B for non-colocalized detection, C and D for nanogap-based colocalization. Compared to the insets A, B, and D, 3D nanogap arrays presented in the inset C keep the SPR curve from broadening due to lower plasmonic damping and longer array period.

3.5.3 Molecular Binding Capacity

To evaluate molecular binding capacity as the sensor performance, Fig. 8 shows the isotherm characteristics measured in two ways using the optimum triangular nanogap arrays in 3D at $\theta_{eva} = 60°$, which was determined to produce the maximum RAS as shown in Fig. 7b. First, the array period was increased when the concentration of target DNA was fixed as 4 µM, which in effect decreases the concentration of probe DNA. Secondly, the isotherm characteritic was also measured by varying the concentration when the array period and the concentration of probe DNA were fixed at 2 µm and 4 µM respectively.

To extract the detection sensitivity of colocalized SPR sensing based on 3D nanogaps, resonance angle shifts ($\Delta\theta_{SPR}$) were measured by varying the array period as shown in Fig. 8a, where the amount of probe DNA is gradually reduced with an increase of array period while the amount of target DNA remains constant. The relation between $\Delta\theta_{SPR}$ and the number of nanopatterns (N_{np}) was presented as

$$\Delta\theta_{SPR} = -0.015 + 0.030\log N_{np}. \tag{5}$$

In Eq. 5, $\Delta\theta_{SPR}$ is proportional to $\log(N_{np})$, which means SPR shifts are linearly increased when the amount of p-DNA is exponentially increased. As N_{np} is reduced, the detection and the nature of resonance angle shifts change from being target-limited to probe-limited with transition taking place at $N_{np} = 3300\sim13,000$ ($\Lambda = 5-10$ µm), i.e., the limit of detection arises in the probe-limited detection of DNA hybridization. If we assume surface probe density to be 4×10^{12} molecules/cm^2 and hybridization efficiency 40%, a total effective nanogap area of 264,000 nm^2 in a 650-µm-diameter beam spot measures approximately 4200 molecules. This corresponds to the molecular binding capacity of 1.6 fg/mm^2 based on the molecular weight of 7.4 kDa for a 24-mer t-DNA. The results suggest that the detection sensitivity of 3D nanogap-based SPR sensing, compared to that of conventional SPR sensors, should be improved by more than three orders of magnitude.

Figure 8b shows the measured angular shifts when the concentration of target DNAs ($C_{t\text{-DNA}}$) was varied. In this case, p-DNA concentration was fixed at 4 µM and other conditions remained the same as in Fig. 8a. Surprisingly, the relation between $\Delta\theta_{SPR}$ and $C_{t\text{-DNA}}$ is similar to what was measured from $\Delta\theta_{SPR}$ and N_{np}, which implies that the amount of DNA hybridization is limited by p-DNAs (N_{np}) or t-DNAs ($C_{t\text{-DNA}}$). On condition that a much smaller number of DNAs participate in hybridization, 160% of enhancement was measured based on the slope in the inset $\Delta\theta_{SPR}/\log(C_{t\text{-DNA}})$.

4 Notes

1. Overall procedure was automatically controlled by computer. The size of a measurement chamber is roughly 10 mm (L) × 1.5 mm (W) × 0.5 mm (D), and ambient temperature was maintained at 19.6 ± 0.2 °C for measurement stability.

2. Two concentric motorized rotation stages were employed to implement angle-scanning of $\theta/2\theta$ measurement. The motorized rotation stages were controlled with a minimum motion increment of 0.0002°, absolute accuracy of 0.008°, uni-directional repeatability of 0.001°, and wobble at 20 μrad.

3. ITO provides chemical and optical stability to SPR sensors in terms of material degradation [20]. During the evaporation, chamber temperature was maintained at 200 °C to enhance adhesion by annealing.

4. The area of nanogap produced by a circular (triangular) nanoaperture with ϕ (L) = 600 nm and d_g = 20 nm was numerically calculated as listed in the table at 2-μm array period, i.e., a unit area of 2 × 2 μm². For 2D gratings, the array period was 400 nm. Area unit is nm².

θ_{eva}	2D grating	3D circular aperture	3D triangular aperture
30°	320,000	24,000	19,000
45°	400,000	28,000	24,000
60°	550,000	35,000	33,000

5. The estimated surface density of immobilized p-DNA is 4.0×10^{12} molecules/cm² [21].

References

1. Kim D (2005) Effect of the azimuthal orientation on the performance of grating-coupled surface-plasmon resonance biosensors. Appl Optics 44:3218–3223

2. Campbell CT, Kim G (2007) SPR microscopy and its application to high-throughput analyses of biomolecular binding events and their kinetics. Biomaterials 28:2380–2392

3. Dasary SSR, Singh AK, Senapati D, Yu H, Ray PC (2009) Gold nanoparticle based label-free SERS probe for ultrasensitive and selective detection of trinitrotoluene. J Am Chem Soc 131:13806–13812

4. Vollmer F, Arnold S (2008) Whispering-gallery-mode biosensing: label-free detection down to single molecules. Nat Methods 5:591–596

5. Oh Y, Lee W, Kim Y, Kim D (2014) Self-aligned colocalization of 3D plasmonic nano-gap arrays for ultra-sensitive surface plasmon resonance detection. Biosens Bioelectron 51:401–407

6. Byun KM, Kim S, Kim D (2005) Design study of highly sensitive nanowire-enhanced surface plasmon resonance biosensors using rigorous coupled wave analysis. Opt Express 13:3737–3742

7. Kim K, Yoon SJ, Kim D (2006) Nanowire-based enhancement of localized surface plasmon resonance for highly sensitive detection: a theoretical study. Opt Express 14:12419–12431

8. Moon S, Kim Y, Oh Y, Lee H, Kim HC, Lee K, Kim D (2012) Grating-based surface plasmon resonance detection of core-shell nanoparticle mediated DNA hybridization. Biosens Bioelectron 32:141–147

9. Yu H, Kim K, Ma K, Lee W, Choi JW, Yun CO, Kim D (2013) Enhanced detection of virus particles by nanoisland-based localized surface plasmon resonance. Biosens Bioelectron 41:249–255

10. Hoa XD, Kirk AG, Tabrizian M (2009) Enhanced SPR response from patterned immobilization of surface bioreceptors on nano-gratings. Biosens Bioelectron 24:3043–3048

11. Ma K, Kim DJ, Kim K, Moon S, Kim D (2010) Target-localized nanograting-based surface plasmon resonance detection toward label-free molecular biosensing. IEEE J Sel Topics Quantum Electron 16:1004–1014

12. Kim Y, Chung K, Lee W, Kim DH, Kim D (2012) Nanogap-based dielectric-specific colocalization for highly sensitive surface plasmon resonance detection of biotin-streptavidin interaction. Appl Phys Lett 101:233701

13. Yang H, Lee W, Hwang T, Kim D (2014) Probabilistic evaluation of surface-enhanced localized surface plasmon resonance biosensing. Opt Express 22:28412–28426

14. Ryu Y, Moon S, Oh Y, Kim Y, Lee T, Kim DH, Kim D (2014) Effect of coupled graphene oxide on the sensitivity of surface plasmon resonance detection. Appl Optics 53:1419–1426

15. Chung K, Rani A, Lee J-E, Kim JE, Kim Y, Yang H, Kim SO, Kim D, Kim DH (2015) A systematic study on the sensitivity enhancement in graphene plasmonics sensors based on layer-by-layer self-assembled graphene oxide multilayers and their reduced anologues. ACS Appl Mater Interfaces 7:144–151

16. Kubatkin S, Danilov A, Hjort M, Cornil J, Brédas J-L, Stuhr-Hansen N, Hedegård P, Bjørnholm T (2003) Single-electron transistor of a single organic molecule with access to several redox states. Nature 425:698–701

17. Elhadj S, Singh G, Saraf RF (2004) Optical properties of an immobilized DNA monolayer from 255 to 700 nm. Langmuir 20:5539–5543

18. Fromm DP, Sundaramurthy A, Schuck PJ, Kino G, Moerner WE (2004) Gap-dependent optical coupling of single "bowtie" nanoantennas resonant in the visible. Nano Lett 4:957–961

19. Haes AJ, Zou S, Schatz GC, Van Duyne RP (2004) Nanoscale optical biosensor: short range distance dependence of the localized surface plasmon resonance of noble metal nanoparticles. J Phys Chem B 108:6961–6968

20. Szunerits S, Castel X, Boukherroub R (2008) Surface plasmon resonance investigation of silver and gold rilms coated with thin indium tin oxide layers: Influence on stability and sensitivity. J Phys Chem C 112:15813–15817

21. Steel AB, Herne TM, Tarlov MJ (1998) Electrochemical quantitation of DNA immobilized on gold. Anal Chem 70:4670–4677

Chapter 3

Two-Dimensional Surface Plasmon Resonance Imaging System for Cellular Analysis

Tanveer Ahmad Mir and Hiroaki Shinohara

Abstract

Optical biosensors based on surface plasmon resonance (SPR) phenomenon have received a great deal of attention in cellular analysis applications. Sensitive and high-resolution SPR imaging (SPRi) platforms are very useful for real-time monitoring and measurement of individual cell responses to various exogenous substances. In cellular analysis, mainstream SPR-based sensors have potential for investigations of cell responses under ambient conditions. Evaluations that account only for the average response of cell monolayers mask the understanding of precise cell-molecular interactions or intracellular reactions at the level of individual cells. SPR/SPRi technology has attracted a great deal of attention for detecting the response of cell monolayers to various substances cultivated on the gold sensor chip. To unleash the full strength of SPRi technology in complex cell bio-systems, the applied SPR imaging system needs to be sufficiently effective to allow evaluation of a compound's potency, specificity, selectivity, toxicity, and effectiveness at the level of the individual cell. In our studies, we explore the utility of high-resolution 2D-SPR imaging for real-time monitoring of intracellular translocation of protein kinase C (PKC), and detection of neuronal differentiation in live cells at the level of individual cells. The PC12 cell line, which is one of the most commonly used neuronal precursor cell lines for research on neuronal differentiation, was chosen as a nerve cell model. Two dimensional SPR (2D-SPR) signals/images are successfully generated. We have found that cells treated with the differentiation factor nerve growth factor (NGF) showed a remarkable enhancement of SPR response to stimulation by muscarine, a nonselective agonist of the muscarinic acetylcholine receptor.

Key words Nerve growth factor, Surface plasmon resonance, Surface plasmon resonance imaging, Muscarine, Protein kinase C

1 Introduction

The ability to assess single-cell activity under physiological conditions in a label-free format would be highly beneficial for evaluating cellular and molecular binding dynamics. Conventional approaches to evaluating single-cell activity with high precision and temporal resolution are limited. The majority of currently available methods for studies of single cells present difficulties such as the need for fluorescent tagging and washing steps prior to readout. Fluorescent

Avraham Rasooly and Ben Prickril (eds.), *Biosensors and Biodetection: Methods and Protocols Volume 1: Optical-Based Detectors*, Methods in Molecular Biology, vol. 1571, DOI 10.1007/978-1-4939-6848-0_3, © Springer Science+Business Media LLC 2017

labeling may also alter membrane receptors and binding sites, thereby compromising measurements of in vivo interactions. By contrast, label-free methods directly provide additional information due to the ability to evaluate the intrinsic biological and physical characteristics of the individual cell.

SPRi biosensors are composed of three main components: optics (lasers or light emitting diodes) that illuminate a gold metal film, a high refractive index glass prism, and a charge-coupled device (CCD) camera detector that records changes in reflectance and captures images [1]. In general, SPR biosensors are based on surface evanescent waves that propagate parallel to a metal surface. The waves are generated under attenuated total reflection conditions when energy from a photon of p-polarized light impinges a thin Au metal sensor chip. Surface plasmons result from coupling of oscillating modes of free electron density at a specific incident angle. The plasmons possess a maximum intensity at the sensor surface and decay exponentially away from the metal/dielectric interface. This, in turn, gives SPR a sensing depth of only few hundred nanometers. SPR generates a dip in the reflection intensity curve in the cell area close to the substrate [2].

SPR experiments can be performed on a variety of commercially available platforms. Several reports have been published implementing one-dimensional SPR system for cell stimulation. In addition, conventional 2D-SPR imagers have been implemented for living cells [3]. However, the resolution of the 2D imaging systems is not sufficient for the observation of cellular responses at the individual cell level. To address this drawback our laboratory applied a high-resolution 2D-SPR imager for real-time and label-free monitoring of mammalian cells [4, 5].

The 2D-SPRi technique is ideally suited for single-cell analysis, as cell stimulation with exogenous molecules and their interactions are monitored by local refractive index changes. This allows real-time, noninvasive, and tag-free evaluation of cellular reactions such as binding dynamics, adhesion, spreading behavior, and morphological changes [6–9].

Figure 1 shows a 2D-SPR apparatus of Kretchmann configuration (A) and its schematic representation (B). The 2D-SPRi (2D-SPR 04A, NTT-AT Co. Corp., Japan) apparatus is equipped with a collimator (to parallelize incident light) and a CCD camera with four kinds of lens ($1\times$, $2\times$, $4\times$, and $7\times$) for image magnification. The optical system consists of a high-refractive index SF6 glass prism (RI $= 1.72$) deposited with a thin (50 nm) gold layer. To avoid external light interference, the measurement system is assembled in the temperature controlled (air conditioned) instrumental box. A light emitting diode (640 nm LED) producing the incident light beam passes through a collimator lens to illuminate the back side of the gold sensor chip through the coupling prism. The light beam from a collimated source passes through a polarizer filter and

a

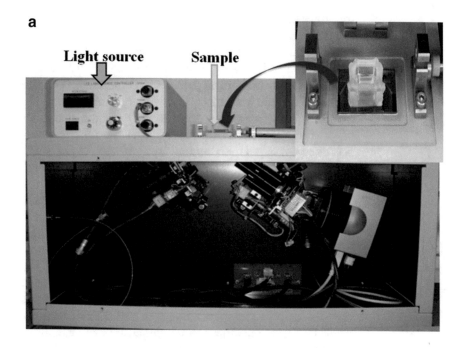

b

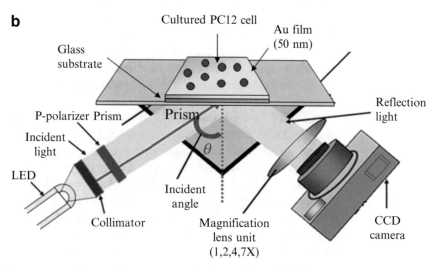

Fig. 1 (**a**) Shows a 2D-SPRi apparatus of Kretchmann configuration. (**b**) Schematic illustration of the used 2D-SPR system equipped with collimator and a CCD camera with four kinds of lens (1×, 2×, 4×, and 7×) for image magnification. A light emitting diode (LED) produces the incident light beam (670 or 770 nm) after passing through a collimator lens that illuminates the back side of the sensor chip through the coupling prism. The light beam from a collimated source passes through a polarizer filter and impinges on the sensor chip containing cells at a specific angle near the SPR angle. The reflected light is passed through a narrow band-pass filter, a 7× microscope objective collects the reflected light and forms reflectivity images that are detected recorded by the digital CCD camera at a fixed angle

impinges on the sensor chip containing cells at a specific angle near the SPR angle. The reflected light passes through a narrow band-pass filter and a magnification objective lens ($1\times$, $2\times$, $4\times$, or $7\times$) collects the reflected light and forms reflectivity images that are captured and stored by the cooled CCD camera. 2D-SPR recording is based on the reflection intensity modulation of incident p-polarized light beam at a given angle with the spatial capabilities of imaging. The polarizing filter allows the recordings of both p-polarized and s-polarized light, but s-polarized light is used only as a reference signal to eliminate artifacts and to improve the image contrast.

For cell-stimulation and intracellular signal analysis by 2D-SPRi, the gold sensor chip surface is modified with poly-L-lysine initially and is incubated overnight in a humid environment, and PC12 cell suspension is inoculated onto the same sensor chip to produce the mammalian cell sensor sample. The gold sensor slide on which the PC12 cells are cultured is placed on the top of the prism, and an index matching fluid is used between the prism and the SPR chip to eliminate unwanted reflection. During measurement, CCD camera visualizes whole sensor surface, and records the reflected light images (SPR images) from the entire observable area. Sequential changes in refractive index in each CCD pixel correspond to the region that is measured (e.g., regions 1, 2, 3, etc.). The interaction between ligands and cell membranes on the SPR chip influences the index of refraction, thereby causing a shift in reflectivity of incident light thus yielding a profile of refractive indexes.

2D-SPRi system has the ability to observe both angle shifts in the reflectivity curve, as well as to measure time-dependent cellular interaction dynamics. In angle shift measurements, when cells are cultivated onto the sensor chip, the SPR response signal is observed as angle shifts from lower to higher angle, as a result of a change in the index of refraction at the sensor surface. For time course measurements the light reflectivity is measured at an optimum angle close to the plasmon angle. The overall SPR response is measured as a change in the light intensity reflected from the sensor chip surface, and SPR images of cells are captured by the CCD camera (Fig. 1). For data analysis, a reference image is collected from the area without cells. Single-cell investigation on sensor chip is done by simply controlling the cell density, and by adjusting the high-resolution magnification lens unit. In our studies, the $7\times$ magnification lens is adjusted for monitoring the activity of individual cells. To analyze reflection intensity response of various cell regions with 2D-SPRi at single cell level, from the total cells in the observation area, a random selection for region of interest is made and the reflection intensity percentage of the selected cells is averaged. The overall reflection intensity change response is determined as the percentage of reflected light intensity of P-polarized light by subtracting cell region SPR response recorded before (reference) from a cell region response recorded after exposing to stimuli. Additionally, a threshold of reflection intensity is used to decide whether the cell has responded or not.

The key to applying 2D-SPR imaging system for the study of cellular analysis is the development of a sensitive, nondestructive, label-free assay of nearly all living cells and tissues. Developing a real-time cell-based sensor system seemed most valuable to cell biologists to provide a technical means to study the dynamics of intracellular signaling and neuronal differentiation rather than just snapshots.

In this chapter, the utility of a 2D-SPR imager for studying mammalian cells at individual cell level is demonstrated with two biological systems: the influence of exogenous substances on intracellular signal transduction response in normal PC12 cells, and the effect of exogenous substances to observe the intensity modulation of intracellular signaling as functional readout for neuronal differentiation associated with intracellular PKC translocation.

2 Materials

1. PC12 cell line (cell no. RCB0009) derived from rat adrenal gland is obtained from RIKEN BioResource Center.

2. Dulbecco's Modified Eagle Medium (DMEM) is obtained from GIBCO (Cat. No. 31600-034).

3. Horse serum is obtained from GIBCO (Cat. No. 16050-122).

4. Fetal bovine serum (FBS) is obtained from (ICN Biomedicals, Cat. No. F4135).

5. Penicillin/streptomycin is obtained from GIBCO (Cat. No. 15070-063).

6. 0.05% Trypsin-EDTA is obtained from GIBCO (Cat. No. 25200).

7. NaCl is obtained from Wako Pure Chemicals Industries (Cat. No. 191-01665).

8. NaOH is obtained from Kanto Chemical (Cat. No. 37184-00).

9. KCl is obtained from Wako Pure Chemicals Industries (Cat. No. 163-03545).

10. $CaCl_2 \, 2H_2O$ is obtained from Wako Pure Chemicals Industries (Cat. No. 031-00435).

11. $NaHCO_3$ is obtained from Wako Pure Chemicals Industries (Cat. No. 191-01305).

12. Glucose is obtained from Wako Pure Chemicals Industries (Cat. No. 041-00595).

13. HEPES is obtained from Nacalai Tesque (Cat. No. 17546-34).

14. PBS without calcium and magnesium is obtained from Wako Pure Chemicals Industries (Cat. No. 168-19323).

15. Dimethyl sulfoxide (DMSO) is obtained from Sigma (Cat. No. D2650).

16. Nerve growth factor is obtained from Sigma (Cat. No. N0513).

17. T-25 tissue culture flasks are obtained from IWAKI (Cat. No. 3103-025).

18. Hanks' balanced salts (HBBS) is obtained from Sigma (Cat. No. H6136).

19. 24-well plate is obtained from Sigma (product number: Z707791).

20. Hemicytometer is obtained from Sigma (product number: Z359629).

21. Bottle top filter is obtained from Corning® bottle-top vacuum filters-(CLS430049).

22. flexiPERM (11 × 7 × 10 mm) is purchased from Greiner Bio One (Cat. No 90034067.

23. 2D-SPR apparatus (2D-SPR 04A) coupled with a collimator, four magnification lenses (1×, 2×, 4×, and 7×), and a cooled charge-coupled device (CCD) camera is provided by NTT Advanced Technology (NTT-AT.Co.Corp) Japan.

24. Light emitting diode (640 nm LED) provided by NTT Advanced Technology (NTT-AT.Co.Corp) Japan, with 2D-SPR 04A apparatus.

25. CCD camera provided by NTT Advanced Technology (NTT-AT.Co.Corp) Japan, attached with 2D-SPR 04A apparatus.

26. 2D-SPR analysis software provided by NTT Advanced Technology (NTT-AT. Co. Corp) Japan, assembled on the computer monitor connected with 2D-SPR 04A apparatus.

27. 50-nm gold layer deposited high-refractive index glass (SF6 chips, 18 × 17 mm) is obtained from BAS, Japan.

28. Index-matching fluid ($n = 1.72$) obtained from Cargille Laboratories.

29. Sterile bench installed in the clean culture room.

30. CO_2 incubator installed in the clean culture room.

31. Inverted microscope (Olympus, cat. no. IX70) equipped with appropriate filter and magnifier (Nikon) installed in the clean culture room.

32. (+)-Muscarine chloride is obtained from Sigma (Cat. No. M6532-5MG).

33. Phorbol-12-myristate-13-acetate (PMA) is obtained from Sigma (Cat. No. P8139-1MG).

34. Staurosporine from Streptomyces sp. is obtained from Sigma (Cat. No. S4400).

3 Methods

All the solution preparations and cell cultures were performed under sterile conditions.

3.1 Cell Culture and Preparation of SPR Sensor Surface

1. A 50-nm gold thin film-coated high-refractive index glass (SF6) chip adhered with a square well of flexiPERM ($11 \times 7 \times 10$ mm) is used (*see* **Note 1**).

2. Prior to cell culture, SPR sensor slides are put into small cell culture dishes and are exposed to ultraviolet (UV) sterilization for 10 min.

3. Cryopreserved nerve model cells (PC12 cells) are harvested and then transferred into a 15 ml centrifuge tube containing 5 ml DMEM medium supplemented with 10% horse serum (HS) and 5% fetal bovine serum (FBS) (*see* **Note 2**).

4. T-25 flasks are filled with 5 ml of cell suspension and incubated in a cell culture incubator for 3–4 days (*see* **Note 3**).

5. After a given incubation period, T-25 flask is taken into a sterile clean bench. Old medium is removed and cells are thoroughly washed with sterile Hanks' balanced salts (HBBS) solution.

6. To detach the cells from T-25 flask surface, 1 ml of 0.05% trypsin-EDTA is added and agitated gently for 1 min. Finally, 4 ml of medium (DMEM, 10% HS, 5% FBS) is added, the total content is transferred into a 15 ml tube to stop the function of trypsin-EDTA and centrifuged.

7. After centrifugation, supernatant is discarded and the pellet aggregate is mechanically separated into single cells by gentle trituration of the suspension using pipette.

8. To prepare a SPR sensor surface of cellular analysis, a homogeneous solution is diluted in culture medium so that the cell suspension becomes 4.0×10^5 cells/ml. A total volume of 300 μl of medium containing cell suspension is poured onto the sensor chip to prepare one sensor sample (*see* **Note 4**).

9. For cellular analysis using NGF-treated neuronal differentiated PC12 cells, a homogeneous solution is diluted in culture medium so that the cell suspension becomes 3.0×10^5. A total volume of 300 μl of medium containing cell suspension is poured onto the sensor chip to prepare one sample (*see* **Note 5**).

10. In neuronal differentiation study, sensor chip surface is first modified with poly-L-lysine, PC12 cells are cultured onto the coated sensor surface. After 24 h, normal DMEM medium is exchanged with new DMEM containing 80 ng/ml NGF and only 1% HS.

3.2 Preparation of Working Solutions

1. First, 1 g Dulbecco's Modified Eagle Medium (DMEM) powder is diluted with 84 ml sterile Millipore water and then added 0.37 g $NaHCO_3$ while stirring gently. Then, 1 ml penicillin–streptomycin (1%), 10 ml horse serum (10% vol/vol), and 5 ml FBS (5% vol/vol) are added using 0.22-mm top bottle filter (*see* **Note 6**).

2. Hanks' balanced salts (HBBS) is prepared by dissolving original HBBS powder pack in cell culture-grade distilled, deionized water at room temperature. The prepared solution is sterilized using a 0.22-μm sterilizing filter, the solution is dispensed into sterile containers, and stored from 2 °C to 20 °C.

3. Just before use, phorbol-12-myristate-13-acetate (PMA) is dissolved as 1 and 0.2 mg/ml solutions in Dimethyl sulfoxide (DMSO) (*see* **Note 7**).

4. Just before use, staurosporine is dissolved as 1 mg/ml solutions in Dimethyl sulfoxide (DMSO) (*see* **Note 8**).

5. Just before use, (+)-Muscarine chloride is dissolved in water (*see* **Note 9**).

6. Just before use, KCl is dissolved in water.

7. Nerve growth factor (NGF) is dissolved in HBBS solution at 50–200 ng/ml and stored at −20 °C.

3.3 2D-SPR Imaging Experimental Setup

2D-SPR experiments are performed using a high-resolution 2D-SPR apparatus, as shown schematically in Fig. 1. A 50-nm gold-coated high-refractive index glass chip, assembled with a square wall-type reusable cell culture chamber system (flexiPERM) made of silicone that can easily stick on the sensor chip due to its naturally adhesive underside to all smooth surfaces without glue creating growth chambers. PC12 cells are maintained on the sensor chips for over 24 h in a humidified atmosphere containing 5% CO_2 at 37 °C. After taking out sensor chip from the incubator, the growth medium is discarded, and cells are rinsed twice with Hanks' solution (pH 7.4, 37 °C). After washing, Hanks' solution (240 μl) is added to the sensor chip. To eliminate the unwanted reflection and recording of reflection image of P-polarized parallel incident light, the sensor chip is adjusted on the top of the prism using index matching fluid. Thereafter, a volume of 30 μl of stimulant solution (KCl and PMA) or reference solution for negative control (Hanks' solution) is carefully injected onto the sensor slides using a manual pipette. During the experiment, SPR curves (reflection intensity changes) are monitored in real time and SPR images are recorded. Offline data is analyzed on a computer connected with the measurement system using 2D-SPR analysis software (2D-SPR analysis software provided by NTT-AT Company with the 2D-SPR apparatus). For inhibitory response, PC12 cells are pre-incubated with 100 nM staurosporine (a PKC inhibitor) solution for 30 min prior to PMA treatment.

In order to evaluate reflection intensity change response of various cell locations with 2D-SPRI at the individual cell level, from the total cell population on the chip observation region a random sorting of cell regions of interest is done and the reflection intensity percentage of the selected cell areas is averaged. The time-dependent reflection intensity change responses are determined as the percentage of reflected light intensity of P-polarized light to that of s-polarized light at a single incident angle as a function of time by simply subtracting cell region SPR responses recorded before (negative control) and after exposing to stimuli. To determine a resonance angle, SPR angles of the sensor surface with cells and the bare surface are measured by changing the incident angle from 49° to 57°. For overall data analysis, a threshold of reflection intensity is used to decide whether the cell has responded to stimuli or not. For example, responded cells are determined by $\Delta RI \geq 15\%$ over reflection intensity increase at 1 min from K^+ stimulation, $\Delta RI \geq 5\%$ over reflection intensity increase at 24 min from phorbol-12-myristate-13-acetate (PMA) injection, $\Delta RI \geq 3\%$ over reflection intensity increase 20 min from muscarine injection, respectively.

3.4 SPR Data Analysis

1. 2D-SPR software (originally developed by NTT-AT. Co, Japan, and incorporated in the computer connected to the 2D-SPR apparatus) was used for experimental SPR analysis. However, other imaging software (e.g., ImageJ) theoretically can be used for the analysis.

2. At the time of cell culturing on the gold chip, the cell number was controlled by checking the samples using phase contrast microscopy before placing on the 2D-SPR measurement platform.

3. During SPR measurement the CCD camera visualizes the entire sensor surface. Selected areas of interest in the viewable regions of the computer screen are encircled and selected for analysis.

4. The circle size is not fixed and the actual size of the circle was not measured.

5. As shown in Fig. 2a, b, areas corresponding to pixels containing cells are marked as regions 1–5. An area without cells (control) is marked as region 6.

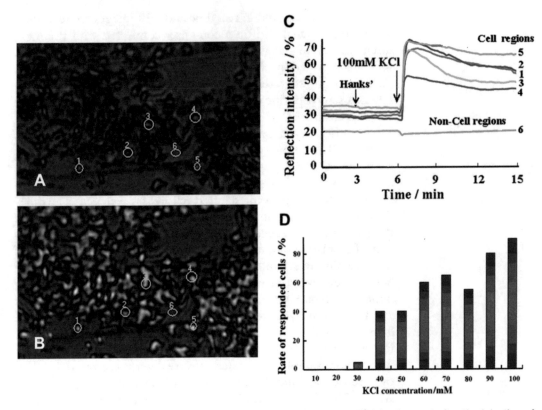

Fig. 2 (**a**) 2D-SPR image of the sensor chip surface is recorded before K⁺ injection and after the injection of Hanks' solution (negative control). (**b**) 2D-SPR image of the sensor chip surface recorded after 100 mM KCl injection, showing a robust increase in reflection intensity in PC12 cell regions. (**c**) Time course of the reflection intensity changes at individual PC12 cell regions on KCl stimulation with the 2D-SPR imager. (**d**) Representative columns indicating the percentages of SPR-responding cells according to concentrations of KCl. Reprinted from ref. 6, with permission from Elsevier

6. The images of the encircled areas as well as signal patterns (SPR curves) correspond to these encircled areas measured in real time (results in Fig. 2c, regions 1–5).

7. The refractive index in each CCD pixel corresponding to the chosen region is recorded and saved.

8. 2D-SPR analysis software automatically records the reflected light images (SPR images), which can be exported to Excel.

9. Finally, the saved Excel data was used for (time course measurement) graph plotting.

10. Because of the low magnification (7×) while the cells are clearly visible, the cell volume/dimensions or cell organelles are not measured, SPR imaging research is still at its infancy, and improvement in the resolution of SPR imaging system is greatly needed.

3.5   High KCl-Induced Response in PC12 Cells Observed with 2D-SPR Imager

1. Prior to experiment, the 2D-SPR imager is assembled. The sensor chip on which cells are adhered is fixed on the prism of the 2D-SPR imager.

2. The cell and blank areas on the sensor chip surface are observed and then the measurement is started. SPR curve that is dependence of reflectivity of the area of interest on incident angle is first evaluated to get exact resonance angle and to obtain suitable measurement angle.

3. SPR curve in terms of time course measurement after exposing to stimulant is observed at the suitable measurement resonance angle.

4. After stabilization of baseline, cells are first exposed to Hanks' solution as reference. As expected, no change in reflection intensity is observed after exposing to negative control solution (*see* **Note 10**).

5. To monitor SPRi sensorgram for intracellular signal transduction pathway (mainly PKC translocation) via depolarization, the cells are exposed to different concentrations of K^+ solution and sequential alterations of average reflected light intensity in terms time course observation for each pixel of the selected areas of interest on the monitor are obtained and analyzed using the 2D-SPR analyzer software and images are captured by the CCD camera.

6. After exposing cell to K^+ solution, 2D-SPR images of PC12 cell areas are gone from dim to bright and variation in cell-to-cell reflection intensity change response is also observed. It is speculated that this variation may be due to differences in the physique of the respective cells (Fig. 2) (*see* **Note 11**).

3.6   PKC Agonist and Antagonist-Induced Response in PC12 Cells Observed with 2D-SPR Imager

1. To confirm the results whether K^+-induced reflection intensity change in 2D-SPR sensing is genuinely implicated to intracellular signal transduction (translocation of PKC) in PC12 cells, cells are exposed to Phorbol-12-myristate-13-acetate (PMA), a specific activator of Protein Kinase C (PKC).

2. SPR images of the PC12 cells recorded with the CCD camera before and after stimulation of varying the concentrations of Phorbol-12-myristate-13-acetate (PMA), the obtained SPR curves showed a slow increase with fluctuation after gaining phorbol-12-myristate-13-acetate (PMA) access to the cell interior (Fig. 3) (*see* **Note 12**).

3. To further validate the results and examine the inhibitory effects, PC12 cells are pretreated with a potent inhibitor of PKC (staurosporine) for 30 min, on the Phorbol-12-myristate-13-acetate-induced response (*see* **Note 13**).

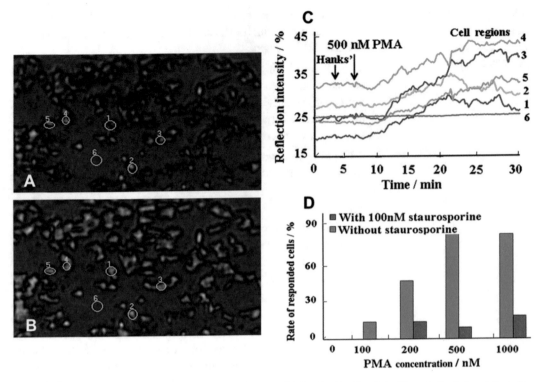

Fig. 3 (a) 2D-SPR image of the sensor chip surface obtained after the injection of Hanks' solution for background observation. (b) 2D-SPR image of PC12 cell regions obtained after phorbol-12-myristate-13-acetate (PMA)stimulation (500 nM), which showed enhancement of the reflection intensity in cell regions. (c) Reflection intensity change corresponding to the intracellular refractive index change, measured with the 2D-SPR imager for PC12 cells adhered to a gold chip on 500 nM phorbol-12-myristate-13-acetate (PMA) injection. After injecting 500 nM PMA, a slow elevation of reflection intensity was seen at cell regions. No reflection intensity change is observed in non-cell regions. (d) Representative columns indicating the percentages of SPR-responded cells in the presence (*red*) or absence (*blue*) of 100 nM staurosporine. Reprinted from ref. 6, with permission from Elsevier

4. The representative columns of comparative graph for the dependence of reflection intensity change induced by stimulation with various concentrations of phorbol-12-myristate-13-acetate (PMA) in the presence or absence of staurosporine showed a clear effect on these activators or inhibitors on SPRi generated signal (Fig. 3).

3.7 2D-SPRi Observation of the Effect of Muscarine on Normal and NGF-Treated Differentiated PC12 Cells

1. 2D-SPR imaging system is further applied to observe the intensity modulation of intracellular signaling as functional readout for differentiation associated with intracellular PKC translocation, PC12 cells are exposed to Muscarine (acetylcholinergic receptor agonist) before and after differentiation (*see* **Note 14**).

2. Compared to non-differentiated cells, NGF-treated differentiated PC12 cells showed a remarkable enhancement in reflection intensity change signal (Fig. 4).

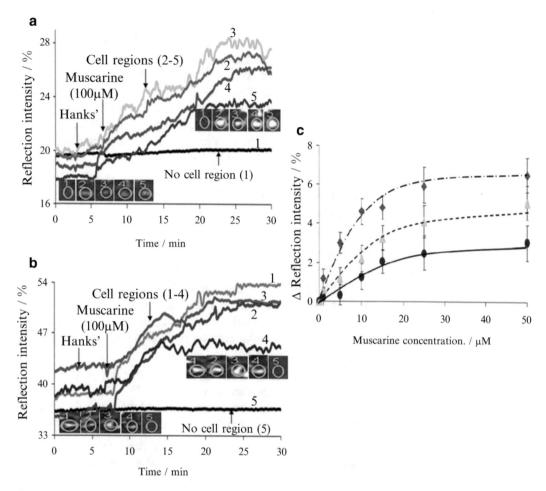

Fig. 4 Time course of the reflection intensity change for undifferentiated (**a**) and differentiated (**b**) PC12 cells at individual cell regions on muscarine stimulation with the 2D-SPR imager. After the application of muscarine, slow and fluctuating elevation of reflection intensity was seen at cell regions. No reflection intensity change is observed at the non-cell region. (**c**) Dependence of reflection intensity change at the undifferentiated cell regions and the differentiated cell regions (2 days treatment, 4 days treatment) on the concentration of muscarine as a stimulant. The PC12 cells are treated with 80 ng/ml NGF for 2 or 4 days. Reprinted from ref. 7, with permission from Elsevier

4   Notes

1. After finishing the experiment, the medium is discarded and the chips are washed carefully and thoroughly to remove suspended dead cells or other material.

2. Cells are centrifuged at $500 \times g$ for 3 min, supernatant is decanted, and the cell pellet diluted in pre-warmed culture medium.

3. Usually, optimum confluency is obtained within this period.

4. The cell suspension is diluted to 4.0×10^5 cells/ml (for experiments before differentiation) with culture medium and poured onto a sensor chip placed in the sterilized culture dish. The cells are evenly distributed by brief agitation of the gold sensor chip, and the cells are allowed to adhere and spread typically overnight to reach a stable baseline before being placed onto the 2D-SPR imaging platform for stimulant stimulation and observation of reflection intensity changes.

5. For the neuronal differentiation study, the cell suspension is diluted to 3.0×10^5 cells/ml with culture medium and poured onto poly-L-lysine coated sensor chips.

6. Bottles are labeled as follows: • DMEM • Sterile • Date • Initials. Dry powder must be free-flowing. Do not use if powder is caked or clumping, indicating accumulation of moisture.

7. Phorbol-12-myristate-13-acetate (PMA) solution is prepared on the day of the experiment. Stock solutions are stored at $-20\ ^\circ\text{C}$. This compound is irritating to eyes, respiratory system, and skin, so avoid contact with skin and eyes by wearing suitable protective clothing.

8. Staurosporine solution is prepared on the day of the experiment. Stock solutions are stored at $-20\ ^\circ\text{C}$. This compound is irritating to eyes, respiratory system, and skin, so avoid contact with skin and eyes by wearing suitable protective clothing.

9. Muscarine solution is prepared on the day of the experiment. Stock solutions are stored at $-20\ ^\circ\text{C}$. This compound is irritating to eyes, respiratory system, and skin, so avoid contact with skin and eyes by wearing suitable protective clothing.

10. Hanks' buffer is used to avoid possible confounding effects of the measuring buffer solution that may stem from the mechanical disturbance caused by the turbulence during solution ejection by manual system.

11. It is shown in Fig. 2c, the reflection intensity of the PC12 cell region is dramatically increased and individual cell regions showed a similar response pattern; however, the reflection intensity was uneven variation that may be borne from differences in physical condition of the particular cell.

12. It is shown in Fig. 3c, the extracellular administration of Hanks' solution stimulation (bottom line) failed to produce any alteration in reflection intensity response curve. However, a few minutes after phorbol-12-myristate-13-acetate (PMA), a potent tumor promoter that can intercalate into membrane and potently activate intracellular PKC by recruiting it to cell membranes, thereby mimicking the mechanism of the natural PKC activator, diacylglycerol exposure, the reflection intensity

response curve increased, accompanied by a conspicuous increase of intracellular signal transduction. It is observed that the signal transduction that might be associated with the translocation of PKC is monitored for individual cells exhibiting a similar slow elevating SPR curve. It is thought that, unlike diacylglycerol, phorbol esters are not readily metabolized, and exposure of cells to these molecules results in prolonged activation of PKC.

13. To further study whether the reflection intensity change signal measured by the 2D-SPR imaging system with the stimulation of these diacylglycerol surrogates indeed associated with the activation of PKC, the effect of treating cells with an antagonist of PKC, staurosporine, on the response to phorbol-12-myristate-13-acetate (PMA) is determined and compared. As shown in in Fig. 3d, Staurosporine-treated cells decreased the response after treating with phorbol-12-myristate-13-acetate (PMA) solution and lead reflective intensity decreases toward the basal level.

14. The effect of nerve growth factor (NGF) on PC12 cell culture is evaluated by observing reflection intensity change responses at the individual cell level under the consideration that NGF tends to decrease cell growth; proliferation; different responses could be recorded in PC12 cells maintained under controlled circumstances in comparison with those incubated in the presence of NGF. It is shown in Fig. 4, after Hanks' solution injection no reflection intensity change response is detected, whereas after exposing the cells to muscarine solution a slow and fluctuating elevation of reflectivity curves is obtained, with a similar response pattern for both non-differentiated and differentiated PC12 cells. However, a significant enhancement in reflection intensity change response is observed in NGF-treated PC12 cells. The obtained data demonstrates that the 2D-SPR imager is not only useful for intracellular signal detection but could also be useful to distinguish differentiated cells from non-differentiated counterparts noninvasively, in real time and independent of any fluorescent or radioactive probes.

References

1. Suzuki M, Ohshima T, Hane S, Iribe Y, Tobita T (2007) Multiscale 2D-SPR biosensing for cell chips. J Robot Mechatron 19:5519–5523

2. Homola J (2008) Surface plasmon resonance sensors for detection of chemical and biological species. Chem Rev 108(2):462–493

3. Yanase Y, Hiragun T, Ishii K, Kawaguchi T, Yanase T, Kawai M, Sakamoto K, Hide M (2014) Surface plasmon resonance for cell-based clinical diagnosis. Sensors 14 (3):4948–4959

4. Horii M, Shinohara H, Irebe Y, Suzuki M (2007) Application of high resolution 2D-SPR imager to living cell-based allergen sensing. In: Viovy J-L, Tabeling P, Descroix S, Malaquin L (eds) The proceedings of ITAS 2007 conference. Chemical and biological microsystems society, Paris, pp 451–453

5. Horii M, Shinohara H, Irebe Y, Suzuki M (2011) Living cell-based allergen sensing using a high resolution two-dimensional surface plasmon resonance imager. Analyst 136:2706–2711

6. Shinohara H, Sakai Y, Mir TA (2013) Real-time monitoring of intracellular signal transduction in PC12 cells by two-dimensional surface plasmon resonance imager. Anal Biochem 441:185–189

7. Mir TA, Shinohara H (2013) Two-dimensional surface plasmon resonance imager: an approach to study neuronal differentiation. Anal Biochem 443:46–51

8. Mir TA and Shinohara H (2012) Real-time monitoring of cell response to drug stimulation by 2D-SPR sensor: an approach to study neuronal differentiation. In: Proceeding of the 14th international meeting on chemical sensors. doi: 10.5162/IMCS 2012/P1.1.18

9. Mir TA and Shinohara H (2012) 2D-SPR biosensor detects the intracellular signal transduction in PC12 cells at single cell level. In: Proceeding of the sixth international conference on sensing technology. doi: 10.1109/ICSensT.2012.6461763

Chapter 4

Immunosensing with Near-Infrared Plasmonic Optical Fiber Gratings

Christophe Caucheteur, Clotilde Ribaut, Viera Malachovska, and Ruddy Wattiez

Abstract

Surface Plasmon resonance (SPR) optical fiber biosensors constitute a miniaturized counterpart to the bulky prism configuration and offer remote operation in very small volumes of analyte. They are a cost-effective and relatively straightforward technique to yield in situ (or even possibly in vivo) molecular detection. They are usually obtained from a gold-coated fiber segment for which the core-guided light is brought into contact with the surrounding medium, either by etching (or side-polishing) or by using grating coupling. Recently, SPR generation was achieved in gold-coated tilted fiber Bragg gratings (TFBGs). These sensors probe the surrounding medium with near-infrared narrowband resonances, which enhances both the penetration depth of the evanescent field in the external medium and the wavelength resolution of the interrogation. They constitute the unique configuration able to probe all the fiber cladding modes individually, with high Q-factors. We use these unique spectral features in our work to sense proteins and extra-cellular membrane receptors that are both overexpressed in cancerous tissues. Impressive limit of detection (LOD) and sensitivity are reported, which paves the way for the further use of such immunosensors for cancer diagnosis.

Key words Surface plasmon resonance, Optical fiber sensors, Bragg gratings, Immunosensing, Cells, Proteins

1 Introduction

Biosensors are by definition a combination of a biological receptor compound and a physical or physicochemical transducer. Optical methods of transduction offer important advantages such as:

- Noninvasive, safe, and multi-dimensional detection (wavelength, intensity, phase, polarization, etc.).

- Well-established technologies readily available from communication and micro-nano technologies industries (lasers, detectors, passive components, etc.).

Avraham Rasooly and Ben Prickril (eds.), *Biosensors and Biodetection: Methods and Protocols Volume 1: Optical-Based Detectors*, Methods in Molecular Biology, vol. 1571, DOI 10.1007/978-1-4939-6848-0_4, © Springer Science+Business Media LLC 2017

– Visible and near-infrared optical frequencies coincide with a
wide range of physical properties of bio-related materials.

Biosensors are continuously developed to respond to the
demand for rapid, accurate, and in situ monitoring techniques in
different areas such as medical diagnosis, genomics, proteomics,
environmental monitoring, food analysis, and security. Label-free
optical biosensors yield direct and real-time observation of molecular interaction, without requiring the use of labels since they sense
binding-induced refractive index changes. Among the different
optical biosensor configurations (measurements of absorbance,
reflectance, fluorescence, refractive index changes, etc.), those
using surface plasmon polaritons (SPPs) have been intensively
developed during the past three decades.

SPPs result from collective excitations of electrons at the
boundary between a metal and a dielectric. The strong sensitivity
of the plasmon-propagation constant to the permittivity of a nanoscale layer of material (bioreceptors) deposited on the metallic
surface has led to the monitoring of surface plasmon resonance
(SPR) to detect (bio)chemical changes occurring between molecules. As a result, biochemical reactions are measured with SPR by
monitoring their effective refractive index. In practice, as sketched
in Fig. 1, the so-called Kretschmann-Raether approach realizes this
by launching light beams from a high refractive index prism to a
thin metallic interface at an angle such that light is totally reflected
[1]. In doing so, an evanescent wave extends in the metal overlay.
When the component of the propagation constant of the light
along the interface matches that of a plasmon excitation of the
other side of the metal layer, part of the light couples to the
plasmon, which decreases the reflection (at θ_0 in Fig. 1 where θ_c
denotes the critical angle of incidence). The interrogation is made
either by varying the wavelength and keeping the incidence angle
constant or by using monochromatic light and modifying the

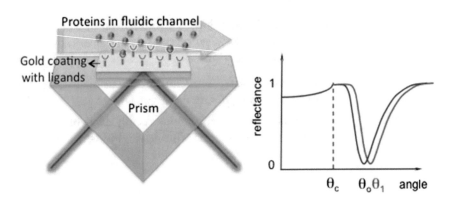

Fig. 1 Sketch of the Kretschmann-Raether prism approach for plasmonic generation and operating principle of
its interrogation to measure binding of biomolecules on receptors grafted on the gold substrate

angle. In both cases, the polarization state of the light has to be parallel to the incidence plane so that the plasmon wave is normally polarized to the interface. Biochemical reactions occurring above the metal surface affect the effective index of the plasmon wave, which is detected through a shift of the SPR (from θ_0 to θ_1). The sensitivity to the surrounding refractive index often ranges in the order of 10^{-6}–10^{-7}.

Numerous transducers have been developed, bringing additional assets with respect to the Kretschmann-Raether prism approach, which is exploited in most commercial systems. In this aspect, optical fiber-based sensors are particularly attractive. With their lightweight, compactness, and ease of connection, they provide remote operation in very small volumes of analyte of the order of microliter. And with their continuous development and optimization, they appear perfectly suited for in situ or even possibly in vivo diagnosis. They also have the potential to assay different parameters simultaneously.

To excite SPR from an optical fiber, light confined in the fiber core has to be locally outcoupled and brought into contact with the surrounding medium. In practice, this is achieved either by a geometrical modification (polishing or etching of the cladding) so as to expose the evanescent wave to the surrounding medium or by using in-fiber gratings (refractive index modulations photo-imprinted in the fiber core along the propagation axis). Hence, various architectures are available [2]: etched multimode optical fibers, side-polished, D-shaped, tapered, or U-bent optical fibers, long period fiber gratings (LPFGs) and tilted fiber Bragg gratings (TFBGs). Configurations based on cladding removal/decrease can be quite easily achieved. The SPR is in this case spectrally manifested by a broadband resonance (full width at half maximum (FWHM) ~20 nm or higher) in the transmitted amplitude spectrum. Operation in reflection mode is possible by using a mirror deposited on the cleaved fiber end face beyond the sensing region. However, these configurations considerably weaken optical fibers at the sensor head and may prevent their use in practical applications, out of laboratory settings. For this reason, large core fibers (unclad 200–400 μm core fibers) are the most spread in practice [3]. These configurations operate at visible wavelengths, which limits the extension of the evanescent wave in the surrounding medium. Indeed, its penetration depth is proportional to the operation wavelength (λ) and usually ranges between $\lambda/5$ and $\lambda/2$, depending on the mode order [4]. Hence, operation at near-infrared telecommunication wavelengths enhances the penetration depth, which in turn improves the overall sensor sensitivity to large-scale targets such as proteins or cells. Such operation can be easily achieved with in-fiber gratings.

Gratings preserve the fiber integrity while providing a strong coupling between the core-guided light and the cladding. LPFGs consist in a periodic refractive index modulation of the fiber core of a few hundreds of μm. They couple the forward-going core mode into forward-going cladding modes. Their transmitted amplitude spectrum is composed of a couple of wide resonances (FWHM ~20 nm) dispersed in a wavelength range of a few hundreds of nm. TFBGs are short period (~500 nm) gratings with a refractive index modulation slightly angled with respect to the perpendicular to the optical fiber axis. In addition to the self-backward coupling of the core mode, they couple light into backward-going cladding modes. Their transmitted amplitude spectrum displays several tens of narrow-band cladding mode resonances (FWHM ~200 pm or even below) located at the left-hand side of the Bragg resonance (or core mode resonance) corresponding to the core mode self-coupling. According to phase matching conditions, every cladding mode resonance possesses its own effective refractive index and the maximum refractometric sensitivity is obtained when this effective index tends to the surrounding refractive index value. TFBGs act as spectral combs and constitute the only optical fiber configuration able to probe simultaneously but distinctively all the cladding modes supported by an optical fiber [5].

SPR optical fiber sensors can be obtained from the above-listed structures surrounded by a noble metal (most often gold or silver). Sheaths of thickness ranging between 30 and 70 nm are most generally used. SPR generation is achieved when the electric field of the light modes is polarized mostly radially at the surrounding medium interface. The orthogonal polarization state is not able to excite the SPR, as the electric field of the light modes is polarized mostly azimuthally (i.e., tangentially to the metal) at the surrounding medium interface and thus cannot couple energy to the surface Plasmon waves. Depending on the configuration, refractometric sensitivities in the range $[10^2$–10^5 nm/RIU (refractive index unit)] have been reported [6].

When comparing the sensor performances between different configurations, it is not sufficient to compare only sensitivities (i.e., wavelength shifts), without considering the wavelength measurement accuracy. It is more convenient to refer to the figure of merit (FOM) of the device. The FOM corresponds to the ratio between the sensitivity and the linewidth of the resonance, taking into account that it is easier to measure the exact location of a narrow resonance than a broad one [7]. Hence, in terms of experimentally demonstrated FOM, TFBGs outclass all other configurations by more than one order of magnitude [2]. This results from their sensitivity close to 500 nm/RIU and the narrowness of their resonances.

Plasmonic generation in gold-coated TFBGs has been pioneered in 2006 by the team of Prof. Jacques Albert at the *Carleton University of Ottawa*, Canada [8]. Since then, our two research groups have worked together on the spectral characterization of these probes that are inherently polarization selective [9, 10] and on their subsequent use for biochemical sensing [11, 12].

In our work, we use the well-acknowledged antibody-antigen affinity mechanism to assess the unique protein detection and quantification capabilities of gold-coated TFBGs and evaluate their performances in terms of both sensitivity and limit of detection (LOD). We have conducted experiments on two kinds of bioreceptors, e.g., proteins and extracellular membrane receptors that are both overexpressed in the case of cancer cells. In particular, we have quantified a certain type of cytokeratins that are secreted by lung cancer tumors and cells suspensions. Using another functional film on the same sensor, we have also detected extracellular membrane receptors in native membranes of different human epithelial cell lines. A differential diagnosis has been demonstrated between two cell lines, with overexpressed membrane receptors (positive control) and with a low number of these receptors (negative control). Such results make an important step forward toward the demonstration of in vivo diagnosis.

In the remaining of this chapter, we will first describe the operating principle of the sensor. Then, we will focus on its performances in the quantification of proteins and cells in vitro.

2   Materials

2.1   Optical System to Produce TFBGs

1. Frequency-doubled Argon-ion laser emitting at 244 nm with a mean output power of 50 mW (Newport SpectraPhysics).

2. Uniform phase mask with a pitch of 1090 nm (Coherent).

3. Automated one-axis translation stage (Physik Instrumente).

4. 5-cm-focal distance cylindrical lens in front of the phase mask (Newport).

5. Diaphragm to spatially filter the UV laser beam (Newport).

6. Telecommunication-grade single-mode optical fiber (Corning SMF-28) .

2.2   Gold Deposition on the Optical Fiber Surface

1. Sputtering chamber (LEICA EM SCD 500).

2. 99.99% purity gold foil used as the target in the sputtering chamber.

2.3   Optical System to Measure the Spectrum of Gold-Coated TFBGs

1. Optical vector analyzer (Luna Technologies OVA CTe).

2. In-line linear polarizer (General Photonics).

3. In-house developed microfluidic chamber to immerse gold-coated TFBGs in controlled liquids.

4. Hand-held Abbe refractometer accurate to 10^{-4} RIU (*Reichert AR200*).

1. Dithiolalkanearomatic PEG6-COOH (thiols) (SensoPath Technologies, N° SPT-0014A6) was prepared at 2 mM in absolute ethanol.

2. Phosphate buffer saline (PBS) (Sigma N° P4417) was prepared to yield 10 mM phosphate (PO_4^-), 2.7 mM potassium chloride (KCl), and 137 mM sodium chloride (NaCl), at pH 7.4 in milliQ water.

3. The mixture N-hydroxysuccinimide (NHS) (Sigma-Aldrich 56480) and N-(3-)dimethylaminopropyl)-N'-ethylcarbodiimide hydrochloride (EDC) (Sigma-Aldrich E7750) was prepared at a ratio 1:5 with a final concentration of NHS at 100 mM and 500 mM of EDC in milliQ water.

4. 1% (w/v) bovine serum albumin (BSA) (Acros Organics 134731000) was made in phosphate buffer solution at pH 7.4.

5. Polyclonal anti-Cytokeratin 7 antibodies (AbCK7) (Biorbyt orb10410) have been diluted in phosphate buffer solution at pH 7.4, at 20 µg/mL.

6. Cytokeratin 7 peptide (CK7pep) (Biorbyt orb72193) made of 23 amino acids, corresponding to a molecular weight of 2.65 kDa, was diluted in phosphate buffer solution at pH 7.14, at different concentrations.

7. Cytokeratin 7 full protein (CK7FP) (abcam 132933) which is characterized by the association of 469 amino acids and a molecular weight equal to 78 kDa, has been prepared at different concentrations in PBS.

8. Fetal bovin serum (FBS) (Sigma F7524) was used to complement the phosphate buffer solution for the measurement in complex media.

3 Methods

3.1 Operating Principle of TFBGs

The sensor is obtained from a single-mode optical fiber in which a photo-imprinted refractive index grating is formed over a short section of the core, and that is further coated with a bilayer coating: a very thin metal coating and a suitable biochemical recognition binding layer.

The glass optical fiber is a 125-µm-thick cylindrical waveguide made of two concentric layers, the core in the middle surrounded by a cladding that is thick enough to prevent light propagating in the core to interact with the fiber surroundings. The refractive index of the core is slightly higher than that of the cladding to allow for the guidance of light in the fiber core. Our experiments are performed on 8 µm core single-mode optical fibers that guide light into a single optical mode at wavelengths between 1300 and 1650 nm. Such fibers are widely available at low cost (less than 100

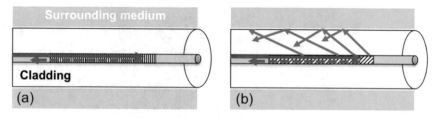

Fig. 2 Sketch of the light coupling mechanism for uniform FBGs (**a**) and tilted FBGs (**b**)

US$ per km) and are telecommunication-grade. In the same way, a large quantity of equipment is available to characterize, manipulate, and use such fibers and devices made from them.

As sketched in Fig. 2a, a uniform FBG is a periodic and permanent refractive index modulation of the fiber core that is imprinted perpendicularly to the propagation axis [13, 14]. It behaves as a selective mirror in wavelength for the light propagating in the core, reflecting a narrow spectral band centered on the so-called Bragg wavelength $\lambda_{\text{Bragg}} = 2n_{\text{eff,core}}\Lambda$ where $n_{\text{eff,core}}$ is the effective refractive index of the core mode (close to 1.45) and Λ is the grating period. In practice, $\Lambda \sim 500$ nm to ensure that the Bragg wavelength falls in the band of minimum attenuation of the optical fiber centered on 1550 nm. The Bragg wavelength is inherently sensitive to mechanical strain and temperature, through a change of both n_{eff} and Λ [13]. In practice, a change of temperature of +1 °C yields a Bragg wavelength shift of ~10 pm (at 1550 nm). Such change is easily measured with standard telecommunication instruments since the full spectral width of the reflected light from a typical 1-cm-long grating is of the order of 100 pm. However, our purpose here is to measure events occurring on or near the fiber cladding surface and the core mode reflection spectrum is inherently and totally insensitive to such changes (because the penetration depth of core-guided light into the cladding does not exceed a few micrometers). We use a small modification of the FBG to couple light from the core to the cladding and still benefit from narrowband spectral resonances that will reveal small changes at the cladding boundary.

A TFBG corresponds to a refractive index modulation angled by a few degrees relative to the perpendicular to the propagation axis (Fig. 2b) [15]. In addition to the self-backward coupling of the core mode at the Bragg wavelength Gold-coated tilted fiber Bragg gratings (TFBGs):, the grating now redirects some light to the cladding whose diameter is so large that several possible cladding modes can propagate, each with its own phase velocity (and hence effective index $n_{\text{eff,clad}}$). These possible modes of propagation correspond to different ray angles in Fig. 2b. Again, there is a one-to-one relationship between the wavelength at which coupling occurs for a given cladding mode and its effective index. This relationship

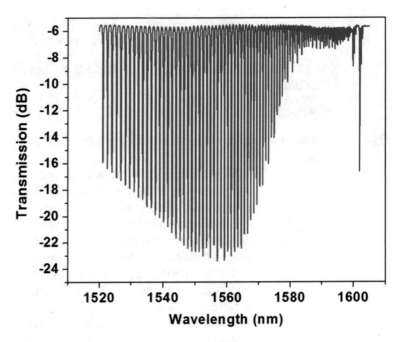

Fig. 3 Typical transmission spectrum Gold-coated tilted fiber Bragg gratings (TFBGs):transmission spectrum of a 10° tilted fiber Bragg grating measured in air (using linearly polarized input light as described in Reference [5])

is expressed by a similar phase matching condition as before: $\lambda^i_{clad} = (n_{eff,core} + n^i_{eff,clad})\ \Lambda$ where the index "i" reflects the fact that the fiber cladding can support many modes. Figure 3 shows a TFBG transmission spectrum where each resonance corresponds to the coupling from the core mode to a group of backward propagating cladding modes. As a result of phase matching, the spectral position of a resonance now depends on the effective index of its associated cladding mode, which in turn depends on the optical properties of the medium on or near the cladding surface.

Therefore, spectral shifts of individual resonances can be used to measure changes in fiber coatings and surroundings. Laffont and Ferdinand were the first to demonstrate SRI sensing with TFBGs in 2001 [16]. They observed a progressive smoothing of the transmitted spectrum starting from the shortest wavelengths as the SRI increased from 1.30 to 1.45. Several data processing techniques can be used to quantitatively correlate the spectral content with the SRI value, either based on a global spectral evolution or a local spectral feature change. The first method involves monitoring the area delimited by the cladding mode resonance spectrum, through a computation of the upper and lower envelopes as resonances gradually disappear when the SRI reaches the cut-off points of each cladding mode [16, 17]. Another technique tracks the wavelength shift and amplitude variation of individual cladding mode resonances as they approach cutoff [18]. Both techniques present minimum detectable SRI changes of ~10^{-4} RIU (refractive index unit). In terms of wavelength tracking, this result is obtained with a sensitivity that peaks between 10 and 15 nm/RIU for the modes near cutoff. In all cases, the Bragg wavelength provides an absolute power and wavelength reference that can be used to remove uncertainties related to systematic fluctuations (such as power level changes from the light sources) and even temperature changes (because it was demonstrated in Chen et al. [19] that all resonances shift by the same amount when the temperature changes). The TFBG thus allows inherently temperature-insensitive SRI measurements and large signal-to-noise ratio. For biochemical experiments however, it is usually necessary to improve the LOD levels to at least 10^{-5} RIU, by increasing the wavelength shift sensitivity while keeping noise level down and spectral features narrow [20]. Fortunately, it has been recently demonstrated that the addition of a nanometric-scale gold coating overlay on the TFBG outer surface considerably enhances the refractometric sensitivity through the generation of surface Plasmon resonances (SPR) [8, 21, 22]. This sensitivity increase can be easily achieved with gold-coated TFBGs.

3.2 Photo-Inscription of TFBGs

1. TFBGs are manufactured in the same way as conventional FBGs, i.e., through a lateral illumination of the fiber core using an interference pattern of ultraviolet (UV) light between 190 and 260 nm. The exposure of the fiber to bright interference fringes for a few minutes locally and permanently increases the mean refractive index of the core. The ultraviolet interference pattern is reproduced inside the fiber much like in a photographic process. Mass production of identical gratings is straightforward when the phase mask technique is used to generate the interference pattern. A phase mask is a diffractive element made in a pure silica substrate that is optimized to maximize the diffraction in the first (+1 and −1) diffraction orders. An interference pattern at half the period of the mask is therefore produced in the fiber core when the optical fiber is located in close proximity behind the mask, as depicted in Fig. 4. In order to obtain a TFBG, we usually rotate the phase mask in the plane perpendicular to the incident laser beam. We routinely produce TFBGs with tilt angles varying between 4° and 10°, the larger angle being preferred to operate when the SRI lies near the index of water and water solutions, which is often the case in biochemical research, because then the strongest cladding mode resonances are located near 1550 nm (when the Bragg wavelength is near 1610 nm), where they are easier to measure. Their physical length is typically 1 cm.

2. Prior to the manufacturing process, the single-mode fibers are hydrogen-loaded to enhance their photosensitivity (capability to change their refractive index when exposed to a light

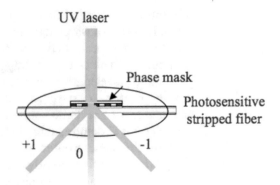

Fig. 4 Sketch of the phase mask technique

interference pattern) to ultraviolet light. This is done in a vessel under pressure (200 atm) and at moderate temperature (70 °C).

3.3 Gold Deposition on the Optical Fiber Surface

1. We use a sputtering chamber to extract gold from a target.

2. Two consecutive depositions are made in the same conditions, with the optical fibers rotated by 180° between both processes, to ensure that the whole outer surface is covered by gold.

3. The optimum gold thickness has been found to be approximately 50 nm, for the narrowest, deepest SPR attenuation. In spite of the two-step deposition that yields a slightly nonuniform thickness around the fiber circumference, we did not find this nonuniformity to be detrimental for SPR generation in practice. Also, as the vacuum is obtained in the chamber starting from ambient air and not from an inert (Argon) starting atmosphere, there is no need for a 2–3-nm-thick adhesion layer of chromium or titanium between the silica surface and the gold coating.

3.4 Surface Functionalization

The gold-coated TFBGs are functionalized for biosensing purposes. The chemistry involved in this process depends on the target application. In our case study, it is based on the antigen/antibody affinity. Whatever the analyte to be detected, a self-assembled monolayer (SAM) is manufactured, prior to the biomolecules immobilization.

1. For this, gold-coated TFBGs were first thoroughly rinsed with ethanol and dried under nitrogen.

2. Gold-coated TFBGs were then immersed in a solution of thiols (2 mM) dispersed in ethanol. Thiols incubations were usually done during 18 h at room temperature in a 1-mm-thick capillary tube sealed at both ends to prevent solvent evaporation. At the end of the incubation, the functionalized gold-coated TFBGs were removed from the tube and again rinsed with ethanol (*see* **Note 1**).

3. The next step of the surface functionalization consisted in the esterification of the carboxylate functions of the SAM. The reaction was realized by an immersion of the gold-coated TFBGs in a mixture of NHS/EDC (1:5 M/M) in H_2O milliQ for 2 h. The surface was then rinsed with milliQ water and dried under nitrogen.

4. The gold-coated TFBGs were immediately immersed in a primary antibody solution (AbCK7 at 20 μg/mL in PBS at pH 7.4). After 2 h of incubation at room temperature, the gold-coated TFBGs were rinsed with PBS (*see* **Note 2**).

5. A blocking step was realized by immersion of the optical fibers in BSA solution 1% (w/v) in PBS at pH 7.4 for 1 h to avoid nonspecific interaction.

6. Bioreceptors were finally rinsed with PBS and ready to be used for Cytokeratin 7 proteins detection (*see* **Note 3**).

These four consecutive steps (grating manufacturing, gold deposition, surface functionalization, and bioreceptors grafting) yield the plasmonic probe sketched in Fig. 5. The illustration below displays the photo-inscribed optical biosensor with antibodies (orange semicircle) immobilized on the gold layer deposited on the optical fiber's cladding. The optical fiber is finally presented here in a packaging (gray cylinder) in the presence of antigens (blue sphere) in a bulk solution.

3.5 Interrogation of Gold-Coated TFBG Immunosensors

1. Transmitted amplitude spectra of gold-coated TFBGs were recorded by means of an optical vector analyzer from *Luna Technologies*, which was chosen for both its high measurement resolution (1.25 pm) and fast scanning rate (~1 s to cover the full TFBG spectrum). This analyzer gathers a narrowband

Fig. 5 Sketch of a gold-coated tilted fiber Bragg grating used for biosensing purposes which displays the functionalized photo-inscribed optical fiber inserted in a packaging

tunable laser source and a detector. It can work either in transmission or in reflection.

2. A linear polarizer was placed behind the optical source to control and orient the state of polarization (SOP) of the light launched into the TFBG. It is worth mentioning that care was taken to avoid polarization instabilities (use of short fiber lengths, strong curvatures avoided and ambient temperature kept constant to within 1 °C).

Figure 6 displays the transmitted amplitude spectrum of a gold-coated TFBG immersed in salted water (refractive index measured close to 1.356 at 589 nm), which was recorded with a linear input SOP optimized to maximize coupling to the SPR, corresponding to the radial polarization, as further explained in the following. This spectrum exhibits the typical SPR signature around 1551 nm, which is due to the maximum phase matching of the cladding mode to the surface plasmon mode of the gold water interface, according to [8]. The Bragg wavelength appears at the right end side, centered at 1602 nm. In the following, the Bragg wavelength is used to remove any effect from surrounding temperature changes by monitoring its shift. This intrinsic feature is very interesting in practice as a change of 0.1 °C induces an SRI change of 10^{-5}, thus susceptible to generate erroneous spectral modifications.

Figure 7 displays a zoom around the SPR signature for radially- (EH and TM modes) and azimuthally polarized (HE and TE modes) amplitude spectra that yield antagonist behaviors in liquids,

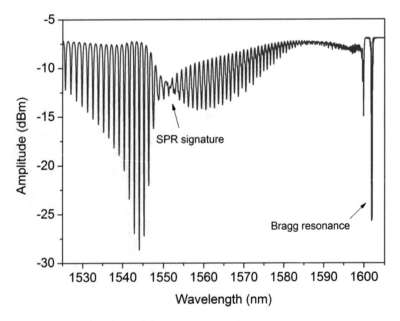

Fig. 6 Transmitted amplitude spectrum of an SPR-TFBG immersed in salted water (radial polarization—SRI = 1.358)

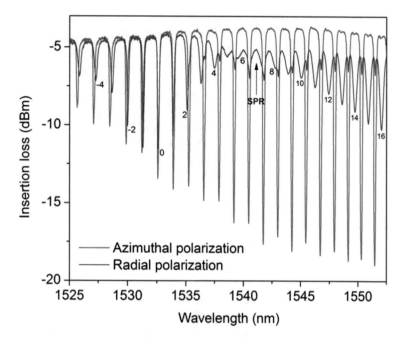

Fig. 7 Transmitted amplitude spectra for two orthogonal SOPs (radial and azimuthal polarizations) of a TFBG immersed in salted water (SRI = 1.338)

as explained in [8]. It is obvious from this figure that EH and HE modes come in pairs and to facilitate the following discussion, the mode resonances are labeled as a function of their position with respect to the most important one in the EH spectrum (mode 0 in our numbering).

Figure 6 reveals that, for short wavelengths (mode -2), the behavior is similar to that of bare TFBGs in air, with EH modes at a longer wavelength than HE ones. Then, getting closer to the SPR mode and for each cladding mode resonances pair, the EH mode wavelength increases less than the HE one. The crossing point occurs for the 0 mode. Past this mode, EH resonances appear on the short wavelength side of HE ones. This peculiar behavior results from the fact that EH modes begin to localize in the gold sheath as they approach the SPR, which lowers their effective refractive index (due to the small value of the gold refractive index). HE modes are tangentially polarized at the gold boundary and hardly penetrate it; therefore, they do not feel this effective index decrease.

Beyond the crossing, the $+2$ EH mode now lags significantly behind the HE mode, which points out to a strong localization of the EH mode field in the gold layer and the strong influence of the SPR (and hence of the SRI) on the mode resonance. The next modes show an even stronger lag of the EH mode but the associated resonance also becomes somewhat wider, because of the loss of energy to the metal. In the case of the $+4$ mode, the wavelength

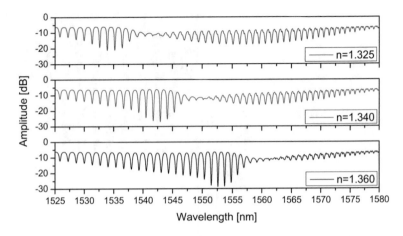

Fig. 8 SPR signature in the transmitted spectrum of gold-coated TFBG for coarse changes in SRI

separation becomes so great that the EH and HE modes are completely dissociated. And further analysis of the modes in the vicinity of the SPR becomes meaningless.

Gold-coated TFBGs were immersed in different refractive index liquids to measure their refractometric sensitivity. Figure 8 shows the transmitted amplitude spectra of a 50-nm gold-coated TFBG measured for three different refractive index values measured by a hand-held Abbe refractometer (*Reichert AR200* accurate to 10^{-4} RIU). For such large SRI changes, the SPR location can be unambiguously located by following the strongest attenuation in each spectrum. Therefore, the interrogation relies on the tracking of the wavelength shift of the center of the envelope of the most attenuated resonances. Using this technique, a linear response is obtained, as depicted in Fig. 9 and the SRI sensitivity is ~550 nm/RIU in the range between 1.32 and 1.42.

For high-resolution refractometric sensing over an SRI range limited to 10^{-3} typically, accurate measurements of the SPR mode are not possible from the radially polarized spectrum taken alone. Indeed, the SPP is only revealed by its absence from the spectrum and it cannot be reliably measured for wavelength shifts limited to a few picometers. Hence, different methods have been developed to track the SPR shift, mainly based on a comparison between both orthogonally polarized amplitude spectra as reported in our previous works [9, 10]. In practice, modes slightly off the SPR are used, because they combine relatively high sensitivity with a narrow spectral width and they can be "followed" by a combination of their changes in amplitude and wavelength. Indeed, as they stand on the shoulder of the SPR envelope, a slight change of the SPP location yields a modification of the peak-to-peak amplitude of these modes, as shown below.

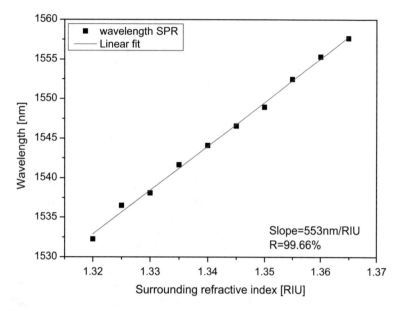

Fig. 9 SPR wavelength shift as a function of the SRI value

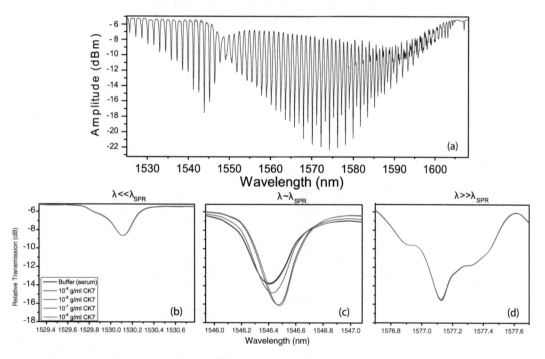

Fig. 10 Spectra measured during a biosensing experiment (**a**) with three spectral regions shown in detail: on the short wavelength side of the SPR (**b**); near the SPR (+2 mode) (**c**); and on the long wavelength side (**d**)

Such high-resolution sensing is demonstrated in Fig. 10 with spectra measured during a biochemical binding experiment, i.e., for an SPR shift close to the detection limit (a full report on these experiments can be found in [12]). By zooming in on resonances near the SPR and a few nm away, it becomes clear that the presence of a comb of resonances with widely different sensitivities allows for very small wavelength shifts to be detected unambiguously with high precision. Most often, the focus is made on the +2 mode among the spectral comb, as it appears to be the most sensitive in terms of both amplitude variation and wavelength shift.

Optical fiber devices have two important advantages over bulky prism SPR configurations: light propagates essentially without loss in short lengths of fibers, resulting in very high signal-to-noise ratios, and interfacing devices to light sources and detectors consists essentially of plugging connectors into widely available fiber optic instrumentation, instead of having to carefully align optical beams through imaging systems. In terms of experimentally demonstrated FOM, TFBGs exhibit a value reaching 5000, which surpasses all configurations by more than one order of magnitude [2]. As shown in Fig. 11, this results from the fact that they exhibit narrow resonance bands (FWHM ~0.1–0.2 nm) compared to even the best possible theoretical value (~5 nm obtained by calculating the reflection from the base of a prism in the Kretschmann-Raether configuration), also keeping in mind that the experimental SPR FWHM from other fiber configurations all exceed 20 nm and more.

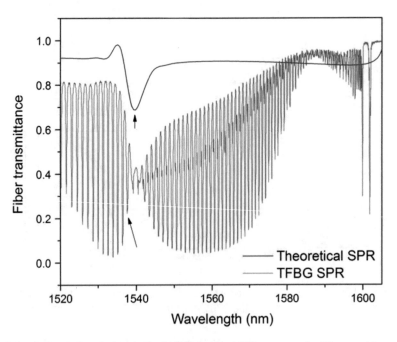

Fig. 11 Comparison between the best theoretical SPR response for 50 nm gold on silica in the Kretschmann-Raether configuration (*thick blue line*) and a measured TFBG-SPR spectrum with the same thickness of gold (*thin red line*). The arrows indicate the resonance to be followed in each case

3.6 Proteins and Cells Quantification with Gold-Coated TFBG Immunosensors

Up to now, and to the best of our knowledge, there have been a few reports where gold-coated TFBGs were used for biochemical sensing. In [23], the probe was associated with aptamers and a measurement of the dissociation constant was demonstrated. In [24], the probe was used to measure the intracellular density of non-physiological cells, namely human acute leukemia cells. In [25], the sensor was integrated into cell culture equipment and was used for real-time monitoring of cellular response to chemical stimuli obtained by adjunction of trypsin, serum, and sodium azide. The corresponding effects—detachment of cells from the surface, uptake of serum, and inhibition of cellular metabolism, respectively—were monitored through a shift of the SPR signature in the transmitted amplitude spectrum of gold-coated TFBGs.

In the following, we summarize our main achievements toward the demonstration of lab-on-fiber devices with gold-coated TFBGs suited for cancer diagnosis [26, 27].

3.7 Sensing Cytokeratins

A diverse range of tumor markers has been associated with lung cancer. In this study, we focused on cytokeratins 7 (CK7) that are particularly useful tools for diagnosis in oncology. In particular, CK7 profile of lung tumor has proved to be a useful aid in the differential diagnosis of carcinomas, since primary and metastatic tumors present different profiles. In fact, primary lung tumors express cytokeratin 7 (CK7+) while secondary tumors are deficient in CK7 (CK7−). Moreover, it has been demonstrated that cytokeratin fragments can be released from malignant cells and consequently CK fragments can be located in blood circulation and are therefore easily accessible with an optical fiber properly modified.

The cytokeratin 7 antigen detection by the bioreceptor is based on the specific chemical reaction with its corresponding antibody (AbCK7) previously immobilized on the optical fiber surface. The operating principle is based on the recognition of the cytokeratin 7 antigen epitope by the fragment antigen-binding (Fab), a region present on the cytokeratin 7 antibody. In addition to the full protein (CK7FP) detection made of 469 amino acids, corresponding to a large biomolecule, we proposed here to monitor a protein fragment called cytokeratin 7 peptide (CK7pep) made of only 23 amino acids to present a comparative study depending on the size of the target. In fact, the detection of small molecule antigens remains a challenge in the SPR immunosensing. Most of the SPR biosensors are based on large molecule detection since SPR response in the presence of small mass suffers from various disadvantages not encountered with full protein, such as low signal-to-noise ratio.

1. Detection of CK7 full protein.
   (a) Gold-coated TFBGs, preliminary functionalized with cytokeratin 7 antibodies, have been first connected to the LUNA and held straight between two clamps, for transmission recording (*see* **Note 4**).

(b) The first measurement consisted in the calibration of the signal. The gold-coated TFBG was immersed in PBS, and then the polarization controller allowed reaching the modes s and p, which were the references.

(c) Detection started with the immersion of the gold-coated TFBG in CK7 full protein solutions diluted in PBS. To reach the sensitivity of the biosensor, initial measurements were realized in a phosphate buffer with increasing CK7FP concentrations from $1E-12$ to $1E-6$ g/mL.

(d) Recording of the transmission spectrum during experiments put in evidence an evolution of the SPR-TFBG starting from $3E-12$ g/mL until it has reached a threshold at $2E-8$ g/mL, as illustrated in Fig. 12a. Such observations indicate that the biosensor was highly responsive since our sensitivity, estimated lower than 1 pM, is very good (*see* **Note 5**).

2. Detection of CK7 peptide.

(a) Gold-coated TFBG functionalized with AbCK7 was plugged in transmission on the LUNA coupled with the polarizer.

(b) Reference measurement (modes s and p) was performed in phosphate buffer solution at pH 7.4.

(c) Then gold-coated TFBG was immersed in CK7 peptide solutions at different concentrations, diluted in PBS at pH 7.4.

(d) The amplitude monitoring has highlighted a similar trend than the detection of CK7FP. Nevertheless, differences between CK7pep and CK7FP detection have been noticed

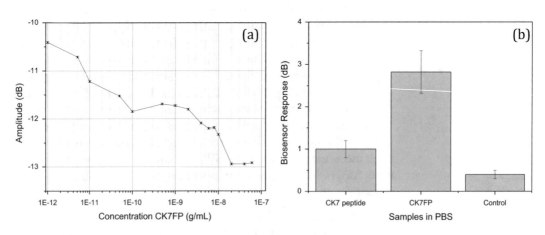

Fig. 12 Amplitude monitoring of the sensitive mode in the presence of CK7FP at different concentrations diluted in PBS (**a**). Biosensor response obtained for optical fiber functionalized with AbCK7, in the presence of CK7 peptide or CK7FP in comparison with OF functionalized without AbCK7 (**b**)

during experiment. First, instead of $3E-12$ g/mL, second the final amplitude measured between the maximal and the minimal threshold has been determined at $1.0 +/- 0.2$ dB against 2.8 dB for CK7FP, as illustrated in Fig. 12b $1E-9$ g/mL.

3. Detection of CK7 peptide in complex media.

   (a) The gold-coated optical fiber biosensor was performed on phosphate buffer complemented with 10% fetal bovine serum (FBS), resulting in a complex buffer full of proteins including albumin, enzymes, and others; as well as vitamins and nutrients. Objectives were to get close to physiological conditions on the one hand, and to guarantee the selectivity of the biosensor on the other hand. In the presence of FBS in the media, the optical fiber ended up with numerous compounds able to attach on the surface.

   (b) The functionalized gold-coated TFBG was tested on CK7 peptide solution diluted in PBS complemented with 10% FBS, and the SPR-TFBG spectra were recorded in transmission.

   (c) The amplitude monitoring, as presented in Fig. 13a, clearly shows the evolution of SPR-TFBG signal as soon as the CK7pep was injected until it has reached a minimum threshold at $1E-7$ g/mL. Figure 13b presents the average output and errors for a three-fold repeated CK7pep detection on distinct optical fibers prepared on the exact same conditions. All three fibers have presented the identical amplitude evolution depending on the CK7 concentration, including an amplitude shift from $1E-9$ g/mL, which identifies that the lowest concentration detected is estimated at 0.4 nM.

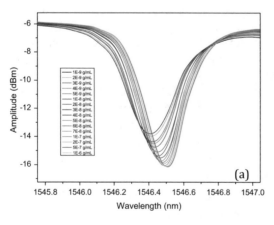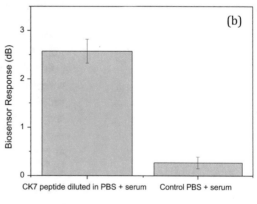

Fig. 13 Shift of the sensitive mode during incubation of the optical fiber functionalized with AbCK7 on CK7 peptide at different concentrations diluted in PBS + 10% FBS (**a**). Biosensor response obtained for optical fiber functionalized with AbCK7 immersed in PBS + 10% FBS, in the presence or in the absence of CK7 peptide (**b**)

3.8 Sensing Transmembrane Receptors

In this section, we report on the use of gold-coated TFBGs for selective cellular detection through membrane protein targeting. The focus is made on the epithelial growth factor receptor (EGFR), which is a transmembrane receptor from the 4-tyrosine kinase receptors family. It is an important biomarker and therapeutic target that it is over-expressed by numerous cancer cells.

1. Surface functionalization.
   The sensor surface selectivity was ensured by bio-functionalization through a two-step approach.

   (a) First, the clean waveguide surface was activated with the Carboxylic acid-SAM formation reagent (cat. n: C488) obtained from *Dojindo* (Japan).

   (b) Then, monoclonal mouse immunoglobulin G (IgG) raised against the human epidermoid carcinoma cell line (*Santa Cruz Biotechnology Inc.* and *American Type Culture Collection* (USA)) were immobilized on the surface through carbodiimide covalent biochemistry using *Dojindo*'s amine coupling kit (cat. n: A515-10). This antibody was diluted in the *Dojindo*'s reaction buffer to 0.01 μg/mL for efficient covalent immobilization during 30 min. The manufacturer's instructions were followed for each step for both processes.

2. EGFR detection by SPR-TFBG.
   Two cell lines were used: the first one with overexpressed human epidermal growth factor receptors (EGFRs)—A431 cell line, (EGFR positive, *EGFR (+)*). The second cell line was EGFR negative – OCM1 cell line, (*EGFR (−)*).

   (a) The culture process was done at 37 °C in a humidified incubator with 5% CO_2 atmosphere. No antibiotic was used and cell cultures were free of mycoplasma and pathogenic viruses. Cells suspensions were then prepared for SPR experiments. For this, A431 and OCM1 cells from a confluent monolayer were detached mechanically by a gentle scraping of cells from the growth surface into ice-cold phosphate buffered saline (PBS). Cells were then washed three times in cold PBS by repeated centrifugation (at 2500 RPM and 4 °C for 3 min). Pellets were then again suspended to a concentration of 2–5 × 10^6 cells/mL. 0.5 mL volumes were used for experiments. A handheld automated cell counter from Millipore (Scepter 2.0, PHCC00000) was used to count the cells.

   (b) For immunosensing experiments, gold-coated TFBGs were hold straight between two clamps and were then approached toward the cells suspensions, using a vertical translation stage. To demonstrate that a differential response can be obtained between both cell lines, TFBGs were first immersed in RPMI-1640 cell culture media (*Roswell Park Memorial Institute* (RPMI) 1640, supplemented with 10% FBS—refractive index = 1.3366), until a stable baseline was reached.

(c) In the first step, sensors were incubated in a suspension of OCM1 during ~10 min. After this assay, a secondary baseline of the media (RPMI-1640) was recorded.

(d) In the second step, sensors were incubated in a suspension of A431 during the same period. Finally, the third media baseline was recorded before, and after an extra rinsing step with the media.

(e) The conditioning in media was 5 min and assays lasted 10 min. Figure 14 shows the processed data (evolution of the peak-to-peak amplitude of the +2 mode) obtained after such binding experiments. It clearly figures out that an important response change is obtained with the A431 cell line (here at a concentration of 3×10^6 cells/mL), for which EGFRs are overexpressed. This is not the case for the other cell line for which the response remains comparable to the background noise level obtained in RPMI-1640 media. Error bars are the standard deviation obtained for three experiments made in the same conditions.

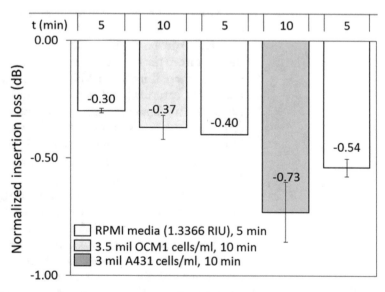

Fig. 14 Different assays presenting specific cell interactions. *Dark gray* bars represent A431 cells, *light gray bars* represent OCM1 cells. *Blank bars* are washing and rinsing steps for RPMI-1640 cell culture media supplemented with FBS

Table 1
Sensor response as a function of the cells concentration, bringing an estimate about the LOD

Cells conc. (10^6 ml1)	Response (%)
0.5	N/A
1.1	N/A
2.0	6.6
2.9	20.0
3.5	23.3
5.1	90.0

(f) Different cells suspensions were used with various concentrations to estimate the LOD of the sensors. Table 1 summarizes the obtained results in terms of sensor response, which is here computed as the relative amplitude change of the +2 mode with respect to its value measured in the reference solution (RPMI medium). It shows that the LOD is ~2×10^6 cells/mL, which is relevant with respect to the target application. Obviously, the sensor response is not linear with the cells concentration. The saturation has not been properly measured but it is beyond ~5×10^6 cells/mL.

Optical fiber-based biochemical sensors fill an increasingly well-defined niche as they do not require elaborate light management schemes to probe molecules and materials. Light remains guided in the fiber from the source to the detector, apart from localized probe regions where it interacts with the immediate surroundings of the fiber. This feature is the dominant factor in making such sensors less expensive to fabricate and to use than conventional biosensors (especially those based on SPR and resonant waveguide gratings). However, this comes with two main inconveniences: a higher (worse) limit of detection and the impossibility to scale up toward massively parallel testing.

Our work features a robust biosensing platform with a simple four-step protocol enabling label-free cells sensing. It addresses the first issue by combining SPR effects with a grating-based approach in a single sensor that keeps all the advantages of the fiber solution. TFBGs provide a dense spectral comb of high Q-factor resonances that probe the SPR envelope and allow an order of magnitude improvement in the determination of the SPR shifts under biochemical interactions at the fiber surface. We have shown that we can separately excite resonances that enable SPR generation and others that do not (which can thus serve as reference channels).

The sensor also provides absolute temperature information, from the wavelength position of Bragg core resonance that does not sense changes of the surrounding refractive index medium. This additional information can be used to average out noise and to eliminate many cross-sensitivity factors. We have also demonstrated that our biosensing platform can be used to investigate the binding of living human epithelial cells through specific interaction of transmembrane proteins and target extracellular membrane receptors in biological matrices.

The strength of our platform arises from its near-infrared operating wavelength, thus allowing improved penetration depth and full compatibility with telecommunication-grade equipment, including switch matrix devices that allow automatic interrogation of several tens of optical fibers in relatively rapid succession. Hence, different fibers coated with different bioreceptors could be easily interrogated to assay several parameters simultaneously, which would solve the second aforementioned limitation. Additionally for practical use, it could considerably enhance the reliability of a diagnosis.

Much of the development required for the exact quantification and deployment of practical sensors based on gold-coated TFBGs is still in progress. Therefore, it is hoped that the achievements presented herein will stimulate further research in this area.

4   Notes

1. Sulfur atoms of the SAM attach within minutes on the gold substrate; nevertheless, carbon chain requires hours to be arranged via van der Waals forces, in the aim to form a packed and stable monolayer.

2. Polyclonal antibodies were used instead of monoclonal to improve the protein detection, since polyclonal antibodies react with multiple epitopes of the protein.

3. SPR-TFBG experiments were realized in a constant volume (500 μL) in a constant temperature.

4. Recording of the amplitude spectra was realized after the stabilization of the signal.

5. Proteins detection measurements require at least triplicates experiments to obtain statistics. Column bars in Figs. 11 and 12 represent the average of the amplitude shift while the error bars correspond to the standard deviation of three experiments realized in the same conditions.

References

1. Kretschmann E, Raether H (1968) Radiative decay of non radiative surface plasmon excited by light. Z Naturforsch 23:2135

2. Caucheteur C, Guo T, Albert J (2015) Review of recent plasmonic fiber optic biochemical sensors: improving the limit of detection. Anal Bioanal Chem 407:3883

3. Pollet J, Delport F, Janssen KPF, Jans K, Maes G, Pfeiffer H, Wevers M, Lammertyn J (2009) Fiber optic SPR biosensing of DNA hybridization and DNA-protein interactions. Biosens Bioelectron 25:864–869

4. Baldini F, Brenci M, Chiavaioli F, Gianetti A, Trono C (2012) Optical fiber gratings as tools for chemical and biochemical sensing. Anal Bioanal Chem 402:109–116

5. Albert J, Shao L-Y, Caucheteur C (2013) Tilted fiber Bragg grating sensors. Laser Photon Rev 7:83–108

6. Sharma AK, Rajan J, Gupta BD (2007) Fiber-optic sensors based on surface plasmon resonance: a comprehensive review. IEEE Sensors J 7:1118–1129

7. Offermans P, Shaafsma MC, Rodriguez SRK, Zhang Y, Crego-Calama M, Brongersma SH, Rivas JG (2011) Universal scaling of the figure of merit of plasmonic sensors. ACS Nano 5:5151–5157

8. Shevchenko YY, Albert J (2007) Plasmon resonances in gold-coated tilted fiber Bragg gratings. Opt Lett 32:211–213

9. Caucheteur C, Shevchenko YY, Shao L-Y, Wuilpart M, Albert J (2011) High resolution interrogation of tilted fiber grating SPR sensors from polarization properties measurement. Opt Express 19:1656–1664

10. Caucheteur C, Voisin V, Albert J (2013) Polarized spectral combs probe optical fiber surface plasmons. Opt Express 21:3055–3066

11. Albert J, Lepinay S, Caucheteur C, DeRosa MC (2013) High resolution grating-assisted surface plasmon resonance fiber optic aptasensor. Methods 63:239–254

12. Voisin V, Pilate J, Damman P, Mégret P, Caucheteur C (2014) Highly sensitive detection of molecular interactions with plasmonic optical fiber grating sensors. Biosens Bioelectron 51:249–254

13. Othonos A, Kalli K (1999) Fiber Bragg gratings: fundamentals and applications in telecommunications and sensing. Artech House, London

14. Erdogan T (1997) Fiber grating spectra. J Lightwave Technol 15:1277–1294

15. Erdogan T, Sipe JE (1996) Tilted fiber phase gratings. JOSA A 13:296–313

16. Laffont G, Ferdinand P (2001) Tilted short-period fiber Bragg grating induced coupling to cladding modes for accurate refractometry. Meas Sci Technol 12:765–772

17. Caucheteur C, Mégret P (2005) Demodulation technique for weakly tilted fiber Bragg grating refractometer. Photon Technol Lett 17:2703–2705

18. Chan CF, Chen C, Jafari A, Laronche A, Thomson DJ, Albert J (2007) Optical fiber refractometer using narrowband cladding-mode resonance shifts. Appl Optics 46:1142–1149

19. Chen C, Albert J (2006) Stain-optic coefficients of individual cladding modes of single mode fibre: theory and experiment. Electron Lett 42:1027–1028

20. White IM, Fan XD (2008) On the performance quantification of resonant refractive index sensors. Opt Express 16:1020–1028

21. Shevchenko YY, Chen C, Dakka MA, Albert J (2010) Polarization selective grating excitation of plasmons in cylindrical optical fibers. Opt Lett 35:637–639

22. Caucheteur C, Chen C, Voisin V, Berini P, Albert J (2011) A thin metal sheath lifts the EH to HE degeneracy in the cladding mode refractometric sensitivity of optical fiber sensors. Appl Phys Lett 99:041118

23. Shevchenko Y, Francis TJ, Blair DAD, Walsh R, DeRosa MC, Albert J (2011) In situ biosensing with a surface plasmon resonance fiber grating aptasensor. Anal Chem 83:7027–7034

24. Guo T, Liu F, Liu Y, Chen NK, Guan BO, Albert J (2014) In situ detection of density alteration in non-physiological cells with polarimetric tilted fiber grating sensors. Biosens Bioelectron 55:452–458

25. Shevchenko Y, Camci-Unal G, Cuttica DF, Dokmeci MR, Albert J, Khademhosseini A (2014) Surface plasmon resonance fiber sensor for real-time and label-free monitoring of cellular behavior. Biosens Bioelectron 56:359–367

26. Ribaut C, Voisin V, Malachovska V, Dubois V, Mégret P, Wattiez R, Caucheteur C (2016) Small biomolecule immunosensing with plasmonic optical fiber grating sensor. Biosens Bioelectron 77:315–322. doi:10.1016/j.bios. 2015.09.019

27. Malachovska V, Ribaut C, Voisin V, Surin M, Leclère P, Wattiez R, Caucheteur C (2015) Fiber-optic SPR immunosensors tailored to target epithetial cells through membrane receptors. Anal Chem 87:5957–5965

Chapter 5

Biosensing Based on Magneto-Optical Surface Plasmon Resonance

Sorin David, Cristina Polonschii, Mihaela Gheorghiu, Dumitru Bratu, and Eugen Gheorghiu

Abstract

In spite of the high analytic potential of Magneto Optical Surface Plasmon Resonance (MOSPR) assays, their applicability to biosensing has been limited due to significant chip stability issues. We present novel solutions to surpass current limitations of MOSPR sensing assays, based on innovative chip structure, tailored measurements and improved data analysis methods. The structure of the chip is modified to contain a thin layer of Co-Au alloy instead of successive layers of homogenous metals with magnetic and plasmonic properties, as currently used. This new approach presents improved plasmonic and magnetic properties, yet a structural stability similar to standard Au-SPR chips, allowing for bioaffinity assays in saline solutions. Moreover, using a custom-designed measurement configuration that allows the acquisition of the SPR curve, i.e., the reflectivity measured at multiple angles of incidence, instead of the reflectivity value at a single-incidence angle, a high signal-to-noise ratio is achieved, suitable for detection of minute analyte concentrations. The proposed structure of the MOSPR sensing chip and the procedure of data analysis allow for long time assessment in liquid media, a significant advancement over existing MOSPR chips, and confirm the MOSPR increased sensitivity over standard SPR analyses.

Key words Magneto-optic surface plasmon resonance, Magnetic alloys, Surface plasmon resonance enhancement, Affinity biosensor, Angle-resolved surface plasmon resonance, Fixed-angle surface plasmon resonance

1 Introduction

Surface Plasmon Resonance is a phenomenon that appears when a plane-polarized light is coupled, under total internal reflection (TIR) conditions, into a metal film at the interface between two media. The TIR conditions are met when the incident light beam intersects the interface between two adjacent media that have different refractive indices ($n_1 > n_2$) at an angle θ greater than the critical angle θ_c defined by Eq. 1 derived from the Snell law,

Avraham Rasooly and Ben Prickril (eds.), *Biosensors and Biodetection: Methods and Protocols Volume 1: Optical-Based Detectors*, Methods in Molecular Biology, vol. 1571, DOI 10.1007/978-1-4939-6848-0_5,
© Springer Science+Business Media LLC 2017

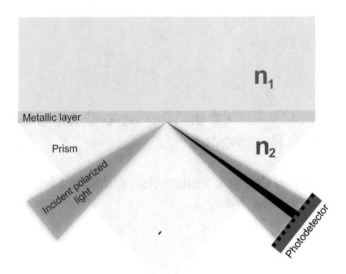

Fig. 1 Principle of surface plasmon resonance in Kretschmann configuration

$$\theta_c = \frac{n_2}{n_1} \qquad (1)$$

where n_2 is the refractive index of the upper medium and n_1 is the refractive index of the lower medium and the light travels from n_1 toward n_2.

The coupling of the light into the metal film is done via a grating (Otto configuration) or a prism (Kretschmann configuration) [1]. In the following, only the Kretschmann configuration is considered.

Depending on the value of the incident light angle θ, the type of the plasmonic metal, and the refractive indices of the two media, the light interacts with the free electrons of the metal film and the energy of the photons is transferred resonantly to these electrons. A drop in the intensity of the reflected light is recorded by a suitable detector (e.g., a photodetector array) placed in the path of the reflected beam—Fig. 1 corresponding to the specific SPR angle. The dependency of the intensity of the reflected light over the incidence angle generates the SPR curve that has a dip minimum corresponding to the SPR angle, θ_{min} (Fig. 1). According to Eq. 2 [2], the SPR angle depends on λ (the wavelength of the incident light), via n^*_m, the complex refractive index of the plasmonic layer and the values of the refractive indices of the media n_1 and n_2.

$$\sin\theta_{min} = \sqrt{\frac{n_m^{*2} \times n_2^2}{(n_m^{*2} + n_2^2)n_1^2}}; \, n_m^* = n_m + i \times k_m. \qquad (2)$$

Considering n^*_m and n_1 constant (as part of the measuring setup, for a given wavelength of the incident light), the SPR angle depends exclusively on n_2—the refractive index of the upper medium. As such, due to their high sensitivity to local changes in the refractive index of the upper media [1], SPR biosensing assays are currently among the methods of choice to assess bioaffinity interactions at interfaces. For example, the minimum resolvable refractive index change for the SPREETA sensor (Texas Instruments Inc.) used in this chapter is 1×10^{-6} RIU (refractive index units) [3].

Nevertheless, there is a continuous strive to improve the sensitivity of the existing SPR analysis. Possible avenues for improvement include optimization of the plasmonic layer into a multilayered structure, improvement in the sensitivity of photodetectors [1], or signal modulation [4, 5]. Promising signal modulation approaches are based on the magneto plasmonic effects [6–8] by employing the transverse magneto-optic Kerr effect (TMOKE). The TMOKE occurs as a result of applying a magnetic field perpendicular to the propagation plane of the incident p-polarized light versus a surface with magneto-optical properties. This leads to modulation of the wave vector resulting in changes in the intensity of the reflected light [9]. The process lays the foundation of the novel magneto-optical SPR (MOSPR) method [10].

In the MOSPR approach the modulation of the reflectivity (i.e., SPR) curve is accomplished by applying an alternating transversal magnetic field, perpendicular to the propagation plane of an p-polarized beam of light [9] incident via a prism coupler, onto a sensor chip exhibiting both magnetic and plasmonic properties (so-called MOSPR chip [4]). MOSPR response measured at a single angle of incidence yields improved sensitivity to the refractive index changes of up to two orders of magnitude, compared to classical SPR assays, as substantiated by both theoretical and experimental studies [2, 9–12].

Important drawbacks of the technique are given by the technologically challenging MOSPR chip fabrication involving metallic multilayers or nanostructures (e.g., nanodiscs, nanocrystals) [6–8] and the intrinsic magnetostrictive effect [5] that causes ferromagnetic materials to change their shape or dimensions during the process of magnetization and adds further stability concerns. Indeed, this complex, multilayered structure of the MOSPR sensing chips is prone to induce a lack of stability when chips are exposed to (saline) liquid samples and, as a result, current highly sensitive MOSPR assays are mostly gas sensors [10, 12, 13].

To overcome this barrier, a multilayered chip comprising a thin film of amorphous Au-Co alloy, capped with a layer of Au (to allow further functionalization) with optimized structure for improved magneto-plasmonic properties, provides a powerful alternative. This structure, recently proposed, exhibits both plasmonic and magnetic properties, and proves excellent stability in (saline) liquid media [14].

In a further improvement, we used the optimized MOSPR structure in an *angle-resolved* MOSPR bioassay (that allows the evaluation of the reflectivity curve in a wider angular range, instead of the reflectivity at a single (fixed) angle of incidence as employed in most of the current SPR/MOSPR experiments [13]) and confirmed experimentally both increased sensitivity in comparison with the current SPR/MOSPR approaches and a high stability (similar to the one achieved with classic, gold-only, SPR chips) when addressing saline liquid samples.

2   Materials

2.1   Surface Chemistry

1. Bovine serum albumin (BSA), human IgG (HIgG), affinity isolated anti-human IgG (AHIgG), N-hydroxysuccinimide (NHS), 1-ethyl-3-(dimethylaminopropyl) carbodiimide (EDC), ethilenediaminetetraacetic acid (EDTA), and ethanolamine were purchased from Sigma–Aldrich (Germany).

2. (1-mercapto-11-undecyl)hexa(ethylene glycol)carboxylic acid was purchased from Prochimia Surfaces, Poland (*see* **Note 1**).

3. Surfactant P20 was provided by GE Healthcare.

4. Matching oil Immersol 518 F, $n_e = 1.518$ from Carl Zeiss.

5. *Aqua regia* HCl:HNO$_3$ 3:1 mixture (*see* **Note 2**).

6. *piranha* solution H$_2$SO$_4$:H$_2$O$_2$ 3:2 mixture (*see* **Note 3**).

2.2   Preparation of Magneto-Optical Sensor Chip

1. Au pellets (99.999%), Co (99.99%) pellets, and Ti sputter target (grade 2) were purchased from Kurt J. Lesker USA.

2. N-BK7 glass wafers were purchased from Sydor Optics.

3. Kurt J. Lesker PVD 75 system with substrate heater, thermal evaporation sources, and reactive sputtering source for metallic layer deposition.

4. PSD–UV–plasma cleaner from Novascan Technologies, Inc.

2.3   Working Solutions

1. immobilization buffer: 10 mM acetate buffer pH 5.5.

2. running buffer: HBS-EP buffer (10 mM HEPES, pH 7.4, 150 mM NaCl, 3 mM EDTA, 0.005% Surfactant P20).

Ultrapure water (Millipore) was used throughout the preparations.

2.4   Measurement Setup

1. Fluidic assembly.

2. SPR module based on SPREETA sensor model TSPR1K23 and custom data acquisition board with appropriate software.

3. Electromagnet with power supply for applying the oscillating magnetic field.

4. PC for data saving and processing.

2.5 Fluidics Assembly

1. Injection pumps—Cavro XLP 600 Tecan Inc.
2. PTFE tubes ID 0.3 mm and fittings from Upchurch Scientific.
3. Polyether ether ketone (PEEK) slab micromachined as flow chamber.
4. PDMS silicone—Sylgard 184 from DowCorning as gasket between the flow chamber and the (MO)SPR chips.

3 Methods

3.1 Fabrication of MOSPR Chips

1. Substrates of BK7 glass (0.3 mm × 4 mm × 11 mm) are cleaned by successive sonication in water, acetone, and isopropanol and dried in a stream of nitrogen.
2. The substrates are exposed to ozone plasma for 10 min using a PSD-UV plasma cleaner.
3. Substrates are loaded in the pressure chamber of the PVD 75 system.
4. The pressure chamber is pumped to a pressure of approximately 2×10^{-6} Torr.
5. Samples are heated to 150 °C for degassing for 30 min using the substrate heating facility of the PVD 75.
6. Samples were rotated at 10 rot/min to ensure uniform coating. All the deposition processes were done at room temperature.
7. Thickness and deposition rate of the deposited layers are constantly monitored by a quartz crystal microbalance (QCM) inside the evaporation chamber of the PVD 75. The QCM works by measuring the changes in the oscillation frequency of a quartz crystal as a result of adsorption of materials on the crystal surface. The PVD 75 QCM is calibrated in terms of adsorbed material (taking into account the elastic properties of the material) and position in relation to the substrate.
8. A layer of Ti (4 nm) is first sputtered on the glass substrates for improving the adhesion of subsequent layers, at 5×10^{-3} Torr Ar pressure and 8 W/cm^2 (*see* **Note 4**).
9. Standard SPR chips are fabricated by evaporating 50 nm of Au onto the Ti modified substrates (*see* **Note 5**).
10. The Au and Co layers are thermally evaporated at rates of 0.5 Å/s and 3 Å/s respectively (*see* **Note 6**).
11. The Au-Co-Au tri-layers consist in a 26 nm Au, 6 nm Co layer, separated from the sensing surface by a 12 nm Au layer. An intermediate layer of 2 nm Ti is inserted between Au and Co to improve adhesion of the two metals.
12. The Au-Co alloy chips are fabricated by evaporating 30 nm layer of alloy Au-Co followed by 15 nm of Au. No intermediate Ti layer is used.

13. The Au-Co alloy is thermally evaporated at a rate of 0.5 Å/s from a single tungsten boat containing corresponding quantities of Au and Co to yield an alloy of 10% Co—volumetric ratio (588 mg Au and 27 mg Co). Alternative techniques for producing thin films of magnetic materials and alloys do exist and may be used for a better control of the film composition and structure. However, they have limitations in respect to costs (*see* **Note 7**), or material type (*see* **Note 8**) .

3.2 Chips Functionalization and Bioaffinity Assays

1. (MO)SPR are cleaned by immersion in *piranha* solution for 10 min and washed with deionized water and ethanol.

2. Cleaned (MO)SPR chips are functionalized by immersion in ethanol solution containing 1 mM of 11-PEG-COOH thiol for 3 h.

3. Functionalized chips are mounted in the flow chamber of the measurement setup and further modified with the ligand (HIgG) by amine coupling [15]:

   (a) A mixture of 100 mM NHS and 400 mM EDC is passed over the chip for 7 min to activate the carboxylic groups of the thiol.

   (b) The ligand (30 μg/ml) is injected for 30 min in 10 mM acetate buffer at pH 5.5 to promote electrostatic pre-concentration.

4. The modified surface is deactivated for 7 min with 1 M ethanolamine pH 8.6.

5. The deactivated surface is blocked and tested for nonspecific adsorption by injecting a solution of 1 mg/ml BSA in HBS-EP (10 min).

6. All the immobilization procedure is performed at a flow rate of 30 μl/min with HBS-EP as running buffer.

3.3 (MO)SPR Measurements and Data Processing

For classical SPR measurements the response is derived from the SPR curve as a function the angle for which the minimum of the reflectivity is recorded [1]. The reflectivity (R_{pp}) is obtained by normalizing the intensity of the measured reflected p-polarized light versus a reference signal (e.g., obtained using an s-polarized light or a solution with an SPR dip outside the measurement range of the device). As pinpointed by Eq. 2, the SPR angle (θ_{min}), depends on the refractive index of the media in the immediate vicinity of the sensing surface [1] and is determined most precisely if the reflectivity dip is the sharpest. Changes at the sensing surface, accompanied by a modification of the refractive index, cause a shift in the SPR angle that is used for label-free assessment of biomolecular surface interactions. This is the preferred mode of operation for real-time measurements having a linear dependence on the refractive index changes and with the amount of molecules accumulated

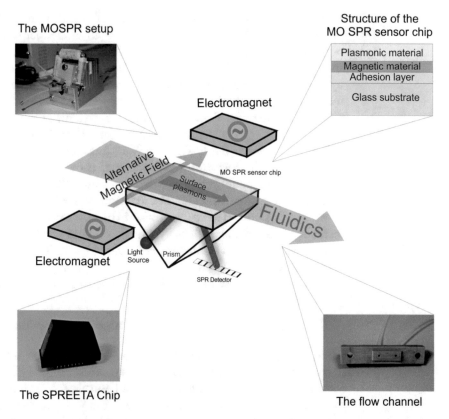

Fig. 2 The magneto optical SPR concept and schematic setup: electromagnet for transverse oscillating magnetic field application; tailored chips with magneto-plasmonic layers; top microfluidics; all system components were custom developed and built around an integrated illumination and detection Kretschman configuration as provided by a Spreeta module

on the surface [1]. Based on the full curve analysis, various data processing methods were developed to improve sensitivity, robustness against noise [16] and foster reduced time of analysis [17].

Figure 2 outlines the MOSPR concept. In brief, the measurement setup, described in more detail elsewhere [18], comprises:

- A SPR module.
- An external electromagnet with power supply for applying the alternating transverse magnetic field.
- A Flow Injection Analysis (FIA) type setup as fluidic assembly.

The SPR measurement module is based on a SPREETA sensor model TSPR1K23 (Texas Instruments USA). The FIA comprises a flow channel connected to an injection pump using PTFE tubes. The flow channel is made by micromachining a polyether ether ketone (PEEK) slab and a gasket of PDMS (Fig. 2). The data acquisition board is designed to give online access to whole curve measurements to allow calculation/extraction of the MOSPR data.

In the Kretschmann configuration [2] of the SPR module, the backside of the chip is illuminated using an embedded divergent p-polarized beam of light in a continuous range of incidence angles, spanning the 61–73° angular domain. The reflected light is assessed by a linear photo detector array comprising 128 pixels, each pixel measuring the intensity of the reflected light corresponding to a different angle of incidence. This allows for virtually simultaneous acquisition of angle specific reflectivity data, the whole SPR curve being measured in ~5 ms. The wavelength used is in the near infra-red region of the spectrum (840 nm) and provides a rather sharp dip and large penetration depths for standard SPR Au chips measured in aqueous media. The relation between the individual pixels of the photo detector array and the corresponding angle of incidence θ, important especially when comparing SPR experimental data with numerical simulations provided by the Transfer Matrix approach, can be derived using the manufacturing parameters of the sensor. In the case of the TSPR1K23 sensor geometry, we find [14, 19]:

$$\theta = U_{max} - (U_{max} - U_{min})/\text{PixelRange} \times \text{PixelPosition}, \quad (3)$$

where $U_{max} = 73.427$ and $U_{min} = 60.425$ are, respectively, the maximum and the minimum angle achieved in the SPREETA sensor, $PixelRange = 128$ is the total number of pixels of the photodetector array and $PixelPosition$ is the value of the position of the individual pixel on the detector array, for which the angle θ is calculated.

For attaching tailored sensor chips to the SPREETA prism, a refractive index matching oil is used on the glass window of the sensor priory cleaned with *aqua regia* [20].

Calibration measurements are performed for each individual chip and relate the SPR responses (i.e., the variations of the measured SPR angle) to the effective refractive index values and involve assessment of water and glycerol solutions of various concentrations with known refractive index. The calibration process comprises injections in the measurement flow channel of solutions of known refractive index. The SPR response is recorded and the SPR angle is related to the refractive index of the calibration solution. Using several solutions with different refractive indices in the domain of interest, one can derive specific calibration curves for the SPR angle as a function of the refractive index of the solution in the flow channel. Further, the SPR analytic responses may be expressed as refractive index variations versus time (i.e., sensorgrams) or as concentrations of analyte (i.e., calibration curves).

MOSPR measurements are performed using tailored MOSPR sensor chips and an electromagnet placed transversal on the outside of the measurement chamber actuated with a sinusoidal voltage. The voltage is applied using a custom-built alternative current source with a frequency domain of 0.1–100 Hz and a maximum current of 1 A. Field strengths of up to 5 mT (demonstrated as being sufficient for saturation of the magneto-optical chips [14]) were achieved using 120 mA, at the frequency of 0.3 Hz, current for electromagnet actuation.

In the typical experiment, reflectivity values are recorded when there is no magnetic field (R_{pp}) as well as for each direction of magnetization ($M+$) and ($M-$) when electromagnet actuation is performed.

Applying the alternating magnetic field did not create significant heating of the electromagnet nor had an effect over the SPR measurement on a classical gold surface, proving that there is no crosstalk between the magnetic field and the electronic and optic components of the SPREETA chip.

3.3.1 Extracting the MOSPR Data

The MOSPR signal is calculated from the reflectivity curves recorded in different states of magnetization of the MOSPR chip and is defined as $\Delta R_{pp}/R_{pp}$ [11]. Here, R_{pp} is the reflectivity (standard for SPR assays), obtained when not applying a magnetic field and ΔR_{pp} represents the difference between the reflectivity measured at one direction of magnetization ($M+$) and the reflectivity measured at the opposite direction of magnetization ($M-$). Figure 3a indicates the reflectivity curves in respect with magnetization as well as the corresponding ΔR_{pp}, R_{pp}, $\Delta\theta$ (the difference between the SPR angle measured at $M+$ and $M-$ respectively) and θ_{min} (the SPR angle measured with no magnetic field) data.

The MOSPR response is derived from the reflectivity values corresponding to a selected angle of incidence, as the variation of the reflectivity R_{pp} while switching the direction of the *in plane* magnetization of the sensor [10]:

$$\frac{\Delta R_{pp}}{R_{pp}(0)} = \frac{R_{pp}(M+) - R_{pp}(M-)}{R_{pp}(0)} \tag{4}$$

where $R_{pp}(M+)$ and $R_{pp}(M-)$ represent the reflectivities for the direction of the magnetization perpendicular to the propagation plane of the incident p-polarized light in both ways and $R_{pp}(0)$ represents the reflectivity without magnetization. In real measurements, ΔR_{pp} is obtained from the amplitude of the oscillations of the measured reflectivity (Fig. 3a). The MOSPR curve is defined as the variation of the $\Delta R_{pp}/R_{pp}$ versus the angle of incidence of the light.

As shown in Fig. 3b, for noise-free data a change in the refractive index of the media that yields a small reflectivity change in the SPR signal, determines a larger change in MOSPR signal. To this effect, an intermediary smoothing step is applied to the raw data.

3.3.2 Fitting Function of SPR Data

Reflectivity data for the entire angular range (SPR curves) are collected and fitted with a rational polynomial equation to achieve smoothing.

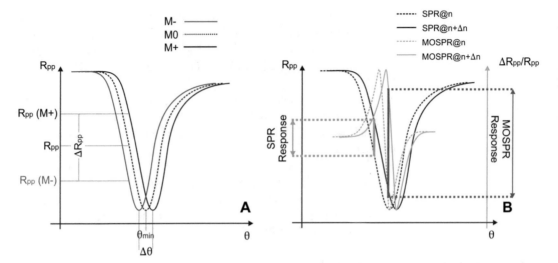

Fig. 3 (**a**) Representative parameters of the measured SPR curves used to derive the MOSPR data: R_{pp} the reflectivity value for a selected incidence angle; ΔR_{pp} corresponds to reflectivity differences between M+ and M− magnetization states, $\Delta\theta$ (the difference between the SPR angles for M+ and M− magnetization states) and θ_{min} (the SPR angle measured with no magnetic field). (**b**) Schematic plot showing ideal SPR (*black*) and MOSPR (*blue*) curves, as well as their variations corresponding to refractive index n (*dotted lines*) and $n + \Delta n$ (*continuous lines*), respectively

For fitting the experimental reflectivity data the following function is proposed [14, 21]:

$$f(x) = \frac{a_0 + a_1 x + a_2 x^2 + a_3 x^3 + a_4 x^4 + a_5 x^5 + a_6 x^6}{b_2 x^4 + b_1 x^2 + 1} \qquad (5)$$

Here, x denotes the pixel position value and is related to the angle of incidence via Eq. 3, and $a_0 \div a_6$, b_1 and b_2 are the polynomial coefficients of the fitting function.

The good match of the rational polynomial function to the theoretical curve is presented in Fig. 4. A theoretical SPR curve is generated using the Transfer Matrix approach considering the literature optical constants of the standard structure comprising a glass prism (BK7 $\varepsilon'=2.2801$), covered with 4 nm Ti adhesion layer ($\varepsilon'= -1.7982$; $\varepsilon''= 20.0613$) and 50 nm gold layer ($\varepsilon' = -30.2270$; $\varepsilon''=2.1235$) corresponding to $\lambda = 840$ nm. The theoretical curve is sampled to 128 points similar to the data received from the SPR detector (128 pixels). A $\chi^2 = 8.9 \times 10^{-9}$ residual error has been obtained for the fit, proving relatively insensitive to noise.

The fitted SPR curves provide corrected reflectivity values for each incident angle and are used to calculate the MOSPR curves. These smoothened MOSPR curves are used for further processing.

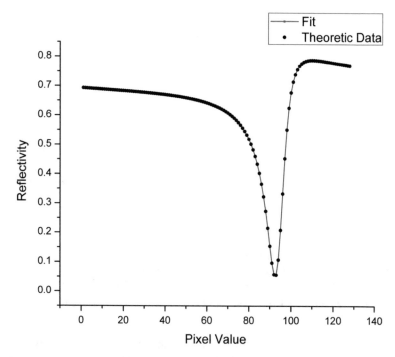

Fig. 4 Theoretical SPR curve for 50 nm layer of gold on glass with water on top (*black dotted line*) and the fit with the phenomenological, polynomial function (*red line*)

3.3.3 Sensitivity Comparison of MOSPR and SPR Sensors

The MOSPR responses are compared with SPR responses in terms of sensitivity and signal-to-noise ratio when analyzing the reflectivity at a given, single angle of incidence.

According to Eq. 4 the MOSPR signal is derived from $\Delta R_{pp}(\theta)/R_{pp}(\theta)$ data using the value of the reflectivity corresponding to a selected angle of incidence θ_s, chosen to yield the highest sensitivity. At this angle, the MOSPR curve (i.e., $\Delta R_{pp}/R_{pp}$ vs. incidence angle) presents the steepest slope [10]. We derive the θ_s angle by analyzing the derivative of the fitted SPR curve. Time series of full SPR curves are recorded, and then fitted with Eq. 5 and the reflectivity values (corresponding to θ_s) are plotted over time. Further, frequency (Fourier) analysis is performed to extract the amplitude of oscillations (corresponding to ΔR_{pp}) and their mean (corresponding to R_{pp}). MOSPR and SPR reflectivity responses derived from raw measurements directly from the detector corresponding to θ_s are presented in Fig. 5a while reflectivity responses calculated from fitted SPR curves are given in Fig. 5b. The SNR is calculated as the ratio between the response (SPR, MOSPR) and the corresponding root mean square deviation (RMSD – calculated on 100 data points). The comparison shows that fitting the curves (and hence reducing the noise) improves greatly the sensitivity (Fig. 5b). The sensitivity increase for the

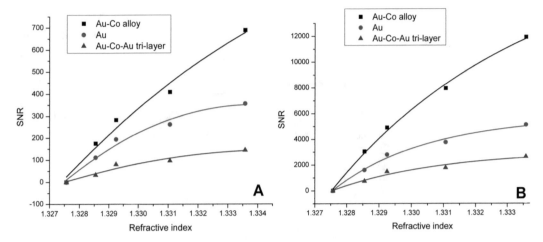

Fig. 5 (**a**) Signal-to-noise ratios for Au—SPR (*red*), Au-Co alloy—MOSPR (*black*) and Au-Co-Au tri-layer—MOSPR (*blue*) derived from raw reflectivity data for different refractive index solutions. (**b**) Signal-to-noise ratio for Au—SPR (*red*), Au-Co alloy—MOSPR (*black*) and Au-Co-Au tri-layer—MOSPR (*blue*) for solutions with different refractive indices—reflectivity values derived from fitted SPR curves

Au-Co alloy chip in respect to gold is 1.5-fold for raw reflectivity approach and reaches 2.5 times when using the reflectivity values derived from fitted SPR curves. In summary, Fig. 5a, b compare the SNR curves derived: (1) directly from the raw reflectivity data, and (2) when using the fitted SPR data. This highlights the advantage of our method in terms of improving on the SNR for both SPR and MOSPR assays.

3.4 MOSPR Chip Stability in Liquid Media

Stability of the MOSPR chips in physiological solutions is tested by exposure to a saline solution (e.g., HEPES buffer solution—HBS) while applying the oscillatory magnetic field. While the alloy shows no sign of destabilization (for virtually indefinite exposure, similar to regular Au sensor chips), in the case of Au-Co-Au tri-layer there is a significant exfoliation of the metallic film (Fig. 6a). Increase in the transparency of the film (observed under transmission microscopy Fig. 6a) further amplified with longer exposure confirms the conclusion of film exfoliation. In the same flow conditions and buffers, no modifications were observed in the cases of plain gold and/or Au-Co alloy (Fig. 6b).

3.5 Bioaffinity Assay

MOSPR assays of bioaffinity interactions are performed for the alloy-based MOSPR chip. Because of the corrosive effect of the saline buffers the Au-Co-Au tri-layers cannot be used even after surface modification and functionalization.

Affinity tests, using a model biomolecular interaction between human immuno-globulin G (HIgG) and Anti-HIgG in a direct assay [1], were performed. IgG is used as model analyte without restricting the generality of the method.

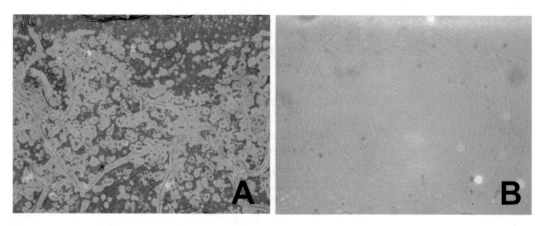

Fig. 6 Optical image of an Au-Co-Au tri-layer chip showing (**a**) pronounced corrosion in the areas exposed to saline solution in contrast with (**b**) non-corroded film

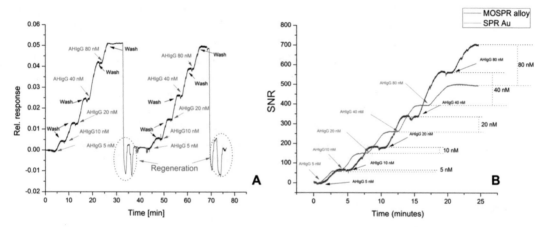

Fig. 7 (**a**) Relative reflectivity for several specific injections of Anti-HIgG (concentration range 5–80 nM) and two regeneration steps on the MOSPR Au-Co alloy chip. (**b**) Signal-to-noise ratio for specific MOSPR (for Au-Co alloy—*black*) and SPR (for Au chip—*red*) responses for several injections of Anti-HIgG

1. The sensor chip surfaces of Au-Co alloy are first functionalized using a carboxylated thiol and then further modified with HIgG via an amine coupling reaction.

2. The bioaffinity assay is performed at 100 μl/min, by injecting successively, for 2 min each, different concentrations of the analyte (A-HIgG) in the range 5–80 nM in HBS.

3. Each injection is followed by a washing step of 2 min to assess the binding level.

4. Surface is fully regenerated with two pulses of 10 mM glycine at pH 2 (2 min).

Several concentrations of Anti-HIgG (ranging from 5 to 80 nM) were injected consecutively over the sensor's surface and the response was recorded (Fig. 7a). The signal shows good stability during the experiment: after regeneration (when the bioaffinity complexes are broken up and Anti-HIgG washed away) the signal always returns to initial baseline values, and comparable signals are obtained for multiple rounds of affinity binding (only two shown in Fig. 7a) demonstrating the good stability of the alloy in buffer and regeneration solutions.

The improvement of the SNR for the Au-Co alloy-based MOSPR assay as compared to the Au-based classic SPR assays (Fig. 7b) is evaluated by comparing the limit of detection (calculated as three times the standard deviation of blank) which is around 0.60 nM for MOSPR and 0.96 nM for the SPR assay, an improvement of 160% of the sensitivity of MOSPR over SPR assays.

3.6 Concluding Remarks

The proposed novel solutions, based on innovative chip structure, and tailored measurement and improved data analysis methods, surpass current limitations of MOSPR sensing assays and have been demonstrated to: (a) enable improved plasmonic and magnetic properties yet a structural stability similar to standard Au-SPR chips, allowing for bioaffinity assays in saline solutions, (b) detection of minute analyte concentrations (IgG is used as model analyte without restricting the generality of the method). The custom-designed MOSPR measurement configuration allows the quasi simultaneous acquisition of the whole SPR curve at high signal-to-noise ratios during long time assessment in liquid media, a significant advancement over existing MOSPR chips, and confirms the MOSPR increased sensitivity over standard SPR analyses.

4   Notes

1. Functionalization reagents are available from other commercial sources.

2. *Piranha* solution reacts violently with many organic materials and should be handled with extreme care.

3. *Aqua regia* preparation produces heat and poisonous vapors; it may react violently with many organic materials and should be handled with extreme care.

4. A chromium underlayer may also be employed.

5. A 50-nm-thick gold layer is optimal for this wavelength of incident light (840 nm). Optimal thickness varies with the wavelength of incident light.

6. Electron beam evaporation may be employed in place of thermal evaporation.

7. The sputtering technique may be used for depositing alloys [22–24]; however, magnetron sputtering magnetic material requires expensive magnetron guns with powerful magnets.

Moreover, manufacturing custom sputtering targets comprising gold alloys may have prohibitively high costs.

8. E-beam epitaxy may be also used for growing crystalline films with well-controlled magnetic properties but this is suitable/limited to substrates with crystalline structure [25].

Acknowledgments

The authors thank the Romanian Executive Unit for Higher Education, Research, Development and Innovation Funding for funding through grants PN II-ID-PCCE-2011-2-0075 and PN-II-RU-PD-2012-3-0467.

References

1. Schasfoort R, Tudos A (2008) Handbook of surface plasmon resonance. The Royal Society of Chemistry, Cambridge

2. Raether H (1988) Surface Plasmons on smooth and rough surfaces and on gratings. Springer, Berlin

3. Incorporated TI (2003) Spreeta on the sensitivity of spreeta. Solutions 2003:1–14

4. Sepúlveda B, Calle A, Lechuga LM, Armelles G (2006) Highly sensitive detection of biomolecules with the magneto-optic surface-plasmon-resonance sensor. Opt Lett 31:1085–1087

5. Regatos D, Sepúlveda B, Fariña D et al (2011) Suitable combination of noble/ferromagnetic metal multilayers for enhanced magneto-plasmonic biosensing. Opt Express 19:8336–8346

6. Armelles G, González-Díaz J (2008) Localized surface plasmon resonance effects on the magneto-optical activity of continuous Au/Co/Au trilayers. Opt Express 16:3112–3114

7. Temnov VV, Armelles G, Woggon U et al (2010) Active magneto-plasmonics in hybrid metal–ferromagnet structures. Nat Photonics 4:107–111. doi:10.1038/nphoton.2009.265

8. Belotelov VI, Kreilkamp LE, Akimov IA et al (2013) Plasmon-mediated magneto-optical transparency. Nat Commun 4:2128. doi:10.1038/ncomms3128

9. Zvezdin A, Kotov V (1997) Modern magnetooptics and magnetooptical materials. IOP Publishing Ltd, Bristol

10. Armelles G, Cebollada A, García-Martín A, González MU (2013) Magnetoplasmonics: combining magnetic and plasmonic functionalities. Adv Opt Mater 1:10–35. doi:10.1002/adom.201200011

11. Regatos D, Fariña D, Calle A et al (2010) Au/Fe/Au multilayer transducers for magneto-optic surface plasmon resonance sensing. J Appl Phys 108:054502. doi:10.1063/1.3475711

12. Manera MG, Montagna G, Ferreiro-Vila E et al (2011) Enhanced gas sensing performance of TiO2 functionalized magneto-optical SPR sensors. J Mater Chem 21:16049. doi:10.1039/c1jm11937k

13. Manera MG, Ferreiro-vila E, Cebollada A et al (2012) Ethane-bridged Zn porphyrins dimers in langmuir-Scha fer thin films: spectroscopic, morphologic, and magneto-optical surface plasmon resonance characterization. J Phys Chem C 116:10734–10742

14. David S, Polonschii C, Luculescu C et al (2015) Magneto-plasmonic biosensor with enhanced analytical response and stability. Biosens Bioelectron 63:525–532. doi:10.1016/j.bios.2014.08.004

15. Johnsson B, Löfås S, Lindquist G (1991) Immobilization of proteins to a carboxymethyldextran-modified gold surface for biospecific interaction analysis in surface plasmon resonance sensors. Anal Biochem 198:268–277. doi:10.1016/0003-2697(91)90424-R

16. Chinowsky TM, Jung LS, Yee SS (1999) Optimal linear data analysis for surface plasmon resonance biosensors. Sensors Actuators B Chem 54:89–97

17. Zhan S, Wang X, Liu Y (2011) Fast centroid algorithm for determining the surface plasmon resonance angle using the fixed-boundary method. Meas Sci Technol 22:025201. doi:10.1088/0957-0233/22/2/025201

18. David S, Gheorghiu E (2011) Towards an advanced magneto-plasmonic sensing platform. Adv Top Electr Eng (ATEE), 2011 7th International symposium 1:12–14

19. Gheorghiu M, David S, Polonschii C et al (2014) Label free sensing platform for amyloid fibrils effect on living cells. Biosens Bioelectron 52:89–97. doi:10.1016/j.bios.2013.08.028

20. Neuert G, Kufer S, Benoit M, Gaub HE (2005) Modular multichannel surface plasmon spectrometer. Rev Sci Instrum 76:054303. doi:10.1063/1.1899503

21. Polonschii C, David S, Tombelli S (2010) A novel low-cost and easy to develop functionalization platform. Case study: aptamer-based detection of thrombin by surface plasmon resonance. Talanta 80:2157–2164

22. Kahn D (1992) Magnetic properties and structure of cobalt hardened gold. SUR/FIN'92 1:305–341

23. Yang K, Clavero C, Skuza J (2010) Surface plasmon resonance and magneto-optical enhancement on Au–Co nanocomposite thin films. J Appl Phys 107:103924

24. Mattox D (2010) Handbook of physical vapor deposition (PVD) processing, 2nd edn. Elsevier, Inc., Berlin

25. Ferreiro-Vila E, Iglesias M, Paz E et al (2011) Magneto-optical and magnetoplasmonic properties of epitaxial and polycrystalline Au/Fe/Au trilayers. Phys Rev B 83:205120. doi:10.1103/PhysRevB.83.205120

Chapter 6

Nanoplasmonic Biosensor Using Localized Surface Plasmon Resonance Spectroscopy for Biochemical Detection

Diming Zhang, Qian Zhang, Yanli Lu, Yao Yao, Shuang Li, and Qingjun Liu

Abstract

Localized surface plasmon resonance (LSPR) associated with metal nanostructures has developed into a highly useful sensor technique. Optical LSPR spectroscopy of nanostructures often shows sharp absorption and scattering peaks, which can be used to probe several bio-molecular interactions. Here, we report nanoplasmonic biosensors using LSPR on nanocup arrays (nanoCA) to recognize bio-molecular binding for biochemical detection. These sensors can be modified to quantify binding of small molecules to proteins for odorant and explosive detections. Electrochemical LSPR biosensors can also be designed by coupling electrochemistry and LSPR spectroscopy measurements. Multiple sensing information can be obtained and electrochemical LSPR property can be investigated for biosensors. In some applications, the electrochemical LSPR biosensor can be used to quantify immunoreactions and enzymatic activity. The biosensors exhibit better performance than those of conventional optical LSPR measurements. With multi-transducers, the nanoplasmonic biosensor can provide a promising approach for bio-detection in environmental monitoring, healthcare diagnostics, and food quality control.

Key words Localized surface plasmon resonance (LSPR), Nanocup arrays (nanoCA), Electrochemistry, Biosensor, BSA (bull serum albumin), Thrombin

1 Introduction

Optical detection is particularly a promising method because it allows remote transduction without any physical connection between excitation sources and detecting elements. In recent years, several nanostructured sensors such as photonic crystal, whispering gallery mode (WGM), and surface plasmon resonance (SPR) were increasingly being employed by optical spectroscopy for DNA detection, antigen-antibody recognition, immunoassays probing, and pathogen identification [1–3]. These sensors can respond to target molecules at low concentrations by observing shifts in resonance wavelength before and after conjugation of biomolecules to the sensor surfaces. A common drawback, however, is that optical

Avraham Rasooly and Ben Prickril (eds.), *Biosensors and Biodetection: Methods and Protocols Volume 1: Optical-Based Detectors*, Methods in Molecular Biology, vol. 1571, DOI 10.1007/978-1-4939-6848-0_6,
© Springer Science+Business Media LLC 2017

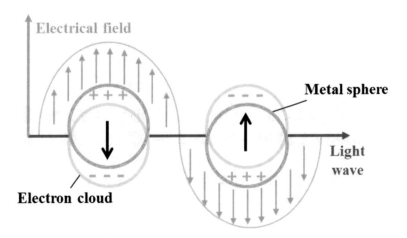

Fig. 1 Schematic of localized surface plasmon resonance (LSPR), where the free conduction electrons in the metal sphere are driven into oscillation due to coupling with incident light

biosensors, such as SPR and WGM, often required a complex system. They should fabricate devices that ensure precise alignment of light coupling to bio-detection volume and generate the desired optical phenomena.

Metal nanostructures were recently reported to modulate localized surface plasmon resonance (LSPR) spectroscopy for probing biomolecular interactions [4–6]. LSPR was often induced by incident light when it interacted with noble metal nanostructures, such as gold nanoparticles, that had smaller sizes than the wavelength of the incident light (Fig. 1). The light can be coupled in direct perpendicular transmission between incident light and spectrometer to obtain LSPR wavelength shifts without complex optical coupling [7–9]. Thus, it can be measured by a robust optical sensing platform minimizing the alignment requirement. However, sizes and positions of single nanostructures such as colloid nanoparticles were always random and difficult to control over a large area. Thus, the wavelength shift may be different for the same analyte because nanoparticles of different sizes gave rise to different optical spectroscopy as biosensors [8]. To generate stable LSPR spectroscopy, periodic nanostructure arrays such as nanohole arrays, nanorod arrays, and nanocone arrays can be fabricated by several nanoelectromechanical systems including electron beam lithography and focused ion beam milling for biosensing [9, 10]. These nanostructures were often fabricated by scanning beam lithographies, such as electron beam lithography (EBL) and focused ion beam (FIB) lithography. Thus, compared to nanoparticles from chemical synthesis, these nanostructure arrays with their more precise sizes and positions provide a promising approach to producing repeatable and stable LSPR spectroscopy for biological and chemical detection. Other detection techniques such as electrochemistry can also be combined with LSPR spectroscopy of nanostructures while providing unique advantages in sensitivity and selectivity. For instance, voltammetry can be used to modulate electrochemical reactions on nanostructures and provide additional selectivity to LSPR monitoring of α-synuclein–small molecule interactions [11].

Nanoimprint lithography is a method of fabricating nanometer scale patterns, and offers advantages including low cost, high throughput, and high resolution. It often creates nanostructure patterns by molding and curing of monomer or polymer resist on nanostructured template, while adhesion between the resist and the template is controlled to allow proper release. Nanoimprint lithography is currently used to fabricate devices for electrical, optical, photonic, and biological applications [12, 13]. In our work, nanocup arrays (nanoCA) were fabricated by nanoimprint and deposited with nanoparticles to modulate a stable LSPR spectroscopy for biosensor applications [14–16]. Nanoplasmonic biosensors using nanoCAs can be employed to monitor small molecule binding to proteins such as odorant binding proteins (OBPs) and peptides, and for odorant and explosive detection. The electrochemical detection technique can also be coupled with LSPR spectroscopy to design an electrochemical LSPR biosensor. This provides multiple sensing information for biosensor applications. The electrochemical LSPR sensor can also be used to quantify immunoreactions and enzymatic activity with demonstrably higher sensitivity.

2   Materials

2.1   Optical and Electrochemical Measurement System

1. Halogen cold light source (DT-MINI-2, Ocean Optics Inc., Dunedin, Florida, USA).

2. Spectrophotometer (USB2000+, Ocean Optics Inc., Dunedin, Florida, USA).

3. Three fiber bundles (QP230-0.25-XSR, Ocean Optics Inc., Dunedin, Florida, USA).

4. Optical fiber attenuator (FVA-ADP-UV, Ocean Optics Inc., Dunedin, Florida, USA).

5. Collimating lens (F230APC-633, Thorlabs Inc., Newton, New Jersey, USA).

6. Absorb sample pool (SPL-CUV-ABS, Pulei Inc., Hangzhou, China).

7. Multi-mode microplate reader (SpectraMax Paradigm, Molecular Devices Co., United Sates).

8. Electrochemical workstation (CHI 660E, CH Instruments, Texas, USA).

9. Ag/AgCl reference electrode (CHI 111, CH Instruments, Texas, USA).

10. Pt counter electrode (CHI 102, CH Instruments, Texas, USA).

2.2 Acquisition and Immobilization of Proteins

1. Phosphate buffer saline (PBS) (*see* **Note 1**).

2. 2-(morpholino)ethanesulfonic acid (MES) buffer (*see* **Note 2**).

3. 1-ethyl-3-(3-dimethylaminopropyl) carbodiimide (EDC) solution (*see* **Note 3**).

4. *N*-hydroxy-succinimide (NHS) solution (*see* **Note 4**).

5. Carboxy-poly(ethylene)-thiol (HOOC-PEG-SH, 2 kDa, Sigma- Aldrich, USA) (*see* **Note 5**).

6. Recombinant expressed plasmid pET-Acer-ASP2 (from Center of Analysis & Measurement, Zhejiang University) (*see* **Note 6**).

7. *E. coli* BL21 (DE3) (from Center of Analysis & Measurement, Zhejiang University).

8. *Bam*H I endonuclease (TaKaR Co., Dalian, China).

9. *Hind* III endonuclease (TaKaR Co., Dalian, China).

10. Isopropyl-β-D-thiogalactopyranoside (Aladdin, Shanghai, China).

11. OBPs, Acer-ASP2.

12. TNT-specific peptides (*see* **Note 7**).

2.3 Fabrication of nanoCA Device

1. UV light-curing flood lamp system (EC-Series, Dymax, USA).

2. E-beam evaporation system (FC/BJD2000, Temescal, USA).

3. Scanning electron microscope (SEM, XL30-ESEM, Philips, Netherlands).

4. Nanocone quartz template fabricated by e-beam lithography.

5. UV curable polymer (NIL-UV-Si, GuangDuo Nano Ltd., Suzhou, China).

6. Dimethyl dichlorosilane (CP, Aladdin, Shanghai, China).

7. Titanium (>99.99%, Sigma-Aldrich, USA).

8. Gold (99.99%, Sigma-Aldrich, USA).

9. (Poly)ethylene terephthalate (PET) substrate.

10. Teflon roller.

11. Transparent epoxy resin adhesive.

3   Methods

3.1 Fabrication of the nanoCA

The nanoCA device can be fabricated for use in LSPR spectroscopy for monitoring binding events occurring on the surface of the device [14, 17]. The nanostructures were imprinted by nanocone template made on quartz substrate and deposited with gold nanoparticles by following steps, as shown in Fig. 2.

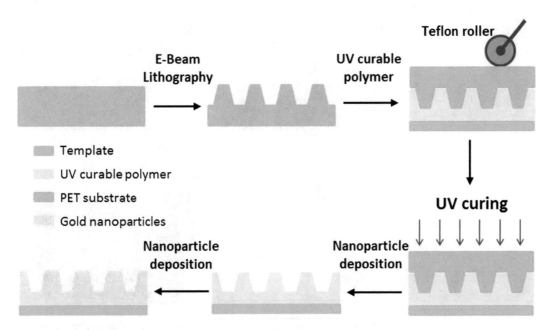

Fig. 2 Flow chart for nanoCA device fabrication by nanoimprint lithography

1. The template had nanocones in arrays, whose height, upper, and base diameter were 300, 200, and 500 nm, respectively. It was passivated with dimethyl dichlorosilane solution for 30 min and then rinsed three times with ethanol and deionized water. This promotes formation of a hydrophobic silane layer on the template, which in turn promotes removal of cured polymer replica with nanocup structures.

2. A 250 μm flexible (Poly)ethylene terephthalate (PET) sheet was used as a supporting substrate, and a Teflon roller was used for evenly distributing the UV curable polymer at the template/PET interface. The depth and opening diameters of the nanocup were 500 and 200 nm respectively, while the diameter of nanoparticles was about 20 nm.

3. To obtain the nanocup array structure, a UV light-curing flood lamp system was used at an average power density of 105 mW/cm^2 to solidify the UV polymer on the template and PET interface. The polymer was cured by UV light for 60 s at room temperature.

4. To deposit gold nanoparticles, 5-nm-thick titanium was deposited first and used as the adhesive layer. Then, gold nanoparticles were deposited on sidewalls and bottom of nanocups. The depositions were both performed with a six pocket e-beam evaporation system. Finally, the nanoCA structure with nanoparticles on sidewalls was successfully fabricated (Fig. 3). Figure 3c shows a scanning electron microscope image of the nanoCA device.

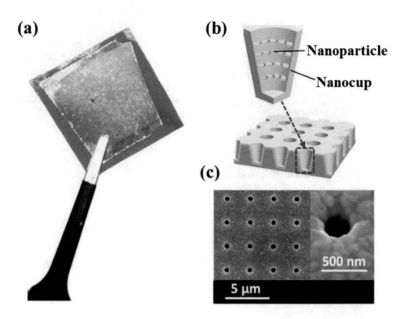

Fig. 3 Fabrication and characterization of nanocup arrays (nanoCA). (**a**) Photograph of nanoCA device in ~ 2 × 2 cm². (**b**) Structure of periodic nanocups on nanoCA chip, with Au nanoparticles deposited along the sidewalls. (**c**) SEM image of the cup arrays

3.2 Optical Detection for LSPR Spectroscopy

Periodic nanostructures have prominent optical features such as absorption and transmission peaks in the visible spectrum. These features were elicited by plasmon resonance and can be utilized in optical sensors. The optical detection can be performed using the following steps.

1. As shown in Fig. 3a, the nanoCA chips were first made into small rounds with diameters of 1 cm by hole puncher. Then the rounds can be immobilized on the bottom of wells by transparent epoxy resin adhesive, leaving 6 h for the adhesive to solidify. LSPR spectroscopy was performed using the microplate reader for multi-mode detection in high throughput.

2. A normal transmission model of the microplate reader was applied to measure the spectra of nanoCA. Light was emitted from the light source on the bottom of wells of the plate, delivered through nanoCA chips and received by the spectrometer. The reader detected LSPR spectroscopy in individual wells of the plate. The lamp and spectrograph can move from one well to other wells by a mechanical driver. The scanning range for spectroscopy was set from 300 to 900 nm and the step was fixed at 1 nm.

3. During measurement, 10 μl sample solutions were often added by pipette on the surface of the chip in the wells of the plate. The small volume of the analyte formed a thin liquid layer and reduced interference from the solution itself.

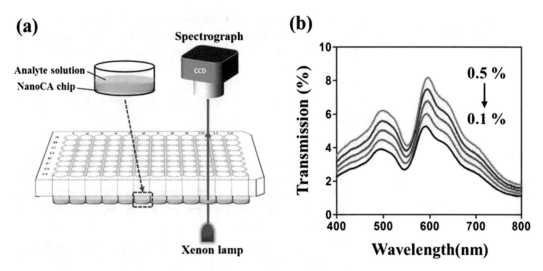

Fig. 4 Nanoplasmonic biosensor system for bio- and chemical detection. (**a**) Schematic diagram of the optical detection using a 96-well plate instrument in transmission mode. (**b**) Transmission spectra of nanoCA in the presence of NaCl at different concentrations (0.1%, 0.2%, 0.3%, 0.4%, and 0.5%). The resonance wavelength shifts with increasing concentration

4. NaCl at different concentrations was employed as increasing refractive index solutions, to determine the refractive-index sensing properties of nanoCA (Fig. 4b). 10 µl NaCl solutions at concentrations of 0.1%, 0.2%, 0.3%, 0.4%, and 0.5% were added by pipette on the surface of the chip. Then, optical detection was performed as described above. The entire transmission spectrum exhibited red shift (shift to right end of the spectrum) with increasing NaCl concentration, indicating the refractive index (RI) sensitivity of nanoCA in LSPR detection. Indeed, transmission peaks around 600 nm were often used as responses of nanoplasmonic sensors. With synergetic LSPR effects of periodical array structure and nanoparticles deposited on the cups, the nanoCA showed a high sensitivity for RI change with ~10^4 nm per refractive index unit (RIU), which was much greater than typical nanoparticle sensors and nanohole sensors based on plasmon resonance [18–20]. Thus, a high-sensitivity LSPR device based on nanoCA is useful as a high-throughput biosensor for monitoring binding of small molecules to proteins.

3.3 Acquisition of Bio-sensitive Material

In sensing property test, the nanoCA device showed a good performance for RIU change on the surface. However, the nanoCA device was only a physical transducer sensitive to refractive index changes, rather than a real sensor with high selectivity to special biochemical analytes. In our work, several olfactory proteins, such as OBPs and peptides, were used to modify nanoCA for bio- and chemical detections.

<table>
<tr><td>

3.3.1 Expression and Purification of OBPs

</td><td>

Oderant binding proteins (OBPs) are an important class of sensing proteins in biological olfactory systems. OBPs have significant binding affinities to various biochemical molecules. Distinct from membrane proteins or receptors, OBPs are soluble, can be expressed at low cost, purified easily, and maintain bioactivity in vitro [21, 22]. Thus, OBPs are useful as biosensors and can be obtained by the following steps:

</td></tr>
</table>

– The recombinant OBPs of *Acer-ASP2* were cloned from full-length cDNA of adult worker bees, *Apis cerana cerana*. The recombinant plasmid *pET-Acer-ASP2* was expressed and transformed into *E. coli* BL21 (DE3) competent cells after 450 bp fragments were excised with BamH I and Hind III from the *pGEM-Acer-ASP2* plasmid.

– The cells were grown in Luria–Bertani broth (including 30 μg/ml kanamycin) at 37 °C with 1.5 mM isopropyl-β-D-thiogalactopyranoside to induce expression of the protein. After 5 h at 28 °C, the bacterial cells were harvested and lysed by sonication and centrifuged into crude cell extracts into pellet and supernatant (3 ml, 1740 × g, 10 min).

– The OBP was then collected from the cells. The pellets of crude cell extracts containing recombinant proteins were precipitated in 1.5 M urea in ddH$_2$O and freeze-dried.

– Finally, the protein was resuspended (500 μg/ml) in phosphate buffered saline (PBS; pH = 7.4) and stored at 4 °C for biosensor experiments. More details can be found in the previous study about expression of the OBPs from honeybee [23].

<table>
<tr><td>

3.3.2 Synthesis of Peptide

</td><td>

Peptides are short protein fragments composed of a chain of amino acids. Peptides can be designed based on known structures of protein binding sites, synthesized with chemical methods, and purified to obtain specific sequences. They are useful in designing artificial receptors to mimic molecular recognition between proteins and analytes. Thus, peptides are ideal candidates for biosensing materials and for biosensor fabrication.

</td></tr>
</table>

– TNT-specific peptide was chemically synthesized based on the reported bio-sensitive sequence (WHWQRPLMPVSI) as olfactory protein for TNT [24].

– The peptide (CLVPRGSC) can be cleaved by thrombin, and was synthesized for use in thrombin detection.

– Chemical synthesis of peptides was performed by standard solid-phase peptide synthesis (SPPS) using BOC chemistry, with stepwise addition of protected amino acids to a growing peptide chain. The synthesis work was performed by Genscript company (Nanjing, China). Peptides were stored as freeze-dried powders (*see* **Note 8**).

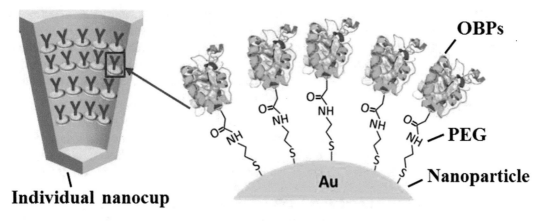

Fig. 5 Self-immobilization of olfactory proteins on sidewalls of nanoCA device

3.4 Self-Immobilization of Proteins

1. The self-immobilization of the proteins on nanoCA was performed by linkage of HOOC-PEG-SH (Fig. 5). The immobilization can be completed by following steps.

2. NanoCA device was first washed with a mixture of 98% H_2SO_4 and 30% H_2O_2 (7:3), and DI water respectively, to remove the organic residues.

3. Subsequently, the nanoCA device was immersed in a petri dish and reacted with 2 ml HOOC-PEG-SH at 1 mg/ml overnight (about 18 h) at room temperature to form Au-S covalent bonding on the surface of the nanoCA device.

4. 2 ml mixing solution of 8 mg/ml EDC and 12 mg/ml NHS in 0.1 M MES buffer (pH = 5, Sigma) was immersed in petri dish for 15 min to activate carboxylate groups of HOOC-PEG-SH after extra HOOC-PEG-SH was washed off with DI water.

5. Finally, the solution was adjusted to pH 8.0 with saturated NaHCO₃, added with 1 ml olfactory proteins at 500 μg/ml, and incubated at room temperature for 2 h. The nanoCA was washed with DI water by gentle shaking for 60 s, followed by drying with nitrogen and stored at 4 °C for further experiments (*see* **Notes 8** and **9**).

3.5 Olfactory Protein Modified Sensor Using LSPR Spectroscopy

The interaction between small molecules and olfactory proteins immobilized on nanoCA surface can affect the refractive index in the metal-dielectric interface of nanocup sidewalls and thus give rise to resonance wavelength shift in LSPR spectroscopy. The wavelength shift can then be used to quantify small molecules involved in binding events, and bio-modified nanoCA can be used as a nanoplasmonic biosensor for bio- and chemical detections [15, 16].

The nanoCA can be used as a biosensor by monitoring interactions between small molecules and olfactory proteins. For instance, several chemical molecules can specifically bind into hydrophobic pockets of OBPs immobilized on the nanoCA surface, which can modulate protein conformation and change relative dielectric constants of the bio-coating [15] (Fig. 6a).

– The OBPs were immobilized on the surface of the nanoCA device by self-assembly to obtain OBP-modified nanoCA for monitoring of molecular binding (*see* Subheading 3.4). For odorant measurement one kind of aromatic odorant, β-ionone, was used as a model molecule to test responses of the bio-nanoCA with OBPs. Ten microliter β-ionone solutions at different concentrations ranging from 10 nM to 10 Mm can be added in wells of the 96-well plates by pipette (*see* **Note 10**).

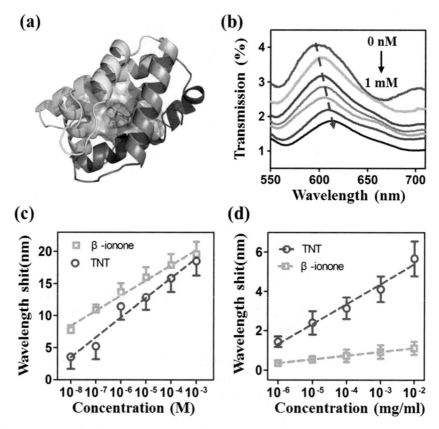

Fig. 6 Bio-nanoCA with olfactory proteins as nanoplasmonic biosensors for odorant and explosive detection. (**a**) Molecular docking of the binding pocket of the OBPs, *Acer-ASP2*. (**b**) Transmission spectrum of OBPs-modified nanoCA for β-ionone. (**c**) Dose-dependent profiles of bio-nanoCA with OBPs in resonance wavelength shift of transmission peak for β-ionone and TNT. (**d**) Dose-dependent profiles of bio-nanoCA with TNT-specific peptide for β-ionone and TNT. (mean ± SD, n = 15)

- Optical detection is performed as described above. The transmission spectra of the bio-nanoCA device can be obtained as illustrated in Fig. 6b. Intensity and location of the transmission peak was notably changed by specific binding of β-ionone and OBPs. Transmission intensity change and LSPR wavelength shift can both be calculated statistically with the OBPs-modified devices. As expected, the wavelength shift showed a good linear dose-dependent response for β-ionone at increasing concentrations (Fig. 6c). These results demonstrate that nanoCA can monitor specific binding of β-ionone to OBPs by the resonance wavelength shift in LSPR spectroscopy.

3.5.2 Nanoplasmonic Biosensor for Explosive Detection

Recent studies reported that trained honeybees are able use their sensing proteins to find nitro-explosives, and various OBPs have been used to modify biosensors for explosive detection [25–27]. We have explored the binding of nitro-explosive molecules to the OBPs of *AcerASP2* by the bio-nanoCA.

- Molecular docking of TNT and the OBP was carried out to show affinity of these two molecules. Like β-ionone, TNT molecules can enter protein cavities, interact with amino acid residues, and form a complex with OBPs. This process might elicit OBP conformation changes like those in β-ionone binding, thereby modulating the LSPR on the nanoCA in the optical measurement.

- The OBPs were immobilized on the surface of the nanoCA device by self-assembly to obtain OBP-modified nanoCA for monitoring of molecular binding (*see* Subheading 3.4). Ten microliter TNT solutions at different concentrations can be added in wells of the 96-well plates by pipette. Then, optical detection can be performed as described above (*see* Subheading 3.2). Figure 6c showed responses of the nanoCA with the OBPs to TNT at increasing concentrations. The nanoCA can monitor the binding of TNT molecules to the OBPs. However, the responses were nonlinear in concentration range from 10^{-7} to 10^{-5} M, and the nanoCA cannot distinguish TNT and β-ionone. It indicated that the nanoCA with OBPs was not a good biosensor strategy for explosive detection of TNT.

- The TNT-specific peptide was immobilized on the surface of the nanoCA device by self-assembly to obtain peptide-modified nanoCA for monitoring of molecular binding (*see* Subheading 3.4). Ten microliter TNT solutions at different concentrations can be added in wells of the 96-well plates by pipette. Then, optical detection can be performed as described above (*see* Subheading 3.2). As shown in Fig. 6d, the nanoCA with the peptides was high selective to TNT, while the nanoCA showed no response to β-ionone.

3.6 Electrochemical LSPR Biosensor for Protein Analysis

LSPR is not only influenced by refractive index change on the nanostructure surface, but also by potential and current density on the surface of a metal film [14]. Thus, LSPR sensors can be combined with electrochemistry to implement synchronized electrical and optical responses. The combination of electrochemistry and optical SPR measurement can provide a multitude of information from different signal transductions and provide novel approaches to elucidate detailed processes for electrochemical reactions.

3.6.1 Electrochemical LSPR Measurement

The apparatus for electrochemical LSPR measurement is depicted in Fig. 7a. Optical measurement was performed in transmission mode. The light was emitted from the source below the nanoCA device, delivered through the device, and finally received by a spectrophotometer. The LSPR signal was measured and analyzed by transmission spectroscopy.

– For electrochemical measurements the potentiodynamic electrochemical measurement of cyclic voltammetry (CV) was used to investigate the electrochemistry properties of nanoCA device and the electrochemical modulation to LSPR. The nanoCA device was used as the working electrode, while platinum

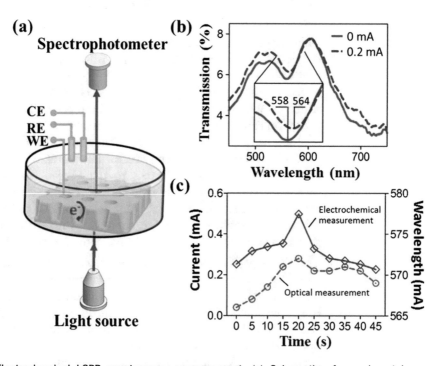

Fig. 7 Electrochemical LSPR spectroscopy measurement. (**a**) Schematic of experimental apparatus for synchronous electrochemical and optical measurement. (**b**) Transmission LSPR spectroscopy of nanoCA with different surface currents. (**c**) Synchronous measurement of LSPR and electrochemistry in CV scanning

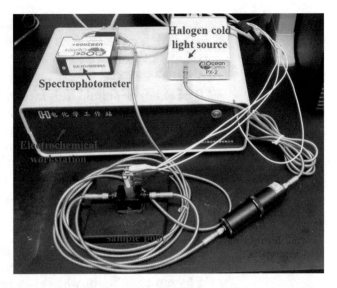

Fig. 8 Photograph of the electrochemical LSPR measurement apparatus

electrode and Ag/AgCl electrodes were used as counter electrode and reference electrodes, respectively.

- In the synchronous measurement, the range of the transmission spectrum was from 300 to 1000 nm with a step of 0.5 nm. The transmission spectrum can be recorded when the surface of nanoCA had RI change and electrochemical current (Fig. 7b). The electrochemical reactions were measured using an electrochemical workstation in a standard three-electrode system. 5 mM $K_4[Fe(CN)_6]/K_3[Fe(CN)_6]$ (1:1) was employed as redox couple in aqueous solution for electrochemical characterization. The scanning rate of CV was fixed at 0.02 V/s and the transmission spectrum was simultaneously recorded and saved every 5 s. Thus, the transmission spectrum was obtained to represent optical LSPR, in 0.1 V steps during voltage scanning with different electrochemical current (Fig. 7c). Figure 8 shows a photograph of the electrochemical LSPR measurement apparatus. For optical detection, there were three fiber bundles, two collimating lenses, an optical fiber attenuator, a light source and a spectrophotometer. Fiber bundles were used to provide optical linkage between the lens, the attenuator, the light source, and the spectrophotometer. The spectrophotometer was also directly connected to the light source by data cable. Thus, the spectrophotometer can synchronously capture light emitted by the light source and delivered through nanoCA devices. For electrochemical detection, the nanoCA device was used as the working electrode, while platinum and Ag/AgCl electrodes were used as counter electrode and reference electrodes, respectively. These three electrodes were all linked to an electrochemical workstation and inserted into a spectrometer measuring cell filled with redox solution. Careful arrangement of these three electrodes should maintain successful delivery through the cell.

– The electrochemical LSPR measurements for nanoCA were further carried out and analyzed with transmission spectrum and synchronous CV scanning. As seen in Fig. 7b, a significant LSPR dip shift was recorded in the transmission spectrum when electrochemical current on the nanoCA surface was changed from 0 to 0.2 mA. The resonance wavelength was shifted from 558 to 564 nm but there was no significant shift observed for LSPR peaks.

– Optical measurements within CV scanning were carried out and the wavelength shifts of LSPR dips were shown in Fig. 7c. During scanning, the wavelength of dip was shifted from 566 nm to 573 nm with an increase of current from 0.25 to 0.5 mA. When the current returned to 0.25 mA, the wavelength also decreased with declining current. These results indicate that the current change on the nanoCA surface can modulate LSPR of nanoCA and produce a wavelength shift in LSPR. The LSPR in transmission dip around 550 nm can be enhanced by electrochemical current and can be analyzed as an indicator in electrochemical LSPR detecting.

3.6.2 Electrochemical LSPR Spectroscopy as Biosensor for Immunoreaction

Utilizing electrochemical enhancement for LSPR on nanoCA, electrochemical LSPR biosensors can be fabricated with higher sensitivity in the synchronous measurement mode. Many biointeractions such as immunoreaction can be probed by electrochemical LSPR spectroscopy on the nanoCA.

– The nanoCA was modified by anti-BSA which can specifically bind to BSA by self-assembly (for methods *see* Subheading 3.4).

– One milliliter BSA solutions at different concentrations can be added in wells of the 96-well plates by pipette.

– The CV scanning can be performed on the nanoCA device (*see* Subheading 3.6). The redox current can be generated from the redox couple and modulated by the binding of BSA to anti-BSA. The CV scanning can provide redox currents around 0.1 and 0.3 V in the voltammogram with electrochemical reactions.

– The transmission spectrum was recorded when CV scanning reached 0.3 V (Fig. 9a, *see* Subheading 3.6). LSPR dip shifts were elicited by the conjunction of anti-BSA and BSA on the surface of nanoCA and enhanced by the CV scanning from 9 to 15 nm. Due to the redox current, the negative protein of HSA also elicited a wavelength shift of about 3 nm. However, this was much lower than the response to BSA of about 15 nm with

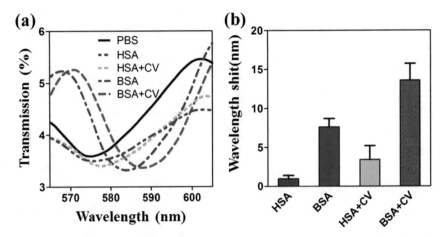

Fig. 9 Electrochemical LSPR detection for specific binding of proteins. (**a**) The transmission spectrum of nanoCA with PBS, 100 μg/ml BSA and BSA plus synchronous CV scanning, 100 μg/ml HAS and HSA plus CV scanning. (**b**) Statistic for shifts in dip wavelength of HSA, HSA plus CV, BSA and BSA plus CV (mean ± SD, $n = 6$)

electrochemical enhancement (Fig. 9b). These results suggest that the electrochemically coupled LSPR measurement had redox currents that modulate LSPR dip shift, in addition to that resulting from RI change. In fact, LSPR measurement with and without CV had detection limits of 13 and 25 μg/ ml, respectively, for BSA detection. Thus, the electrochemically enhanced biosensor fabricated with electrochemical LSPR had higher response and sensitivity than that of conventional optical methods.

3.6.3 Electrochemical LSPR Spectroscopy as Biosensor for Enzymatic Activity

The electrochemical LSPR spectroscopy can also be used in thrombin detection [28]. The detection can be performed by the following steps.

- As shown in Fig. 10a, polyethylene glycol (PEG), peptide (CLVPRGSC), and bovine serum albumin (BSA) were immobilized on surface of nanoCA with the self-assembly (methods, *see* Subheading 3.4), while peptides were used as probes for thrombin.

- One milliliter thrombin solutions at different concentrations can be added in wells by pipette. Thrombin can cleave specific sites of peptide sequences and remove BSA from the surface of nanoCA, thereby modulating LSPR transmission spectra.

- Spectrographic measurement was coupled with electrochemical voltage at 3 V to dissociative BSA from the nanoCA surface and increase LSPR wavelength shift responses. The measurement apparatus was the same as described above (*see* Subheading 3.6).

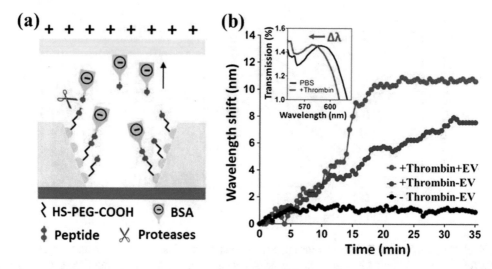

Fig. 10 Electrochemical LSPR sensor for thrombin detection. (**a**) Design of electrochemical LSPR sensor for thrombin using self-immobilization of protein on nanoCA. (**b**) LSPR responses of wavelength shifts of the nanoCA for thrombin detections with (+) and without (−) electrochemical enhancement, electrochemical voltage (EV)

The results for thrombin detection are shown in Fig. 10b. The presence of thrombin can selectively catalyze cleavages at Arg-Gly bonds of peptides, remove BSA from the nanoCA sidewall, and generate time-dependent shifts in resonance wavelength of nanoCA (*see* **Notes 10** and **11**).

– The transmission spectrum can be analyzed by LSPR wavelength shifts with electrochemical enhancement. As expected, as the electrochemical voltage is increased LSPR wavelength shifts significantly. There was stronger voltage enhancement in the presence of thrombin at high concentrations. Indeed, the electrochemically enhanced LSPR biosensor had higher sensitivity and shorter response time than the sensor utilizing LSPR alone. This provides a promising approach for designing more sensitive and rapid LSPR sensors.

4   Notes

1. Phosphate buffer saline (PBS) was used as the solvent for proteins. The solution was prepared by dissolving 137 mM NaCl, 2.68 mM KCl, 1.47 mM KH_2PO_4, and 8.10 mM Na_2HPO_4 in pure water and pH was fixed at 7.4.

2. 2-(morpholino)ethanesulfonic acid (MES) buffer was used as buffer solution for protein self-immobilization. It was prepared by dissolving 100 mM MES and 500 mM NaCl, and its pH was fixed at 6.0.

3. 1-ethyl-3-(3-dimethylaminopropyl) carbodiimide (EDC) solution was prepared at 2 mM, while MES buffer was used as the solvent.

4. *N*-hydroxy-succinimide (NHS) solution was prepared at 5 mM, while MES buffer was used as the solvent.

5. Carboxy-poly(ethylene)-thiol (HOOC-PEG-SH, 2 kDa) was used as linkage between proteins and nanoCA device. The solution was prepared at 1 mg/ml, while ethyl alcohol and pure water with rate 7:3 were used as solvent of HOOC-PEG-SH.

6. OBPs, Acer-ASP2, were cloned from the full-length cDNA of adult worker bees, *Apis cerana* cerana. The protein was resuspended in PBS solution at 500 μg/ml and stored at 4 °C for biosensor experiments (*see* **Note 1**).

7. Peptides used in our study were synthesized by the standard solid phase method. Synthesized peptides were tested by high-performance liquid chromatography (HPLC) and mass spectrometry (MS) to verify amino acid sequence and purity. The peptides were stored in the form of freeze-dried powders before experiments and dissolved in PBS at 500 μg/ml.

8. The proteins involved in the study, such as OBPs, peptides, and BSA, should all be dissolved in PBS solution for use during experiments. Before experiments, proteins should be stored at 4 °C to maintain bio-activity.

9. In designing the electrochemical LSPR biosensor for thrombin, the protein immobilization on nanoCA was completed by two steps of self-assembly of the peptide and BSA. The steps are both as described for linking proteins to HS–PEG–COOH in the above Method section.

10. Protein immobilization on nanoCA should be tested to verify efficient combination of protein and the nanoCA device. Generally, electrochemical impedance spectroscopy and optical spectroscopy can be used to verify the combination.

11. In the electrochemical LSPR detection for thrombin, the electrochemical voltage can range from 0.1 to 5 V. The voltage at 3 V used in our study was optimized to obtain better sensitivity and shorter sensor response times.

Acknowledgments

This work was supported by the National Natural Science Foundation of China (Grant No. 81371643), the Zhejiang Provincial Natural Science Foundation of China (Grant No. LR13H180002).

References

1. Homola J (2008) Surface plasmon resonance sensors for detection of chemical and biological species. Chem Rev 108:462–493

2. Vollmer F, Arnold S (2008) Whispering-gallery-mode biosensing: label-free detection down to single molecules. Nat Methods 5:591–596

3. Nair RV, Vijaya R (2010) Photonic crystal sensors: an overview. Prog Quant Electron 34:89–134

4. Saha K, Agasti SS, Kim C et al (2012) Gold nanoparticles in chemical and biological sensing. Chem Rev 112:2739–2779

5. Willets KA, Van Duyne RP (2007) Localized surface plasmon resonance spectroscopy and sensing. Annu Rev Phys Chem 58:267–297

6. Mayer KM, Hafner JH (2011) Localized surface plasmon resonance sensors. Chem Rev 111:3828–3857

7. Gao HW, Henzie J, Odom TW (2006) Direct evidence for surface plasmon-mediated enhanced light transmission through metallic nanohole arrays. Nano Lett 6:2104–2108

8. Anker JN, Hall WP, Lyandres O et al (2008) Biosensing with plasmonic nanosensors. Nat Mater 7:442–453

9. Stewart ME, Anderton CR, Thompson LB et al (2008) Nanostructured plasmonic sensors. Chem Rev 108:494–521

10. Cao J, Sun T, Grattan KT (2014) Gold nanorod-based localized surface plasmon resonance biosensors: a review. Sens Actuators B 195:332–351

11. Cheng XR, Wallace GQ, Lagugné-Labarthet F et al (2015) Au nanostructured surfaces for electrochemical and localized surface plasmon resonance-based monitoring of α-synuclein–small molecule interactions. ACS Appl Mater Interfaces 7:4081–4088

12. Boltasseva A (2009) Plasmonic components fabrication via nanoimprint. J Opt A Pure Appl Opt 11:114001

13. Pimpin A, Srituravanich W (2011) Review on micro-and nanolithography techniques and their applications. Eng J 16:37–56

14. Zhang D, Lu Y, Jiang J et al (2015) Nanoplasmonic biosensor: coupling electrochemistry to localized surface plasmon resonance spectroscopy on nanocup arrays. Biosens Bioelectron 67:237–242

15. Zhang D, Lu Y, Zhang Q et al (2015) Nanoplasmonic monitoring of odorants binding to olfactory proteins from honeybee as biosensor for chemical detection. Sensor Actuat B Chem 221:341–349

16. Zhang D, Zhang Q, Lu Y, et al. (2016) Peptide functionalized nanoplasmonic sensor for explosive detection. Nano-Micro Lett 8:36–43

17. Gartia MR, Hsiao A, Pokhriyal A et al (2013) Colorimetric plasmon resonance imaging using nano lycurgus cup arrays. Adv Opt Mater 1:68–76

18. Stewart ME, Mack NH, Malyarchuk V et al (2006) Quantitative multispectral biosensing and 1D imaging using quasi-3D plasmonic crystals. Proc Natl Acad Sci U S A 103:17143–17148

19. Kuznetsov AI, Evlyukhin AB, Goncalves MR et al (2011) Laser fabrication of large-scale nanoparticle arrays for sensing applications. ACS Nano 5:4843–4849

20. Kee JS, Lim SY, Perera AP et al (2013) Plasmonic nanohole arrays for monitoring growth of bacteria and antibiotic susceptibility test. Sensor Actuat B Chem 182:576–583

21. Lu Y, Li H, Zhuang S et al (2014) Olfactory biosensor using odorant-binding proteins from honeybee: Ligands of floral odors and pheromones detection by electrochemical impedance. Sensor Actuat B Chem 193:420–427

22. Lu Y, Yao Y, Zhang Q et al (2015) Olfactory biosensor for insect semiochemicals analysis by impedance sensing of odorant-binding proteins on interdigitated electrodes. Biosens Bioelectron 67:662–669

23. Li H-L, Zhang Y-L, Gao Q-K et al (2008) Molecular identification of cDNA, immunolocalization, and expression of a putative odorant-binding protein from an Asian honey bee, *Apis cerana* cerana. J Chem Ecol 34:1593–1601

24. Jaworski JW, Raorane D, Huh JH et al (2008) Evolutionary screening of biomimetic coatings for selective detection of explosives. Langmuir 24:4938–4943

25. Smith RG, D'Souza N, Nicklin S (2008) A review of biosensors and biologically-inspired systems for explosives detection. Analyst 133:571–584

26. Ramoni R, Bellucci S, Grycznyski I et al (2007) The protein scaffold of the lipocalin odorant-binding protein is suitable for the design of new biosensors for the detection of explosive components. J Phys Condens Matter 19:395012

27. Kuang Z, Kim SN, Crookes-Goodson WJ et al (2009) Biomimetic chemosensor: designing peptide recognition elements for surface functionalization of carbon nanotube field effect transistors. ACS Nano 4:452–458

28. Zhang D, Zhang Q, Lu Y et al (2015) Electrophoresis-enhanced nanoplasmonic biosensor with nanocup arrays for protease detection in point-of-care diagnostics. China Nanomed 2015:202

Chapter 7

Plasmonics-Based Detection of Virus Using Sialic Acid Functionalized Gold Nanoparticles

Changwon Lee, Peng Wang, Marsha A. Gaston, Alison A. Weiss, and Peng Zhang

Abstract

Biosensor for the detection of virus was developed by utilizing plasmonic peak shift phenomenon of the gold nanoparticles and viral infection mechanism of hemagglutinin on virus and sialic acid on animal cells. The plasmonic peak of the colloidal gold nanoparticles changes with the aggregation of the particles due to the plasmonic interaction between nearby particles and the color of the colloidal nanoparticle solution changes from wine red to purple. Sialic acid reduced and stabilized colloidal gold nanoparticle aggregation is induced by the addition of viral particles in the solution due to the hemagglutinin-sialic acid interaction. In this work, sialic acid reduced and stabilized gold nanoparticles ($d = 20.1 \pm 1.8$ nm) were synthesized by a simple one-pot, green method without chemically modifying sialic acid. The gold nanoparticles showed target-specific aggregation with viral particles via hemagglutinin-sialic acid binding. A linear correlation was observed between the change in optical density and dilution of chemically inactivated influenza B virus species. The detection limit of the virus dilution (hemagglutinination assay titer, 512) was shown to be 0.156 vol% and the upper limit of the linearity can be extended with the use of more sialic acid-gold nanoparticles.

Key words Biosensor, Sialic acid, Gold nanoparticle, Viral detection, Colorimetric measurement

1 Introduction

Metallic nanoparticles absorb and scatter with great efficiency when interacting with light. This strong interaction between metallic nanoparticles and light occurs because the oscillating electromagnetic field of light initiates the coherent oscillation of the free electrons of the metallic nanoparticles. This oscillation is termed the surface plasmon resonance (SPR). SPR results in dipole oscillation along the direction of the electric field of light (Fig. 1) . The amplitude of the oscillation reaches the maximum at certain frequency of light and it is dependent on the particle size, shape, and refractive index of the solution. For noble metallic nanoparticles,

Avraham Rasooly and Ben Prickril (eds.), *Biosensors and Biodetection: Methods and Protocols Volume 1: Optical-Based Detectors*, Methods in Molecular Biology, vol. 1571, DOI 10.1007/978-1-4939-6848-0_7, © Springer Science+Business Media LLC 2017

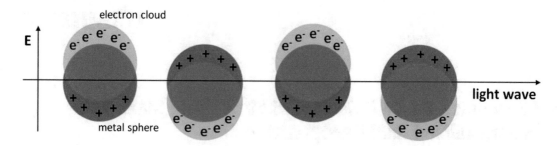

Fig. 1 Schematic illustration of surface plasmon resonance of gold nanoparticles due to collective oscillation of surface electrons by incident light of appropriate wavelengths

the SPR-induced strong absorption or scattering of light can be easily measured by UV-Vis absorption spectrometer.

The SPR band position relies strongly on the size and shape of individual nanoparticles and interparticle distance [1, 2]. Gold nanoparticles typically display an SPR band at around 520 nm. The assembly or aggregation of the metallic nanoparticles usually results in a red shift of the plasmonic band. This SPR band shift can be easily observed by naked eyes with the color of the colloidal gold solution changing from red to purple to blue. Kreibig et al. showed the relationship of SPR band position and interparticle distance decades ago [3]. The interparticle distance-dependent SPR band shift is attributed to the electric dipole-dipole interactions and coupling between plasmons of the neighboring particles in the aggregates. When the interparticle distance is greater than the average particle diameter (dispersed state), the SPR band appears to be red, whereas the color turns to blue when the interparticle distance decreases to less than the average particle diameter (aggregated state).

This phenomenon has been well adopted in various detection schemes with specific targeting elements decorated on the nanoparticle surface for the targets of interest, such as polynucleotides [4, 5], enzymes and proteins [6, 7], cells [8], and heavy metals [9]. In this chapter, we describe a method to develop the gold nanoparticles for the colorimetric detection of influenza virus based on the interaction between hemagglutinin, a protein expressed on the viral surface, and sialic acid, utilized as a surface stabilizing ligand on gold nanoparticles through a simple one-pot synthesis [10].

Sialic acid is a surface ligand presented on the surface of lung epithelial cells and it is recognized as the primary binding site for influenza virus. Hemagglutinin is a surface protein on various viral species that binds to sialic acid on targeted cell surface for the viral infection process [11]. The binding of viral proteins to sialic acid has been studied for various pathogenic viruses, such as influenza virus [11–13], human parainfluenza virus [14], human coronavirus [15, 16], and a specific serotype of rhinovirus [17]. Based on the

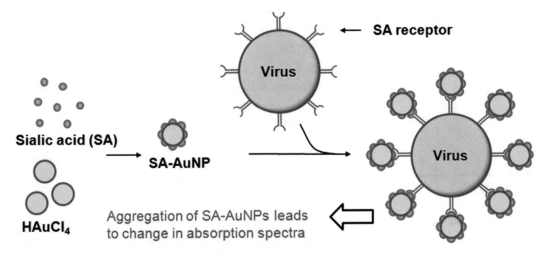

Fig. 2 Detection scheme of viral particles based on the aggregation of SA-AuNPs on the virus surface through sialic acid-hemagglutinin binding

binding capability of sialic acid to certain viruses, it is possible to design a colorimetric sensor for detecting virus using self-reporting plasmonic nanoparticles, which display plasmon shift upon binding to viral particles (Fig. 2). Such a quick and user-friendly detection method based on the binding between hemagglutinin on the influenza virus surface and sialic acid on the gold nanoparticle surface provides an effective means to identify individuals or animals infected with virus, and can help with early therapeutic intervention and reduce the spread of infection.

2 Materials

2.1 Equipment

IKAMAG® Safety Control heating magnetic stirrer (IKA).

Eppendorf bench-top centrifuge (Model 5424).

Nanotrac particle size analyzer (Microtrac).

Ocean Optics USB 4000 UV-Vis spectrometer.

Millipore Milli-DI System.

2.2 Sialic Acid and Gelatin Gold Nanoparticle Synthesis

1. 1.25 mM *N*-acetylneuraminic acid (SA) (Catalogue No. A2388, Sigma-Aldrich, St. Louis, MO) was prepared in DI water.

2. 0.1 wt% gelatin (Catalogue No. 53028, Sigma-Aldrich, St. Louis, MO) was prepared in DI water.

3. 1 M sodium hydroxide (Catalogue No. S5881, Sigma-Aldrich, St. Louis, MO) was prepared in DI water.

4. 0.02 M chloroauric acid (HAuCl$_4$) (Catalogue No. 254169, Sigma-Aldrich, St. Louis, MO) was prepared in DI water.

<table>
<tr><td>

2.3 Influenza Virus Deactivation

</td><td>

1. Dulbecco's Modified Eagle Media (DMEM) (Catalogue No. 11965, Life Technology, Grand Island, NY) was used for the culture of virus.

2. Fetal Bovine Serum (FBS) (Catalogue No. 16000-044, Life Technology, Grand Island, NY) was added to 10% in DMEM.

3. β-propiolactone (Catalogue No. P5648, Sigma-Aldrich, St. Louis, MO) was added to 0.05% of final concentration in the viral solution.

</td></tr>
</table>

3 Methods

<table>
<tr><td>

3.1 Sialic Acid Gold Nanoparticle (SA-AuNP) Synthesis

</td><td>

1. Ten milliliter of 1.25 mM sialic acid (SA) solution was prepared in DI water at room temperature. The SA solution was mixed with 250 μL of 0.02 M HAuCl$_4$ followed by the addition of 50 μL 1 M NaOH in 20 mL glass vial (*see* **Note 1**).

2. The mixture was then stirred and heated for 80 °C at $20 \times g$ for 15 min on a heating magnetic stirrer. The color of the solution changed from yellow to a dark red wine.

3. After the solution was cooled to room temperature, the gold nanoparticles were washed twice by centrifugation at $4500 \times g$) using a bench-top centrifuge (Eppendorf 5424) for 20 min. After the centrifugation, supernatant was removed and the dark red pellet was suspended in DI water and stored until further use.

</td></tr>
<tr><td>

3.2 Influenza Virus Inactivation

</td><td>

1. Virus solution was provided in 10% FBS containing DMEM and it was inactivated by the addition of β-propiolactone (final concentration of 0.05%) for 1 h at 37 °C.

2. Inactivated virus was stored at −20 °C until further use.

</td></tr>
<tr><td>

3.3 Particle Size Analysis

</td><td>

Particle size analysis was performed on a Nanotrac particle size analyzer (Microtrac) with a built-in liquid sample holder. The concentration of the sample was adjusted for the optimal measurement condition. Three 30-s measurements were averaged for particle size determination, with results shown in Fig. 3c.

</td></tr>
<tr><td>

3.4 Determination of SA-AuNP Concentration

</td><td>

To determine the concentration of the SA-AuNPs solution, the volume-density method was used based on the assumption of face-centered cubic (fcc) gold structure and 100% reaction yield. First, the following equation was used to calculate the number of gold atoms in a gold nanoparticle.

$$N = \frac{4\pi\rho r^3}{3} \times \frac{N_A}{M}$$

(N = number of atoms in a gold nanoparticle, ρ = density of fcc gold (19.3 g/cm^3), r = radius of nanoparticle (10.05 nm), N_A is

</td></tr>
</table>

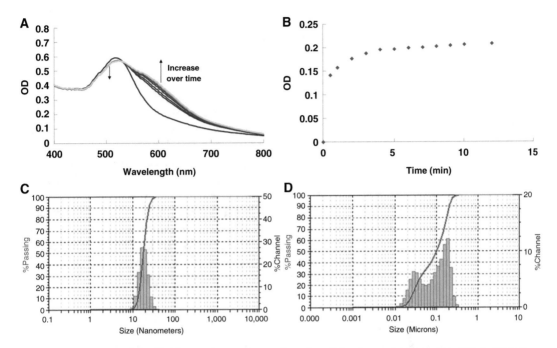

Fig. 3 Typical results of the UV-Vis spectroscopic results and particle size analysis results of SA-AuNP before and after virus incubation. UV-Vis absorption spectra measured at 0, 0.5, 1, 2, 3, 4, 5, 6, 7, 8, 9, 10, and 12 min after the addition of 1.25 vol% influenza B/Victoria. (**a**) Typical UV-Vis spectral change of SA-AuNP solution after the addition of the virus, (**b**) change of OD_{610} over time, (**c**) particle size analysis result measured before the virus addition, and (**d**) particle size analysis result measured after the addition of 1.25 vol% influenza B/Victoria

Avogadro's number, and M is gold atom's atomic weight (196.97 g/mol).

The calculation result yielded ~250,000 gold atoms in 20.1 nm spherical-shaped gold nanoparticles. Since the initial concentration of $HAuCl_4$ is 0.5 mM, the final SA-AuNP concentration is 2 nM when resuspended in the same volume of DI water.

3.5 UV-Vis Measurements

Five hundred microliter of the solution containing SA-AuNP with or without virus was added into the 1 cm path length quartz cuvette and placed in the cuvette holder of the UV-Vis spectrometer (Ocean Optics USB 4000 UV-Vis spectrometer). Then the UV-Vis spectrum was measured immediately. Results are shown in Fig. 3a, b.

3.6 Particle Size Analysis Measurements of SA-AuNP with Virus

Particle size of the SA-AuNP with virus was measured in the same way as described in Subheading 3.3. In short, the solution containing virus incubated SA-AuNP was placed in the built-in liquid sample holder of the particle size analyzer (Microtrac). The sample concentration was adjusted for the optimum result, as shown in Fig. 3d.

3.7 Colorimetric Detection of Viral Particles

1. Five hundred microliter of SA-AuNP solution was mixed with different dilutions of influenza virus solution and immediately added to a 1-cm path length quartz cuvette.

2. The cuvette was then placed in UV-Vis spectrometer (USB 4000, Ocean Optics) and absorption spectra were measured every minute until a plateau was reached.

In a typical experiment, it was observed that the absorption spectrum of SA-AuNPs solution changed immediately after the addition of influenza B/Victoria solution, with increasing OD value at 600–610 nm and decreasing OD value at 510 nm over time. The spectrum becomes stabilized after several minutes, as shown in Fig. 4. The absorbance of 0.4 nM SA-AuNPs solution was measured upon the addition of different dilutions of influenza B/Victoria. We observe that, when the virus concentration was <1.25 vol%, there was a linear correlation between the change of OD value (ΔOD) and the amount of added virus solution, with the detection limit estimated as 0.09 vol%. The final solutions after each test illustrated the gradual change of color from red to purple as the amount of virus solution increased (photograph in Fig. 4) .

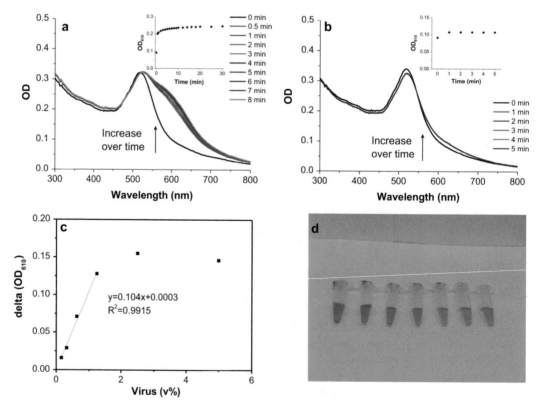

Fig. 4 (**a**) Absorption spectra of 0.4 nM SA-AuNPs over time after the addition of 2.5 vol% influenza B/Victoria solution; *inset*, OD_{610} value change. (**b**) Absorption spectra of 0.4 nM SA-AuNPs over time after the addition of 0.156 vol% virus solution; *inset*, change in OD_{610}. (**c**) Change in OD_{610} value after reaching plateau for different dilutions of virus. A linear correlation is observed for dilutions <1.25 vol% virus. (**d**) Photograph of SA-AuNPs with different dilutions of virus, 5, 2.5, 1.25, 0.625, 0.3125, 0.156, and 0 vol% from *left* to *right*, respectively

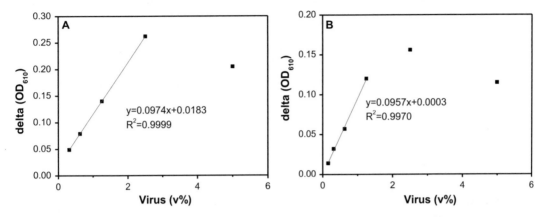

Fig. 5 (**a**) Change in OD_{610} measured for 0.8 nM SA-AuNP solution after the addition of influenza B/Victoria. A linear correlation is observed for dilutions <2.5 vol% virus. (**b**) Change in OD_{610} measured for 0.4 nM SA-AuNP solution after the addition of influenza B/Yamagata. A linear correlation is observed for dilutions <1.25 vol% virus

3.8 Increasing the Linear Dynamic Range of Virus Measurements

To increase the linear range of ΔOD vs. concentration of added virus solution, we used a higher concentration of SA-AuNP solution (0.8 nM) to conduct the similar tests, with results shown in Fig. 5a. It was observed that the linear range of ΔOD vs. concentration of added virus solution was extended to 2.5 vol% of virus in the solution. These results illustrate the importance of using a large amount of SA-AuNPs in the solution for virus detection when applying this scheme. We note that this method can also detect another virus subtype, influenza B/Yamagata. Results obtained from the influenza B/Yamagata solutions are similar to those of influenza B/Victoria and shown in Fig. 5b.

In conclusion, we demonstrate a simple method to synthesize gold nanoparticles using sialic acid as both the reducing agent and the stabilizing agent. Experimental results support the hypothesis that sialic acid molecules on SA-AuNP surface interact with viral envelope protein hemagglutinin, causing a colorimetric change of the SA-AuNP solution. The resulting SA-AuNPs can readily detect influenza B virus (HA titer of 512) diluted to 0.156 vol%. The method is effective for different influenza B lineages (Victoria and Yamagata) .

4   Notes

1. The amount of 1 M NaOH can be varied in the range of 20–100 μL to achieve the best results, as the local deionized water may have slightly different pH values.

Acknowledgment

P.Z. acknowledges support from the National Science Foundation (CBET-0931677/1065633). A.A.W. acknowledges National Institute of Health (R01AI089450 from NIAID) for support.

References

1. Zhong Z, Patskovskyy S, Bouvrette P, Luong JHT, Gedanken A (2004) The surface chemistry of Au colloids and their interactions with functional amino acids. J Phys Chem B 108:4046–4052

2. Link S, El-Sayed MA (1999) Size and temperature dependence of the plasmon absorption of colloidal gold nanoparticles. J Phys Chem B 103:4212–4217

3. Kreibig U, Genzel L (1985) Optical absorption of small metallic particles. Surf Sci 156:678–700

4. Elghanian R, Storhoff JJ, Mucic RC, Letsinger RL, Mirkin CA (1997) Selective colorimetric detection of polynucleotides based on the distance-dependent optical properties of gold nanoparticles. Science 277:1078–1081

5. Storhoff JJ, Marla SS, Bao P, Hagenow S, Mehta H, Lucas A, Garimella V, Patno T, Buckingham W, Cork W, Müller UR (2004) Gold nanoparticle-based detection of genomic DNA targets on microarrays using a novel optical detection system. Biosens Bioelectron 19:875–883

6. Thanh NTK, Rosenzweig Z (2002) Development of an aggregation-based immunoassay for anti-protein a using gold nanoparticles. Anal Chem 74:1624–1628

7. Pavlov V, Xiao Y, Shlyahovsky B, Willner I (2004) Aptamer-functionalized Au nanoparticles for the amplified optical detection of thrombin. J Am Chem Soc 126:11768–11769

8. Medley CD, Smith JE, Tang Z, Wu Y, Bamrungsap S, Tan W (2008) Gold nanoparticle-based colorimetric assay for the direct detection of cancerous cells. Anal Chem 80:1067–1072

9. Darbha GK, Singh AK, Rai US, Yu E, Yu H, Chandra Ray P (2008) Selective detection of mercury (II) ion using nonlinear optical properties of gold nanoparticles. J Am Chem Soc 130:8038–8043

10. Lee C, Gaston MA, Weiss AA, Zhang P (2013) Colorimetric viral detection based on sialic acid stabilized gold nanoparticles. Biosens Bioelectron 42:236–241

11. Sauter NK, Hanson JE, Glick GD, Brown JH, Crowther RL, Park SJ, Skehel JJ, Wiley DC (1992) Binding of influenza virus hemagglutinin to analogs of its cell-surface receptor, sialic acid: analysis by proton nuclear magnetic resonance spectroscopy and X-ray crystallography. Biochemistry 31:9609–9621

12. Varghese JN, Colman PM, van Donkelaar A, Blick TJ, Sahasrabudhe A, McKimm-Breschkin JL (1997) Structural evidence for a second sialic acid binding site in avian influenza virus neuraminidases. Proc Natl Acad Sci U S A 94:11808–11812

13. Weis W, Brown JH, Cusack S, Paulson JC, Skehel JJ, Wiley DC (1988) Structure of the influenza virus haemagglutinin complexed with its receptor, sialic acid. Nature 333:426–431

14. Suzuki T, Portner A, Scroggs RA, Uchikawa M, Koyama N, Matsuo K, Suzuki Y, Takimoto T (2001) Receptor specificities of human respiroviruses. J Virol 75:4604–4613

15. Vlasak R, Luytjes W, Spaan W, Palese P (1988) Human and bovine coronaviruses recognize sialic acid-containing receptors similar to those of influenza C viruses. Proc Natl Acad Sci U S A 85:4526–4529

16. Kunkel F, Herrler G (1993) Structural and functional analysis of the surface protein of human coronavirus OC43. Virology 195:195–202

17. Uncapher CR, DeWitt CM, Colonno RJ (1991) The major and minor group receptor families contain all but one human rhinovirus serotype. Virology 180:814–817

Chapter 8

MicroRNA Biosensing with Two-Dimensional Surface Plasmon Resonance Imaging

Ho Pui Ho, Fong Chuen Loo, Shu Yuen Wu, Dayong Gu, Ken-Tye Yong, and Siu Kai Kong

Abstract

Two-dimensional surface plasmon resonance (2D-SPR) imaging, which provides a real-time, sensitive, and high-throughput analysis of surface events in a two dimensional manner, is a valuable tool for studying biomolecular interactions and biochemical reactions without using any tag labels. The sensing principle of 2D-SPR includes angular, wavelength, and phase interrogation. In this chapter, the 2D-SPR imaging technique is applied for sensing a target microRNA by its corresponding oligonucleotide probes, with sequence complementarity, immobilized on the gold SPR sensing surface. However, the low SPR signal due to intrinsic properties such as low molecular weight and quantity (pico-nanomolar) of the microRNA in clinical samples limits the direct detection of microRNA. Therefore, we developed a biosensing technique known as MARS (MicroRNA-RNase-SPR) assay, which utilizes RNase H to digest the microRNA probes enzymatically for fast signal amplification, i.e., in order to increase both the SPR signal and readout speed without the need for pre-amplification of target cDNA by polymerase chain reaction (PCR). Practically, we targeted microRNA hsa-miR-29a-3p, whose signature correlates to influenza infection, for rapid screening of influenza A (H1N1) patients from throat swab samples.

Key words Biosensing, Disease screening, Influenza, MicroRNA, MicoRNA-RNase-SPR, Two-dimensional surface plasmon resonance, Signal amplification

1 Introduction

Surface plasmon resonance (SPR), a physical phenomenon of plasmonic propagation change near the surface of a nanometer-scale thin metallic (gold, silver) sensing layer, is a sensitive method to detect localized refractive index change on the sensing surface [1]. Two-dimensional surface plasmon resonance (2D-SPR) biosensing, sharing the same principle of SPR, has been extensively used in the last 10 years to study bimolecular interactions [2–5]. The main reason behind this growth is that 2D-SPR imaging provides the advantages of real time, label free, and high throughput for rapid, sensitive detection of target biological binding reactions such

Avraham Rasooly and Ben Prickril (eds.), *Biosensors and Biodetection: Methods and Protocols Volume 1: Optical-Based Detectors*, Methods in Molecular Biology, vol. 1571, DOI 10.1007/978-1-4939-6848-0_8, © Springer Science+Business Media LLC 2017

as antibody–antigen or protein–DNA interactions. There is also a trend toward detection of low-molecular-weight biomolecules such as small peptides and oligonucleotides. Researchers have used 2D-SPR sensing technology for rapid DNA and protein profiling [4, 5]. In this section, we give a brief overview of the principle of SPR and the development of 2D-SPR for biomedical applications.

The SPR phenomenon is an excited charge density oscillation induced by optical light propagating along the boundary between a metal film, such as a gold or silver layer, and a dielectric medium. The incident optical light is composed of p-polarized (transverse magnetic polarization) and s-polarized (transverse electric polarization) light. Only p-polarized light is normal to the boundary and propagation along the x-axis of a metal film, so it can couple into a surface plasmon mode for electron oscillation on the film. The energy of electron oscillation becomes heat and disappears, leading to the SPR phenomenon. However, the plasmonic propagation change in SPR occurs on the nanometer scale, so direct observation of the propagation is impossible with present technologies. Therefore various techniques, namely angular, wavelength, or phase interrogation schemes, are used for SPR signal extraction. Angular interrogation is the measurement of the reflectivity variation with respect to the incident angle under monochromatic light illumination. The SPR coupling angle is the reflected light intensity at its minimum point, or absorption dip (*see* Fig. 1a, b). It is correlated to the change in the dielectric constant value of the sample medium, or the refractive index of the sensing surface. In practice, measuring angular shift of SPR absorption dip or intensity shift at SPR absorption dip at a fixed optimal reflected angle can reveal any biomolecular event on the sensing surface. The wavelength interrogation technique, unlike angular interrogation, uses a fixed illumination angle and observes the reflectivity variation across the wavelengths to address the spectral absorption dip, indicating the presence of SPR, by spectrometer. The spectral location of the adsorption dip, namely resonant wavelength, is determined by the combined effect of the spectral dispersion of the prism and sensing layer, as well as sample medium. In this case, measuring the wavelength shift of SPR absorption dip or intensity shift at SPR absorption dip at a fixed optimal *wavelength* can detect any biomolecular interaction on the sensing surface. Phase interrogation is based on a resonant effect in the SPR system in which in or out of resonance signals cause a massive phase jump of the incident optical wave. The absorption dip in intensity-angle or wavelength diagram corresponds to the maximum resonance slope in the phase-angle or wavelength diagram (*see* Fig. 1c) . The steepness and extent of the resonance slope, caused by massive phase jump, depends on both the material and the thickness of the sensing surface. As the phase jump at the absorption dip leading to a steep phase change across resonance (which can be readily monitored by an optical

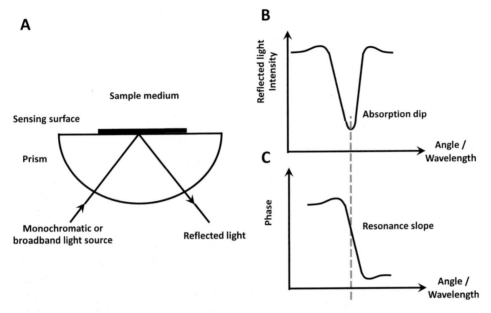

Fig. 1 Angular, wavelength, and phase interrogation schemes are illustrated. SPR sensing surface exhibits the SPR phenomenon under optical wave using monochromatic or broadband light source. The SPR phenomenon is observed from the analysis of reflected light (**a**). In angular interrogation, incident angle of monochromatic light (fixed wavelength) is scanned to identify the absorption dip of the reflected light. In wavelength interrogation the spectrum of incident light at a fixed angle results in an absorption dip at a particular wavelength of the spectral reflected light (**b**). In phase interrogation, SPR signal extraction of reflected light search for maximum resonance slope, which corresponds to absorption dip in above two schemes (**c**). Any bimolecular event on the sensing surface will lead to a shift in the absorption dip, and analysis of the shift in absorption dip generates the SPR signal readout

interferometer), the SPR signal has a better sensitivity factor than angular and wavelength methods using refractive index units (RIU) to represent SPR sensitivity, i.e., 10^{-8}–10^{-9} RIU, in which sensitivity factor is 10^{-6}–10^{-7} RIU. In practice, measuring the phase change of SPR resonance at a fixed optimal angle and wavelength can detect any biomolecular reactions on the sensing surface. Meanwhile, 2D-SPR, also called SPR imaging (SPRI), is the simultaneous sensing of multiple regions of a SPR surface. The optical setups of SPR and SPRI are very similar except for two components: illumination and detection. SPRI requires the detection area to be evenly illuminated so that different regions of sensing probes on the sensing surface are subjected to the same incident light intensity. To achieve this it is more common to use LED than laser illumination. 2D imaging detection is achieved either by 2D photodiode arrays or CCD camera. The advantage of SPRI over SPR for biological applications is mainly the ability of detecting multiple targets, such as cancer biomarkers in one sensing surface. Thus the speed of overall detection is increased allowing for rapid diagnosis, especially in clinical use. However, imaging capability also results in a notable decrease in sensitivity. The sensitivity is usually one-order magnitude lower in SPRI than in SPR, and a typical angular interrogation scheme has a sensitivity of 10^{-5} RIU in SPRI mode compared to the single-point SPR value of 10^{-6} RIU.

MicroRNA, the small noncoding RNA encoded from the human genome to regulate many biological functions, has been found to play an important roles the immune response during influenza virus infection [6]. The microRNA expression pattern thus acts as signature in identifying disease, and rapid detection of microRNA helps in rapid disease diagnosis. 2D-SPR is used to detect microRNA by the sequence complementary hybridization with the oligonucleotide probes immobilized on the SPR sensing surface. The small size and the concentration of microRNA lead to small SPR signals, preventing effective detection. Therefore, we describe the biosensing technique *MicroRNA-RNase-SPR* (MARS) using the RNase H enzymatic reaction to improve SPR signal amplification and readout speed without pre-amplification by PCR [7]. The workflow for microRNA sensing in the MARS assay is show in Fig. 2. Designated microRNA probes with spacers at the 3′end and biotin at the 5′end are immobilized on the SPR sensing

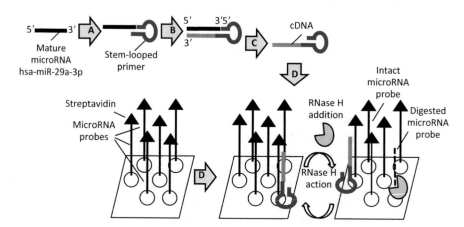

Fig. 2 Workflow for microRNA sensing in the MARS assay. First, mature microRNA is converted into cDNA by stem-loop primers (*top, A:* Stem-loop primer hybridization; *B:* Reverse transcription; *C:* Heat separation/denature; *D:* cDNA addition for sensing). On the SPR gold sensing surface (*bottom*), biotinylated-microRNA probes with spacer C6 are immobilized through thio-linkage. Unbound region on the sensing surface is covered with the blocker PEG 1000-SH. Streptavidins are then attached to the immobilized probes through the biotin-avidin interaction. Target cDNAs with complementary nucleotide sequence are hybridized to the probes to generate RNA–cDNA hybrids that increase the SPR signal. RNase H is added to digest the RNA probes and to release the hybridized cDNAs from the RNA–cDNA hybrids, The released cDNAs then binds to the new probes for further RNase H digestion, leading to the repeated cycle of RNase H reaction for signal amplification as indicated by *forward* and *backward semicircular arrows*. Depletion of the RNA probes on the sensing surface causes a decrease in SPR signal. Reprinted with permission from ref. 8

surface by 3′-thiolated linkage. Since our small microRNA probe leads to insignificant SPR signal change upon association or dissociation, streptavidin - a 60 kDa protein used to increase SPR signal, is added to the RNA probes as a cap on top of the probe through biotin–avidin interaction. This sensing surface is used to capture target cDNAs. RNase H enzymatic digestion removes a streptavidin-conjugated microRNA probe from the SPR sensing surface. Since the cDNA is not degraded, it repeatedly anneals to another microRNA probe leading to further degradation of probes as SPR signal amplies isothermally. We demonstrate the detection of microRNA hsa-miR-29a-3p from throat swab samples for influenza A (H1N1) patient screening. The screening is based on the analysis of hsa-miR-29a-3p, which is downregulated after influenza A H1N1 infection. We demonstrate use of an angular interrogation 2D-SPR imager with fixed angle to determine the intensity change of SPR absorption dip. As mentioned previously, different SPR interrogation systems, namely wavelength or phase, can be used without changing the sensing surface MARS assay platform.

2 Materials

2.1 Two-Dimensional Surface Plasmon Resonance System

1. The GWC SPRimager II system (GWC Technologies) as a SPR imager with temperature set at 25 °C to preheat for at least 1 h before sensing application.

2. External pump (EP-1 Econo Pump; Bio-Rad) with connection tubes of inner diameter 1.6 mm for sample loading site.

3. SF10 right-angle prism is used for SPR prism.

2.2 Surface Chemistry

1. SPRchip™ (GWC Technologies), the uniform gold-coated SF10 glass chip is used as SPR sensing chip, with the ability for 25 dots of 1 mm-diameter sensing probe to create 5×5 arrays.

2. Probe immobilization buffer is prepared sodium acetate at a final concentration of 5 mM (optional with 10% glycerol) in RNase- and DNase-free distilled water and stored at 4 °C.

3. Blocking buffer PEG-1000-SH (Takara Biotechnology) is prepared by dissolving the powder in absolute ethanol to 1 mM and stored at −20 °C.

4. Synthesized microRNA probes (microRNA probe hsa-miR-29a-3p: 5′-(biotin)-uagcaccaucugaaaucgguuauuuuuuuu-(C6)-SH-3′; microRNA probe hsa-miR-181-5p: 5′-(biotin)-aacauucaacgcugucggugaguuuuuuuuu-(C6)-SH-3′) are dissolved individually in the RNase- and DNase-free distilled water at a final concentration of 100 μM, aliquoted and stored at −20 °C for single use.

5. MicroRNA probes at working concentration are freshly prepared by dilution to 1 μM with probe immobilization buffer and stored at 4 °C until use.

2.3 Sensing Solutions

1. SPR working buffer is prepared at 100 mM KCl, 50 mM Tris, 10 mM MgCl2, 10 mM DTT in distilled water, and pH is adjusted to 7.8 by HCl or KOH. This buffer is stored at room temperature (*see* **Note 1**).

2. Synthesized cDNA (Takara, Inc.) (cDNA of hsa-miR-29a-3p: TAACCGATTTCAGATGGTGCTA; cDNA of hsa-miR-181-5p: ACTCACCGACAGCGTTGAATGTT) are dissolved in distilled water at 100 uM and stored in single-use aliquots at −20 °C.

3. Human microRNA sample is collected by throat swabs and immersed into Hank's balanced salt solution. Total RNA in sample solution is extracted by miRNeasy Mini Kit (Qiagen) according to manufacturing protocol. RNA yield is determined by NanoDrop ND-1000 spectrophotometer (Thermo Scientific). Target microRNA hsa-miR-29a-3p from 5 ng total RNA is reversed transcribed into cDNA by TaqMan MicroRNA Reverse Transcription Kit (Applied Biosystems) with mature-microRNA-specific stem-loop primer according to the manufacturer's protocol. The cDNA is stored at −20 °C until use (*see* **Note 2**).

4. Working solutions of microRNA probes are prepared by dilution with probe immobilization buffer to a final concentration of 1 μM.

5. Recombinant streptavidin (Bioss) is stored at −20 °C. Working streptavidin solution is prepared freshly by dilution with the SPR working buffer to 50 μg/mL.

6. Recombinant RNase H from *Escherichia coli* (Takara Biotechnology) is diluted from stock glycerol solution in SPR working buffer at 60 U/mL (*see* **Note 3**) .

3   Methods

3.1 2D SPR Imager Setup and Detection Workflow

1. SPRchip (GWC Technologies) from the sealed package is freshly opened, rinsed with RNase-free distilled water three times, and then air dried before use.

2. MicroRNA probes (final concentration 1 μM, 0.2 μL each) are blotted on the designated position of the SPR gold sensing plate (optional with an injector) and incubated at 4 °C for 16 h (*see* **Note 4**).

3. Uncoated site of the SPR gold sensing plate is blocked with PEG 1000-SH (final concentration 1 mM) at 25 °C for 2 h.

4. The prism is cleaned with absolute ethanol three times and then air dried. Pump tubing and connection tubing are pre-rinsed with RNase AWAY (Life Technologies) decontamination solution and then with RNase- and DNase-free distilled water (Life Technologies) at 200 μL/min before use.

5. The gold sensing plate is inserted with the prism with refractive index matching oil and placed into the SPR machine for SPR signal detection.

6. The external pump is connected to SPR machine, sensing surface is rinsed with SPR working buffer with a flow rate of 200 μL/min for 10 min (*see* **Note 5**).

7. Streptavidin working solution is added at 200 μL/min for 3 min.

8. The sensing surface is rinsed with SPR working buffer at 50 μL/min until a steady baseline with intensity fluctuation less than 0.01 arbitrary unit is obtained.

9. SPR intensity reading is set to zero.

10. Sample cDNA or artificial ssDNA is added at 50 μL/min for 10 min.

11. Biosensing surface is rinsed with SPR working buffer at 50 μL/min for 10 min.

12. RNase H working solution is added at 50 μL/min for 15 min.

13. Biosensing surface is rinsed with SPR working buffer at 50 μL/min until a steady baseline is obtained.

3.2 SPR Sensorgram and Detection Specificity

Specificity is important for this biomedical application since we aim to apply it for rapid screening of multiple microRNA signatures. This relies on the ability of the recognition element microRNA probe to specifically capture the target, as well as the specificity of RNase H to digest the probe for signal output. To evaluate this specificity, microRNA probes of hsa-miR-29a-3p and hsa-miR-181-5p, and single-stranded DNA (ssDNA) probe for miR-29a-3p 3p, are immobilized on different regions of the sensing array, as illustrated in Fig. 3. cDNAs of hsa-miR-29a-3p and hsa-miR-181-5p, RNase H are added subsequently at the time points indicated in the sensorgram of Fig. 3. There is a small increase in signal of microRNA probe and ssDNA hsa-miR-29a-3p only after the injection of cDNA of miR-29a-3p, indicating the hybridization between the microRNA probe and target cDNA occurs with high specificity. A large decrease in signal of microRNA probe hsa-miR-29a-3p after RNase H addition indicates that RNase H acts only on the RNA probes from the RNA–DNA hybrid due to the interaction between cDNA and microRNA probe, but not DNA–DNA hybrid, due to the interaction between cDNA and ssDNA probe. Subsequent injections of cDNA of hsa-miR-181-5p and RNase H cause a

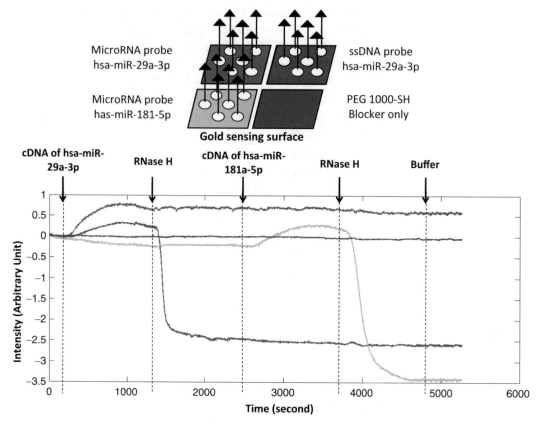

Fig. 3 Specificity of microRNA probe on target cDNA detection. The microRNA probes hsa-miR-29a-3p (*blue*), hsa-miR-181-5p (*green*), ssDNA probe miR-29a-3p (*red*), or PEG 1000-SH blocker only (*purple*) are immobilized on the surface of a SPR chip as shown in the *upper panel*. The cDNA of hsa-miR-29a-3p and hsa-miR-181-5p (100 nM) and RNase H (60 U/mL) are added at the time points indicated in the *lower panel*. The signal change is compared with the blank control (*purple*) to reveal the specificity for probe–target binding and RNase H action on RNA–cDNA and ssDNA–cDNA hybrids. Reprinted with permission from ref. 8

change in the signal of the microRNA probe hsa-miR-181a-5p, which like the pattern of hsa-miR-29a-3p, further supports the specificity of this MARS assay. Saturating the sensor surface with the blocker PEG 1000-SH causes no change in signal after cDNA and RNase H addition, indicating the absence of nonspecific binding of cDNA and RNase H on the sensing surface.

3.3 SPR Detection of microRNA in Throat Swab Sample

We focus on detecting microRNA hsa-miR-29a-3p from throat swab samples to identify influenza A H1N1 infection. The cDNA of a throat swab sample, instead of artificial cDNA, is added to the sensing surface for MARS assay. Figure 4 shows the SPR signal of healthy control and an H1N1-infected sample, with 10 nM artificial cDNA for easy comparison. The larger decrease in signal in the healthy control indicates that more hsa-miR-29a-3p is present. Our published finding showed hsa-miR-29a-3p is correlated to

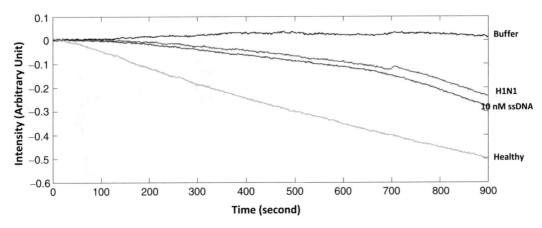

Fig. 4 Screening throat swab samples by MARS assay. The cDNA from healthy control and H1N1-infected samples are added to the SPR sensing surface for hsa-miR-29a-3p detection. Synthesized ssDNA (final concentration at 10 nM) and buffer are also used in the MRSA assay as the standard and reference. The concentration of hsa-miR-29a-3p in a patient sample could be determined from the standard curve of known cDNA concentration of hsa-miR-29a-3p

influenza infection, where low expression level of has-miR-29a-3p has been found in patients with influenza A H1N1 infection compared with healthy individuals [8]. Using a standard curve having standard concentrations of artificial cDNA, it is possible to quantify the microRNA concentration in those samples for more accurate disease screening [7].

4   Notes

1. Solutions must be prepared with RNase- and DNase-free distilled water, as minute amounts of RNase in solutions already lead to SPR signal change. Also, detection of the MARS response may lead to erroneous results if the pH of the reaction solutions deviates from the range of pH 7.0–9.0. The reaction may be halted and bimolecular elements may get denatured when the pH of the reaction solution is out of this range. Figure 5 represents the variation of SPR signals of the RNase H reaction step due to the variation of pH of reaction solutions. The optimal pH of the RNase H reaction is 7.8.

2. The microRNA extraction from human samples must be conducted in a biosafety level-3 culture hood to prevent human infection. The human sample immersed in Hank's balanced salt solution is stored at −80 °C if microRNA extraction is not performed immediately, in order to prevent nonspecific microRNA digestion by RNase present in the human sample.

3. High concentrations of glycerol (above 10% v/v) inhibit enzymatic reaction and causes sudden changes in SPR signal. It is

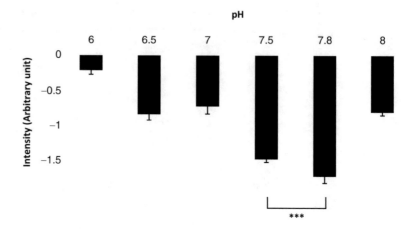

Fig. 5 pH effect on RNase H activity in MARS assay. The RNase H activity under different conditions is measured in the MARS assay by detecting the SPR intensity signal change after addition of the cDNA of hsa-miR-29a-3p (100 nM) on sensing surface with immobilized microRNA probes. SPR signal change is recorded under different pH values in SPR working buffer. Results are mean $\pm$ SEM from three independent experiments and analyzed by Student's *t*-test, where ***p-value <0.001

advisable to dilute the glycerol to a final concentration of 0.1% or lower.

4. The microRNA probe is carefully dotted on the sensing surface using a pipette to prevent accidental mixing with the neighbor probes. After this procedure, the sensing chip can be stored at 4 °C for up to 2 weeks without affecting the SPR signal (*see* Fig. 6).

5. To prevent false SPR signals, rinse thoroughly to ensure that bubbles are removed from the sensing surface, otherwise the bubbles will generate false signals.

Acknowledgment

The research is sponsored by ITF Grant (GHX/002/12SZ), CRF grant (CUHK1/CRF/12G), AoE (AoE/P-0/12) funding from the Hong Kong Special Administrative Region as well as the fund (SGLH20121008144756945) from Shenzhen, China.

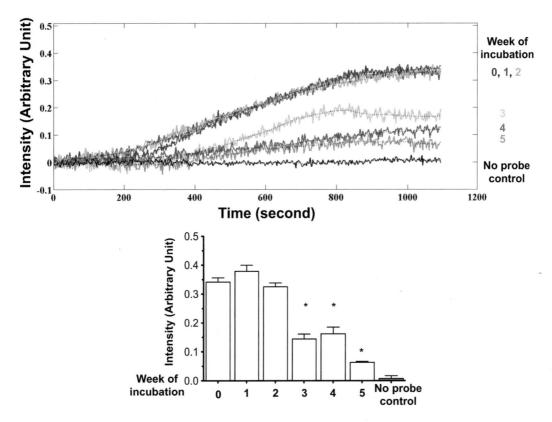

Fig. 6 Stability of the microRNA probes on SPR surface. The microRNA probes immobilized on the SPR gold sensing surface are stored at 4 °C for the times indicated and subjected to SPR detection on target cDNA hybridization, which increase intensity of the SPR signal. Results in the *upper panel* are illustrated as bar charts in the *lower panel*, with mean ± SEM from three independent experiments, where $*p < 0.05$ compared to SPR intensity signal at 0 weeks of incubation. Reprinted with permission from ref. 8

References

1. Kabashin AV, Patskovsky S, Grigorenko AN (2009) Phase and amplitude sensitivities in surface plasmon resonance bio and chemical sensing. Opt Express 17:21191–21204

2. Wong CL, Chen GCK, Ng BK et al (2011) Multiplex spectral surface plasmon resonance imaging (SPRI) sensor based on the polarization control scheme. Opt Express 19:18965–18978

3. Shao Y, Li Y, Gu D et al (2013) Wavelength-multiplexing phase-sensitive surface plasmon imaging sensor. Opt Lett 38:1370–1372

4. Wong CL, Ho HP, Suen YK et al (2007) Two dimensional biosensor arrays based on SPR phase imaging. Appl Optics 46:2325–2332

5. Wong CL, Ho HP, Suen YK et al (2008) Real-time protein biosensor arrays based on surface plasmon resonance differential phase imaging. Biosens Bioelectron 24:606–612

6. Tambyah PA, Sepramaniam S, Ali JM et al (2013) MicroRNAs in circulation are altered in response to influenza A virus infection in humans. PLoS One 8:e76811

7. Loo JFC, Wang SS, Peng F et al (2015) A non-PCR SPR platform using RNase H to detect MicroRNA 29a-3p from throat swabs. Analyst 140:4566–4575

8. Peng F, He H, Loo JFC et al (2016) Identification of microRNAs in throat swab as the biomarkers for diagnosis of influenza. Int J Med Sci 13:77–84

Chapter 9

Gold Nanorod Array Biochip for Label-Free, Multiplexed Biological Detection

Zhong Mei, Yanyan Wang, and Liang Tang

Abstract

Gold nanorod (GNR) based label-free sensing has been attractive due to its unique property of localized surface plasmon resonance (LSPR). Compared to bulk gold, the SPR of GNRs is more sensitive to the refractive index change caused by biological binding in the close proximity. Numerous studies have reported biological detection in solution based GNR probes. However, the biosensing has the intrinsic problems of fluctuating readings and short storage time due to nanoparticle aggregation. In contrast, a chip-based nanorod biosensor is a more robust and reliable platform. We have developed a nanoplasmonic biosensor in a chip format by immobilizing functionalized GNRs on a (3-mercaptopropyl)trimethoxysilane modified glass substrate. The covalent Au-S bond ensures a strong GNR deposition on the substrate. This biochip exhibits a high sensitivity and stability when exposed to physiological buffer with high ionic strength. Another advantage of GNR as optical transducer is its LSPR peak dependence on the aspect ratio, which provides an ideal multiplexed detection mechanism. GNRs of different sizes that exhibit distinct SPR peaks are combined and deposited on designated spots of a glass substrate. The spectral shift of the respective peaks upon the biological binding are monitored for simultaneous detection of specific analytes. Coupled with a microplate reader, this spatially resolved GNR array biochip results in a high-throughput assay of samples as well as multiplexed detection in each sample. Since most biological molecules such as antibodies and DNA can be linked to GNR using previously reported surface chemistry protocol, the label-free nanosensor demonstrated here is an effective tool for protein/DNA array analysis, especially for detection of disease biomarkers.

Key words Gold nanorods, Surface plasmon resonance, Biochip, Multiplex

1 Introduction

When the dimension of a matter is reduced to the nanometer scale (1–100 nm), the material can exhibit distinct properties from its bulk form, including optical, structural, electrical, and chemical properties. These changes make nanoparticles promising for a wide range of biomedical applications such as biosensing, imaging, and drug delivery [1–3]. Among diverse nanomatrials, plasmonic nanoparticles (Au, Ag) are attractive because of their unique optical

Avraham Rasooly and Ben Prickril (eds.), *Biosensors and Biodetection: Methods and Protocols Volume 1: Optical-Based Detectors*, Methods in Molecular Biology, vol. 1571, DOI 10.1007/978-1-4939-6848-0_9,
© Springer Science+Business Media LLC 2017

property, surface plasmon resonance (SPR) which arises from free electron oscillation induced by electromagnetic radiation (i.e., light) [4]. The SPR results in a strong absorption of the incident light at specific wavelength region which can be measured by a UV–Vis spectrophotometer. The SPR band intensity and wavelength depends on the metal type, particle size, shape, composition, and the dielectric constant of the surrounding medium, as theoretically depicted by Mie theory [5]. For a given nanosphere, the SPR condition is $\varepsilon_r = -2\varepsilon_m$, where ε_r is the dielectric function of the metal, and ε_m is the dielectric constant of the medium. This indicates an increase in ε_m requires an increase in ε_r to satisfy the SPR condition. Practically, when the refractive index (n) of the surrounding medium increases, the extinction spectrum shifts to longer wavelengths (known as red-shift). Since most biomolecules ($n \approx 1.4$–1.45) have a higher refractive index than water ($n \approx 1.33$), a red shift of the SPR peak is observed when the biomolecules are absorbed to the nanoparitlce surface in aqueous medium. If the nanoparticle is functionalized with special probe (i.e., antibody), the shifts can be induced by the specific target (i.e., antigen) binding to the probe. Thus, by monitoring the SPR shifts using UV–Vis spectrophotometer, biological events on the surface of plasmonic nanoparticles can be quantitatively detected. This is the mechanism of a label-free SPR biosensor.

Recently, gold nanoparticles (GNPs) have been widely used in biosensing because of their facile surface modification, chemical stability and excellent biocompatibility [6, 7]. GNPs with various shapes including nanosphere, nanarod, nanostar, nanoshell, and nanocage are reported for different sensing purposes [8]. Both theoretical study and experimental results have shown that nanorods have a greater performance than nanospheres, which is characterized by the sensitivity (i.e., plasmon peak shift per unit change in the effective refractive index of the surrounding medium, RIU^{-1}). Given its simple synthesis and large tunability of surface plasmon, gold nanorods (GNRs) have been a good alternative to other shapes in SPR-based biosensing [9]. Different from bulk gold films utilized in conventional SPR techniques, the anisotropic shape and nanosize effect enable gold nanorods an extremely intensified local electromagnetic field along its longitudinal direction, which is highly sensitive to changes in the local refractive index [10]. Once biological events occur at the surface of GNRs, the change in the local refractive index will cause a longitudinal peak shift of the extinction spectra of GNRs, which can be measured by a UV–Vis spectrophotometer. Compared to the instrumental setup in traditional SPR techniques [11], this nanosensing method is more sensitive [12], user-friendly, and cost-effective.

In addition, the longitudinal SPR wavelength of GNR is tunable in a broad range from UV–visible to far infra-red region by varying its aspect ratio (length to width ratio). With advance in the

(A)

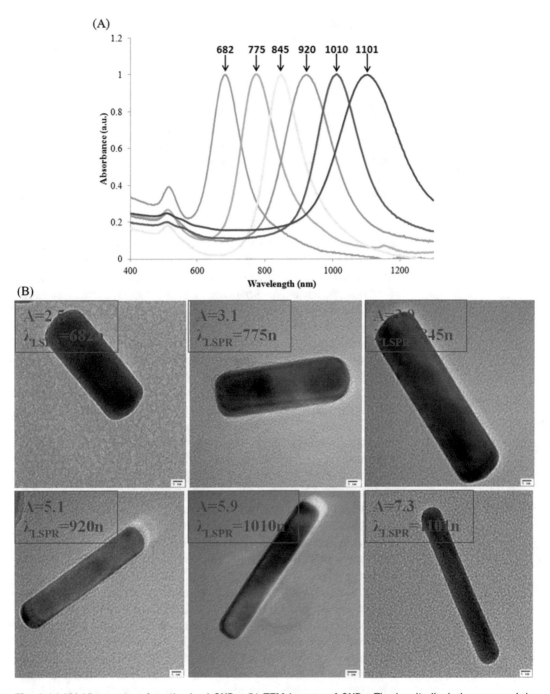

Fig. 1 (**a**) UV–Vis spectra of synthesized GNRs; (**b**) TEM images of GNRs. The longitudinal plasmon peak is tunable by varying rod aspect ratio

seed-mediated growth approach [13, 14], high-quality GNR can be synthesized with precise control of the size and shape. Figure 1 shows the extinction spectra of six GNRs and their respective aspect ratios. As the nanorod size increases, the longitudinal plasmon peak exhibits proportional red shifts. The correlation between aspect ratio (A) and the longitudinal SPR peak (λ_{LSPR}/nm) can be expressed as follows

$$\lambda_{LSPR} = 84.4A + 497.8 \qquad (1)$$

This unique property facilitates a multiplexed biosensing. GNRs of different aspect ratios can be mixed to fabricate multiplexed bioprobes to detect specific analytes simultaneously. For example, if one sample contains two different antigens (antigen 1 and 2) to be detected/quantified simultaneously, GNR 1 (AR = 2.5, LSPR λ = 682 nm) can be functionalized with antibody 1 for antigen 1 detection while GNR 2 (AR = 5.1, LSPR λ = 920 nm) is functionalized with antibody 2. These two functionalized GNRs are then mixed and there will distinct plasmon bands for the specific GNR on the absorption spectrum. The first peak at ~520 nm represents the characteristic gold resonance in transverse direction which is less sensitive to local refractive index changes. The second peak at 682 nm and the third peak at 920 nm reflect the longitudinal peaks from GNR1 and GNR2, respectively. These two dominate peaks are distinctly separated, thereby enabling a simultaneous monitoring of plasmonic shift at both positions independently for the dedicated target antigen detection in a single sample. However, solution-based GNR probe has its intrinsic problems. First, multiple washes cannot be avoided in the process of fabricating GNR bioprobes. These cause extinction intensity change and fluctuate readings. Second, the stability of a GNR suspension depends on temperature and surfactant concentration, which results in short storage time of the GNR probes. As such, it is highly desired to develop a GNR nanosensor in a chip format which significantly strengthen the utility of this label-free biosensing modality as a powerful bioanalytical device [15].

Given that antibodies have primary amine group, it is feasible to chemically modify antibody molecules with thiolation reagent to enable gold–antibody linkage [16]. Thus, these GNR-conjugated antibodies can work as biological receptors or bioprobes for detecting specific antigens. Since many antibodies or antigens are disease-representing biomarkers, rapid detection with a low-cost, specific and sensitive biosensor is of great significance for clinical diagnosis. Therefore, we develop a chip-based GNR biosensor for antigen detection and then we extend this biochip into an array format, which enables high-throughput screening as well as multiplexed detection. Figure 2 is a schematic of three processes involved in a label-free, multiplexed LSPR biochip. Human IgG and rabbit IgG were used as a general analyte for demonstration. First, GNRs of varying sizes are functionalized with anti-human IgG and anti-rabbit IgG, respectively. Next, the functional GNRs are immobilized to a thiolated glass substrate to construct chip-based LSPR

A. Thiolation of antibody, followed by GNR functionalization

B. Fabrication of multiplex GNR biochip

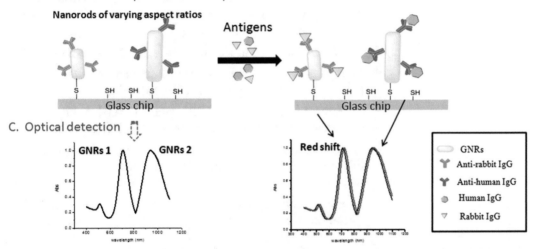

Fig. 2 Schematic of GNR biofunctionalization and design of label-free, multiplexed biosensing by GNRs of different aspect ratios. (**a**) Thiolation of antibody, followed by GNR functionalization. (**b**) Fabrication of multiplex GNR biochip. (**c**) Optical detection

sensor. When the antigens (human IgG and rabbit IgG) are present in the test samples, they will specifically bind to their respective antibodies (i.e., anti-human IgG and anti-rabbit IgG). Upon this binding, pronounced red shifts at the respective dominate LSPR bands are monitored by UV–Vis absorbance measurements. The magnitude of the red shift in the distinct LSPR peak wavelength is directly proportional to the amount of antigen binding onto the GNR probes.

2 Materials

2.1 Chemicals (See Note 1)

1. Gold(III) chloride trihydrate ($HAuCl_4$: 99%).
2. Sodium borohydride ($NaBH_4$:99%).
3. L-ascorbic acid (AA: 99%).
4. Cetyltrimethylammoniumbromide (CTAB).
5. Sodium oleate (NaOL).
6. Silver nitrate ($AgNO_3$:99%).
7. (3-mercaptopropyl)trimethoxysilane (MPTMS).

8. Poly (ethylene glycol) methyl ether thiol (SH-PEG, Mw~5000).

9. 2-iminothiolane hydrochloride (Traut's reagent: 98%).

2.2 Working Solutions

1. $HAuCl_4$ is dissolved in MilliQ water at 10 mM and stored at 4 °C in dark.

2. Bisurfactant solution: a mixture of 7.0 g CTAB and 1.237 g NaOL is added into 250 mL distilled water and sonicated at 50 °C for completely dissolving.

3. $NaBH_4$ is dissolved in ice-cold water at 10 mM before use.

4. Piranha solution: a 3:1 mixture of concentrated sulfuric acid (98%) and hydrogen peroxide (30%) solution.

5. Phosphate buffered saline (PBS, pH 7.4, 0.01 M) is prepared by dissolving one pouch of dry powder in 1 L deionized water.

6. Goat anti-human IgG and anti-rabbit IgG (2 mg/mL) are diluted at 10 μg/mL with PBS and stored at −20 °C before use.

7. Human serum IgG and rabbit serum IgG are dissolved at various concentrations (10, 20, 30, 40, 50, 60 nM) with PBS.

2.3 Instruments

1. Beckman-Coulter UV-NIR spectrophotometer ($DU^@720$, Fullerton, CA).

2. BioTek Multi-Detection Microplate Reader (Synergy™ 2, Winooski, VT).

3. MilliQ@ Direct Water Purification System instruments (EMD Millipore Corporation, Germany).

3 Methods (All Procedures Are Performed at Room Temperature Unless Notified Otherwise)

3.1 Chip Based Nanoplasmonic Biosensing for Single Antigen

3.1.1 Preparation of Gold Nanorods

1. Gold seed solution is prepared by adding 0.025 mL of 10 mM $HAuCl_4$ into 1 mL of aqueous 0.1 M CTAB, followed by adding 1 mL of 10 mM fresh-made ice-cold $NaBH_4$. The resulting solution is incubated at 27 °C for 30 min.

2. To prepare gold growth solution, 4 mM $AgNO_3$ is added to 20 mL of bisurfactant solution. The mixture is kept undisturbed at 30 °C for 15 min. Then 10 mL $HAuCl_4$ is added, followed by 90 min of stirring. 12.1 M HCl and 100 uL AA (64 mM) is added.

3. A specific amount of the seed solution is added to the growth solution. The resulting mixture is incubated at 29 °C overnight for full nanorod growth. To synthesize nanorods of varying aspect ratios, the amount of $AgNO_3$, HCl and seed solution are adjusted following the recipe (*see* Table 1).

Table 1
Growth conditions for GNRs with respective longitudinal SPR wavelengths

AgNO$_3$ (mL)	HCl (µL)	Seed (µL)	λ_{LSPR} (nm)
0.6	120	32	630
1.44	120	32	775
1.92	120	32	840
1.92	140	64	950

4. The suspension is centrifuged twice at 8500 rpm for 30 min. 5 mL MilliQ water is then added to resuspend the solid pellets and centrifuged again at 13000 rpm for 3 min. The resulting pellets are then redispersed in 5 mL solution (*see* **Note 2**).

3.1.2 Fabrication of Nanoplasmonic Biochip

1. Microscopy glass slides (ITO-coated, 7 mm × 50 mm × 0.7 mm; Delta Technologies, Loveland, CO) are cleaned with the piranha solution and heated at 70 °C for 40 min (*see* **Note 3**).

2. Rinse the glass slides with distilled water, ethanol and dry with nitrogen gas.

3. Immerse the cleaned slides in a MPTMS (10% v/v in ethanol) solution.

4. After 2.5 h of incubation, the slides are thoroughly rinsed with ethanol, water and dried with nitrogen gas.

5. To functionalize GNRs with antibodies, 20 µL of Traut's reagents (5 mg/mL) are first added to 0.1 mL of anti-human IgG for 2 h [16]. Then 200 µL of GNR solution and 100 µL of SH-PEG (10 mg/mL) are added for incubation overnight (*see* **Note 4**).

6. The anti-IgG conjugated GNRs are separated from the excess antibody by centrifugation at 8000 rpm for 10 min and finally redispersed in PBS.

7. Immerse the MPTMS-treated glass in the functionalized GNR colloids. After 2 h of shaking at 700 rpm, the glass substrates are rinsed several times with water and dried with nitrogen gas (*see* **Note 5**) .

3.1.3 Antigen Detection

1. Position the GNR-modified glass chip into a UV cuvette (2.5 mL, 12.5 × 12.5 × 45 mm; BrandTech Scientific, Wertheim, Germany) and insert the cuvette to the cuvette holder of the Beckman-Coulter UV-NIR spectrophotometer, as shown in Fig. 3. Measure the UV–Vis absorption spectra of GNRs on glass as the baseline.

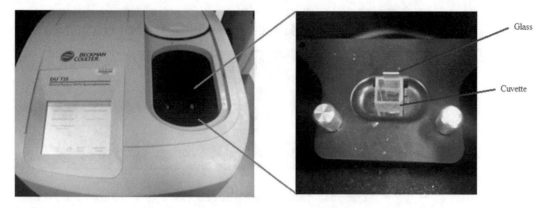

Fig. 3 Schematic of the measurement of UV–Vis spectrum using Beckman spectrophotometer

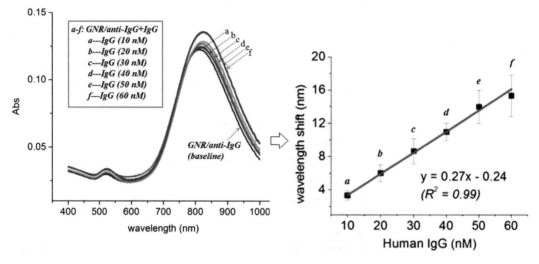

Fig. 4 Detection of human IgG targets by anti-human IgG modified GNRs biochip. The *black curve* is the absorption spectrum of anti-IgG modified GNRs biochip (baseline). The other six *colored curves* are absorptions spectra of the GNR/anti-IgG chip incubated in the target human IgG samples of varying concentrations (*a–f*: 10–60 nM). The peak wavelength at ~840 nm shows a red shift after IgG binding to the GNR/anti-IgG chip. A higher concentration of IgG resulted in a larger red shift. Calibration curve of the red-shift versus IgG concentration shows the longitudinal plasmon shift as a function of human IgG concentration

Dilute human IgG at a series of concentrations from 10 to 60 nM with PBS, then apply these samples on the GNR biochip surface and incubate for 30 min.

2. After washing the chips with water, the UV–Vis absorption spectra are measured again.

3. By comparing the spectra before and after human IgG conjugation, the red shift of longitudinal plasmon peak is observed in proportional to the target concentration (Fig. 4, *see* **Note 6**).

3.2 Nanoarray Based Multiplexed Biosensing for Multiple Antigens

The methods for gold nanorod synthesis and functionlization are similar as described above. To achieve multiplexed biosensing, GNRs with varying sizes are mixed and deposited on one substrate. Here, we take two-antibody detection for example.

1. GNRs with a LSPR wavelength at 840 nm are functionalized with anti-human IgG, while GNRs with a LSPR at 630 nm are functionalized with anti-rabbit IgG.

2. Microscopy glass slides ($3'' \times 1'' \times 1.0$ mm; Fisher Scientific, Pittsburgh, PA) are treated with Piranha and MPTMS solution as described above.

3. 10 μL of antibody-modified GNRs with mixed sizes are dropped onto designated spots on the glass slides and incubated for 2 h. The detection spots are arranged to mimic the layout of a 96 well plate to accommodate an absorption reading by the BioTek plate reader in a high-throughput fashion (Fig. 5).

4. Wash the substrate with water three times and dry it with nitrogen gas.

5. To perform a multiplexed detection, a sample solution (10 μL) of mixed human IgG and rabbit IgG are dropped onto the nanosensing spots and incubated for 30 min until equilibrium.

6. The glass substrate is attached to the bottom of a 96-well plate for absorbance measurement. The UV–Vis absorbance mode is chosen with a scanning range from 400 to 999 nm. As the light source scans the spots from well to well, the corresponding UV–Vis spectra are displayed in the same layout and ready for analysis. Figure 6 is the result of multiplexed biosensing of human and rabbit IgG samples at various concentrations (**Note 7**).

4   Notes

1. All the materials are available from Sigma-Aldrich (St. Louis, MO, USA) unless otherwise noted.

2. The washing steps are crucial for GNR functionalization and immobilization. Three times' washing is optimal to remove the unreacted chemicals and most surfactants. Because CTAB bilayer absorbed on GNR surface could block reactions with gold, it is necessary to decrease its concentration to the critical micelle concentration (CMC) at which GNR is stabilized. Further washing could break the stability of GNR and causes aggregation.

3. ITO-coated glass is chosen because of it conductivity, which makes the GNR on glass observable under scanning electron

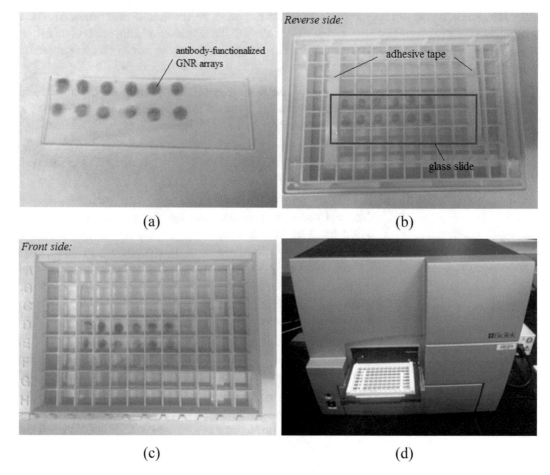

Fig. 5 Representative pictures of (**a**) antibody functionalized GNR array on glass substrates which is expanded to a high-throughput biosensing chip. Each *dark spot*, containing two sizes of functionalized GNRs, works as an individual sensing platform for detecting two antigens simultaneously in one sample. Thus, this biochip can detect 12 samples at one time; (**b**) Reverse side and (**c**) front side of attaching the chip to the bottom of a 96-well plate frame for accommodating the reading on microplate reader (**d**)

microscope (SEM). The ITO coating doesn't influence MPTMS treating.

4. As CTAB is replaced by antibodies, SH-PEG is added 10 min after antibody for stabilizing GNRs.

5. Weakly absorbed (not via covalent bonding) GNR nanoparticles onto the glass can be washed off to ensure a stable and robust biochip [17].

6. The nanorod biochip shows a linear response to the analyte in the range from 10 to 60 nM with high sensitivity of 0.27 nm/nM ($R^2 = 0.99$) for human IgG, which is more than 300% comparison with the reported literature with the sensitivity of 0.0607 nm/nM [17].

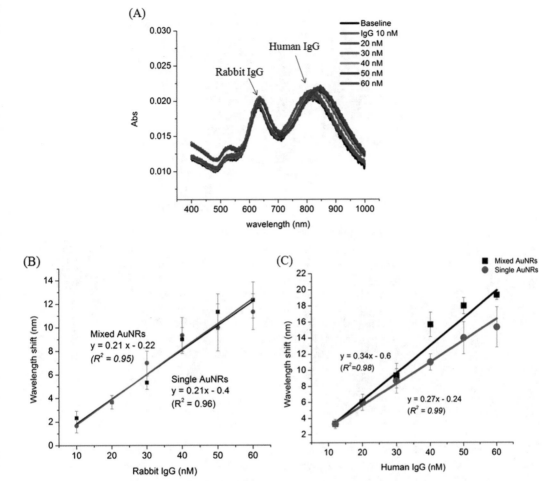

Fig. 6 Multiplexed biosensing of human and rabbit IgG samples at various concentrations. (**a**) UV–Vis spectra of the GNR biochip before and after detection of six various concentrations of mixed IgG. The shorter wavelength around 630 nm is dedicated for rabbit IgG assay, while the longer one around 840 nm for human IgG assay; (**b**) Plotting curve (*black*) of wavelength shift around 630 nm versus the concentration of rabbit IgG; (**c**) Plotting curve (*black*) wavelength shift around 840 nm versus the concentration of human IgG. Both curves suggest that the nanorod chip shows a linear response to the analyte (rabbit or human IgG) in the range from 10 to 60 nM. For comparison, the calibration curves of single GNR based biochip mentioned in Subheading 3.1 are also included (*red*). Interestingly, the multiplexed biosensor has an increased sensitivity than single biosensor for human IgG detection

7. Nonspecific binding studies indicate minimal cross-reactivity of the antibody moieties for high specificity in the multiplexed biosensing. It is observed that the sensitivity is increased as compared to individual detection, especially for longer GNR probes.

8. The biochip achieved by this method demonstrated a random assembly of GNRs on the substrate. Theoretical study has proved that the local electromagnetic field on the tips of a

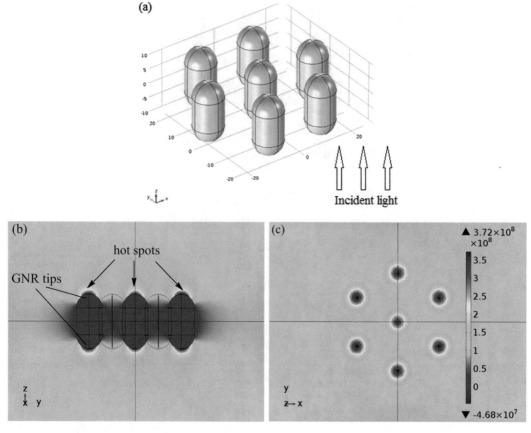

Fig. 7 Local electromagnetic field simulation of a vertical GNR array. (**a**) model of GNR array; (**b**) local field of side view in the cross section along the *y–z* plane; (**c**) local field of the top view along the z direction. The *rainbow bar* shows the intensity. A *redder color* indicates a more intensified local field

nanorod is much stronger than that on the side facets. And the CTAB density of tips is lower than the sides, which ensures a less steric hindrance for the tip modification [18]. For the future perspective, controlling the orientation of the GNR array assembly on the substrate is the current focus. For instance, if GNRs are orderly organized to form a vertically standing array, a large amount of homogeneous hotspots will be created at the GNR surface due to the plasmon coupling effect. Figure 7 is the local electromagnetic field simulation of a vertical nanoarray consisting of seven GNRs using COMSOL 5.0. The incident light with an intensity of 1.0×10^8 V/m propagates along z-direction, passing through the GNR array. The simulation results show that the GNR array exhibit highly uniform and enhanced local field on the tips. This unique property is desired for biomedical applications such as nanoreactor and ultrasensitive SERS sensor [19].

References

1. Giri S, Trewyn BG, Lin VS (2007) Mesoporous silica nanomaterial-based biotechnological and biomedical delivery systems. Nanomedicine (Lond) 2(1):99–111
2. Lee SH, Sung JH, Park TH (2012) Nanomaterial-based biosensor as an emerging tool for biomedical applications. Ann Biomed Eng 40(6):1384–1397
3. Oyelere AK et al (2007) Peptide-conjugated gold nanorods for nuclear targeting. Bioconjug Chem 18(5):1490–1497
4. Chylek P (1986) Absorption and scattering of light by small particles. Appl Opt 25(18):3166
5. Bennett HS, Rosasco GJ (1978) Resonances in the efficiency factors for absorption: Mie scattering theory. Appl Optics 17(4):491–493
6. Upadhyayula VK (2012) Functionalized gold nanoparticle supported sensory mechanisms applied in detection of chemical and biological threat agents: a review. Anal Chim Acta 715:1–18
7. Yu L, Andriola A (2010) Quantitative gold nanoparticle analysis methods: a review. Talanta 82(3):869–875
8. Dreaden EC et al (2012) The golden age: gold nanoparticles for biomedicine. Chem Soc Rev 41(7):2740–2779
9. Anker JN et al (2008) Biosensing with plasmonic nanosensors. Nat Mater 7(6):442–453
10. Jain PK et al (2006) Calculated absorption and scattering properties of gold nanoparticles of different size, shape, and composition: applications in biological imaging and biomedicine. J Phys Chem B 110(14):7238–7248
11. Chou SF et al (2004) Development of an immunosensor for human ferritin, a nonspecific tumor marker, based on surface plasmon resonance. Biosens Bioelectron 19(9):999–1005
12. Sim HR, Wark AW, Lee HJ (2010) Attomolar detection of protein biomarkers using biofunctionalized gold nanorods with surface plasmon resonance. Analyst 135(10):2528–2532
13. Ye X et al (2013) Using binary surfactant mixtures to simultaneously improve the dimensional tunability and monodispersity in the seeded growth of gold nanorods. Nano Lett 13(2):765–771
14. Gulati A, Liao H, Hafner JH (2006) Monitoring gold nanorod synthesis by localized surface plasmon resonance. J Phys Chem B 110(45):22323–22327
15. Wang Y, Tang L (2015) Multiplexed gold nanorod array biochip for multi-sample analysis. Biosens Bioelectron 67:18–24
16. Wang X et al (2015) Gold nanorod biochip functionalization by antibody thiolation. Talanta 136:1–8
17. Wang Y, Tang L (2013) Chemisorption assembly of Au nanorods on mercaptosilanized glass substrate for label-free nanoplasmon biochip. Anal Chim Acta 796:122–129
18. Orendorff CJ, Murphy CJ (2006) Quantitation of metal content in the silver-assisted growth of gold nanorods. J Phys Chem B 110(9):3990–3994
19. Lim DK et al (2011) Highly uniform and reproducible surface-enhanced Raman scattering from DNA-tailorable nanoparticles with 1-nm interior gap. Nat Nanotechnol 6(7):452–460

Chapter 10

Resonant Waveguide Grating Imager for Single Cell Monitoring of the Invasion of 3D Speheroid Cancer Cells Through Matrigel

Nicole K. Febles, Siddarth Chandrasekaran, and Ye Fang

Abstract

The invasion of cancer cells through their surrounding extracellular matrices is the first critical step to metastasis, a devastating event to cancer patients. However, in vitro cancer cell invasion is mostly studied using two-dimensional (2D) models. Three-dimensional (3D) multicellular spheroids may offer an advantageous cell model for cancer research and oncology drug discovery. This chapter describes a label-free, real-time, and single-cell approach to quantify the invasion of 3D spheroid colon cancer cells through Matrigel using a spatially resolved resonant waveguide grating imager.

Key words Adhesion, Colorectal cancer cell, Dynamic mass redistribution, Extracellular matrix, Invasion, Migration, Multicellular spheroid, PTEN, Resonant waveguide grating

1 Introduction

Tumor metastasis is the most devastating aspect of cancer and has been an active area for developing targeted therapeutics [1, 2]. Several in vitro assays have been developed to determine the migratory and invasive capacities of tumor and stromal cells, to elucidate the underlying biochemical and cellular mechanisms of metastasis, and to screen compounds that can inhibit distinct steps of metastasis [3–7]. These assays differ not only in operational and detection principles but also in cell models. Cell migration is often characterized based on the healing rate of a physically wounded two-dimensional (2D) cell monolayer, the recovery rate of a cell monolayer having an exclusion zone, or the number of cells adherent on the underside of a Transwell porous membrane after passing through the membrane. Cell invasion is often examined based on the number of individual cells invading through an extracellular matrix (ECM) (e.g., Matrigel) coated Transwell porous membrane. Chemotaxis, the movement of cells in response to a chemical

Avraham Rasooly and Ben Prickril (eds.), *Biosensors and Biodetection: Methods and Protocols Volume 1: Optical-Based Detectors*, Methods in Molecular Biology, vol. 1571, DOI 10.1007/978-1-4939-6848-0_10,
© Springer Science+Business Media LLC 2017

stimulus, is often studied by introducing a chemoattractant in the lower chamber of a Transwell device. Common to these assays is that they only permit narrowly focused snapshots into complex biological phenomena, and have limited ability to recapitulate in vivo process as primary tumors often display vast structural heterogeneity and microenvironments, and cancer cells can migrate in all directions [7, 8].

In the past years, there is an increase in using three-dimensional (3D) cell models including multicellular spheroids for oncology drug discovery. Compared to 2D cell monolayers, these 3D models are believed to more closely recapitulate the in vivo biology of solid cancers, thus enabling more accurate prediction of cancer behaviors [9–11] and in vivo drug efficacy, potency, and safety [12–14]. Concurrent to the increasing use of 3D models for cancer research is the development of advanced invasion and migration assays [15–20]. For instance, spheroid sprouting assay is used to examine the invasion of cells in a spheroidal structure into surrounding 3D ECM gel [21, 22]; however, this assay mostly detects late invasion events. On the other hand, cell invasion into spheroid assay is useful for studying the invasion of individual cells into a spheroidal cluster of another cell type [23], or the cell invasion between spheroids of two different cell types [24, 25]; however, this assay often uses fluorescently labeled cells, has low temporal resolution, and is cumbersome in quantification [7].

Recently, we have developed a label-free, real-time, single-cell, and quantitative assay, termed label-free single cell $3D^2$ invasion assay, to track the invasion of 3D spheroid cells through a 3D matrix [26, 27]. This chapter describes the protocols to determine the migratory and invasive capacities of three colon cancer cell lines, HT29, HCT116 wild-type (WT), and HCT16 PTEN$-/-$ using wound healing, Transwell invasion, and label-free single cell $3D^2$ invasion assays. PTEN (phosphatase and tensin homolog) is a tumor suppressor negatively regulating the PI3K signaling pathway, and PTEN deletion is correlated with colorectal cancer metastasis and poor patient survival [28].

1.1 Sensor Configurations and Detection Schemes

For the label-free single cell $3D^2$ invasion assay, there are four critical components, including resonant waveguide grating (RGW) biosensor microplate, Matrigel, spheroids of cancer cells, and RWG imager (Fig. 1).

The biosensor microplate, commercially known as Epic® 384well cell culture compatible microplate from Corning Incorporated, is used directly. Within the bottom of each well of the microplate, there is a RWG biosensor. The biosensor itself consists of three layers—a glass substrate, a grating structure, and a waveguide thin film with high refractive index (Fig. 1a). Under illumination with a light source light at a specific wavelength is coupled into the waveguide through diffraction, creating a surface

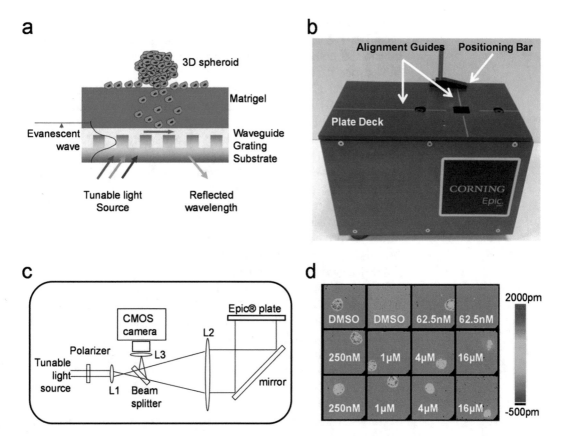

Fig. 1 Spatially resolved resonant waveguide grating (RWG) imager for label-free single cell monitoring. (**a**) The principle of RWG biosensor for monitoring the invasion and adhesion of cancer cells. (**b**) The instrument. (**c**) Optical setup of the imager, which consists of three components: a tunable light source to sweep the wavelength range, an array of optical components to guide and expand light path leading to illuminate at a normal incident angle all biosensors within a plate, and a CMOS camera to record the resonant wavelength image. (**d**) A false-colored DMR image of a 3 × 4 biosensor array. All biosensors were pre-coated with 10 μL 0.1 mg/mL Matrigel. The image was obtained 24 h after spheroids of HT29 cells in the absence and presence of vandetanib at different doses were placed on the top Matrigel surface. False color bar scale: −500 (*blue*) to 2000 pm (*red*). The pattern and area of adherent cells is indicated by the green-to-red spot, while the signal intensity is indicated by the false color scale—*red*: high positive signal; *green*: moderate positive signal; *blue*: negative signal. The well corresponding to the *second left* in the *top row* did not receive any spheroid as the negative control. Condition of spheroid formation: 40K per well seeding density, 4 days culture. The adhesion area and signal was found to be decreased as the dose of vandetanib increases

bound electromagnetic field (also known as evanescent wave) that has a characteristic penetration depth (~200 nm) extending into the medium or cell layer. After propagating within the waveguide for a short distance due to total internal reflection, the light eventually leaks out the waveguide and reflects back. The optical content (e.g., wavelength, angle, intensity) of the reflected light is monitored and recorded in real time, depending on the detection systems. The wavelength of the reflected light, termed as the resonance wavelength, is identical to that of the coupled light and is sensitive to local index of reflection, which is proportional to the density of cellular matters within the penetration depth of the biosensor [29, 30].

Matrigel is used to create a 3D extracellular matrix on the biosensor surface by placing equal volume Matrigel solution into each well of a biosensor microplate at 4 °C followed by overnight gelation at 37 °C (Fig. 1a). Matrigel is a gelatinous protein mixture secreted by Engelbreth-Holm-Swarm (EHS) mouse sarcoma cells, and is the most widely used basement membrane due to its reliability and reproducibility. Matrigel consists of approximately 60% laminin, 30% collagen IV, and 8% entactin. Entactin is a bridging molecule that interacts with laminin and collagen IV, and contributes to the structural organization of these ECM molecules. Matrigel generally contains the transforming growth factor-β (TGF-β), epidermal growth factor, insulin-like growth factor, fibroblast growth factor, tissue plasminogen activator, and other growth factors which occur naturally in the EHS tumor. There is also residual matrix metalloproteinases derived from the tumor cells. However, the Matrigel used here has a reduced level of a variety of growth factors except for TGF-β. Matrigel is a liquid at 4 °C and starts to form a gel above 10 °C [31]. The thickness of Matrigel film after gelation is determined by the volume of the solution used for coating. Coating with 10 μL Matrigel solution results in a thickness of ~630 μm for the 384-well microplate [26]. On the other hand, the density and porosity of Matrigel is determined by the concentration of total protein concentration used for coating. The higher concentration the Matrigel solution is the denser 3D matrix is formed, resulting in slower invasion [27].

Spheroidal clusters of human cancer cells are generated by seeding equal number of cells into each well, followed by culturing in Corning ultra-low attachment (ULA) round-bottom 96-well plates. The ULA plate enables self-assembly of cancer cells into a single 3D spheroidal cluster in the bottom of each well mediated by ultralow attachment surface chemistry and gravitational force [32].

A spatially resolved RWG imager (Fig. 1b) is used to achieve single cell monitoring of cancer cell invasion through the Matrigel. This imager is based on a swept wavelength interrogation scheme, wherein a tunable light source combining a broadband light source with a high precision, narrow-band optical filter is used to illuminate simultaneously a 3 × 4 biosensor array within a 384-well microplate (Fig. 1c), and a complementary metal oxide semiconductor (CMOS) digital camera is used to record the resonance images with a spatial resolution of 12 μm (Fig. 1d) [33, 34]. The wavelength sweeping from 825 to 840 nm is done stepwise in 100 pm every 20 ms. Owing to the fine difference in resonance condition among different pixels as well as the use of wavelength sweeping technique for illumination, the imager permits location-dependent time resolved resonance within a sensor, thus enabling high resolution imaging, although the coupled light tends to travel within the waveguide for a short distance [33]. The time series resonance images are recorded and used to calculate wavelength shifts (in picometer, pm) at each pixel with a temporal resolution of 3 s, resulting in spatially resolved kinetic responses, one per pixel. Since the resonance wavelength is sensitive to the change in local mass density, the kinetic response is often referred to dynamic mass redistribution (DMR) signal [35].

The label-free single cell 3D^2 invasion assay employs the RWG imager to noninvasively track in real time the adhesion event of cells after dissociating from a spheroid and invading through the matrix. The invasion is initiated after a single spheroid of cancer cells is placed on the top surface of the Matrigel film. Since the thickness of Matrigel is far greater than the penetration depth of the biosensor, a cell has to dissociate from the spheroid, invades through the matrix, and finally adheres onto the sensor surface in order to generate a detectable signal. Since adhesion is a rapid process at single cell level [26, 27], the DMR signals obtained can be used to study the behavior and mechanism of the invasion of cells in spheroid through the 3D matrix.

Of note, the RWG biosensor has been shown to permit DMR assay [29, 30, 35], which enables real-time investigation of a wide range of cell phenotypes [36], receptor signaling [37–39], and cellular processes [40, 41] (reviewed in [42–44]).

2 Materials

2.1 Tissue Culture Medium and Cell Line

1. Human colorectal adenocarcinoma HT-29 cell line (ATCC® HTB-38™, American Type Cell Culture, Manassas, VA, USA).

2. The isogenic colorectal carcinoma (CRC) cell lines, HCT116-WT and HCT116-PTEN−/− cells (Horizon Discovery Ltd., Cambridge, UK) (see **Note 1**).

3. McCoy's 5A medium (Catalog #16600, Life Technologies, Carlsbad CA, USA).

4. Complete medium for HT29: McCoy's 5A modified medium supplemented with 10% fetal bovine serum, 4.5 g/L glucose, 2 mM glutamine, and 1× Pennstrep.

5. Complete medium for the isogenic cell lines: McCoy's 5A medium supplemented with 10% fetal bovine serum, 4.5 g/L glucose, 1.5 mM glutamine, and 1× Pennstrep.

6. Serum-free medium for the isogenic cell lines: McCoy's 5A medium supplemented with 4.5 g/L glucose, 1.5 mM glutamine, and 1× Pennstrep.

7. Pennstrep antibiotic solution 100×: 10,000 units penicillin/mL, 10,000 μg streptomycin/mL.

8. Trypsin–ethylenediaminetetraacetic acid (EDTA) solution 10×: 2.5% Trypsin, 0.2% 4Na$^+$-EDTA.

9. Corning® Matrigel Growth Factor Reduced (GFR) Basement Membrane Matrix (Catalog #354230).

10. LY294002 and wortmannin (Tocris Chemical Company, St. Louis, MO, USA).

11. Vandetanib (LC Laboratories, Woburn, MA, USA).

12. Dimethyl sulfoxide (DMSO) .

2.2 Microplates and Instruments

1. Epic® BT high-resolution RWG imager (Corning Incorporated, Corning, NY, USA).

2. Nikon Eclipse TS100 inverted light microscope equipped with NIS Elements F3.0 Software (Nikon Instruments Inc., Melville, NY, USA).

3. Matrix 16-channel electronic pipettor (Thermo Fisher Scientific, Hudson, NH).

4. Corning 75 cm^2 tissue-culture treated polystyrene flask with vented cap (T75).

5. Corning 384-well polypropylene compound storage plate.

6. Corning® Epic® 384-well cell assay microplate (Catalog #5040).

7. Corning UltraLow Attachment (ULA) 96-well round bottomed plate (Catalog #7007).

8. Corning® HTS Transwell®-24 well permeable supports, 6.5 mm Transwell® with 8.0 μm pore polycarbonate membrane inserts, sterile (Catalog #3422) .

2.3 Data Analysis Software

1. Corning Epic® BT Processor software.

2. GraphPad Prism 5 Software (Graph Pad Software Inc., La Jolla, CA, USA).

3. NIH Image J (http://imagej.nih.gov/ij/).

4. Microsoft Excel (Microsoft Inc., Seattle, WA, USA).

3 Methods

3.1 3D Spheroid Cell Culture

This section describes the protocol to form uniform spheroids of three CRC cell lines by directly culturing an equal number of cells in each well of a 96-well ULA round bottomed microplate (Fig. 2a, b). The spheroids formed can be transitional (meaning that it can grow in size over time) or stable (meaning that it stops growing). For the 3D^2 invasion assay, mechanically stable spheroids are preferred since it is important to minimize potential interference with the assay results of mechanically dissociated cells from the

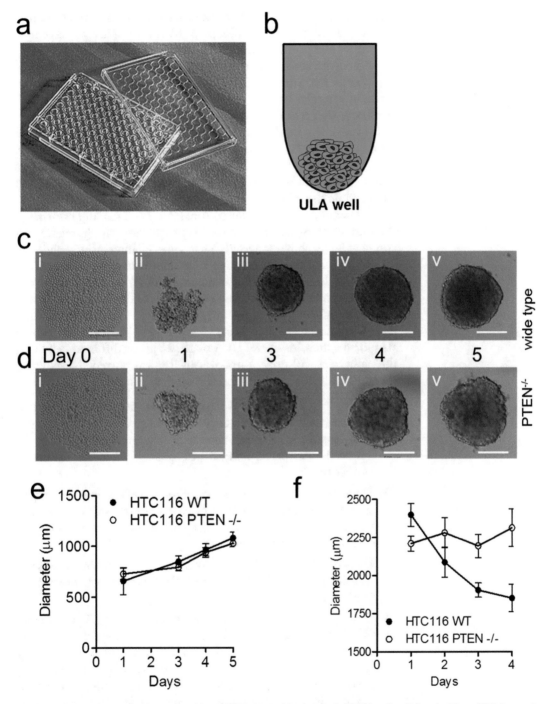

Fig. 2 Growth pattern of spheroidal cells of HCT116 and its isogenic PTEN null cell line in 96-well ULA round bottomed plate. (**a**) ULA microplate. (**b**) schematic drawing showing how a spheroid is formed within a ULA round bottomed well. (**c, d**) Daily light microscopic images of HCT116 WT cells (**c**), or HCT116 PTEN−/− cells (**d**) at Day 0 to Day 5 (*i–v*, respectively). Seeding density: 1000 cells per well; scale bar: 500 μm. (**e, f**) The diameter of spheroidal clusters of both cell lines as a function of culture duration when seeding density is 1000 (**e**) or 20,000 (**f**) cell per well. This figure is adapted from ref. 27 with permission

spheroidal structure. In addition, spheroids that are much smaller than the dimension of the biosensor (2×2 mm) are preferred for their effective positioning onto the biosensor surface.

When the seeding density is 1000 cells per well, the two isogenic cells become loosely attached right after seeding, started forming irregular clusters at day 1, and spheroids at day 3, which continuously grown in size (Fig. 2c–e). This suggests that under transitional spheroid formation condition PTEN knockout has little impact on the formation and growth of spheroids. However, when the seeding density is 20,000 cells per well, the wild-type cells rapidly formed a spheroid whose size continuously decreased till day 4, while the PTEN−/− cells formed a spheroid whose size was barely changed over the 4 day culture (Fig. 2f). This suggests that at the high seeding density the wild-type cell can self-aggregate and compact into a stable spheroid due to cell–cell interaction, while the PTEN null cell can self-aggregate but fail to undergo compaction due to the loss of epithelial characteristics [45]. For a given cell type, the spheroids formed can reach a stable dimension likely due to the limitation of nutrition and oxygen supply and/or accumulation of cellular metabolic waste (e.g., lactate) [46]. HT29 cell can also form spheroids in the ULA plate [26]. For different cell lines, the spheroid culture protocol can be optimized by altering medium, culture duration, and cell seeding numbers.

1. Passage the cells in T-75 flask using $1\times$ trypsin–EDTA when approaching 90% confluency. Generally, cells are split every 5 days, and a 1:10 split is used to provide maintenance culture. The cells between 3 and 25 passages are preferred for assays (*see* **Note 2**).

2. Once approaching confluency, harvest cells using $1\times$ trypsin–EDTA solution. Centrifuge the cells down, remove the supernatant, and resuspend the cell pellet in freshly prepared complete cell culture medium.

3. Dispense 200 μL cell suspensions at varied numbers of cells per well into each well of a 96-well ULA round bottomed plate (*see* **Note 3**).

4. Culture in the complete cell culture medium at 37 °C with 5% CO_2 for 4–8 days. For HT29, partially exchange the media by removing 100 μL old media at day 4 or 7 and then replacing once with 100 μL fresh media. For the two isogenic cell lines, partially exchange the media at day 3. To minimize disruption of the spheroids during media changes, hold the pipette tip at approximately a 45° angle away from the center of the well bottom.

5. Collect time series light microscopic images for each well at specific time points.

6. Calculate the diameter of each spheroidal structure at a specific time using NIH ImageJ software.

3.2 Matrigel Coating

This section describes the protocol to coat RWG biosensor plates or Transwell inserts with Matrigel for invasion assays.

1. Precool Matrigel, biosensor plates, Transwell inserts, pipets, and tips at 4 °C before coating. This step is essential to prevent premature gelation, and to ensure uniform coating.

2. Dilute Matrigel using the complete media to 0.1 mg/mL or 0.2 mg/mL (1 × 50 or 1 × 25 dilution, respectively) at 4 °C.

3. Add 10 or 20 μL the diluted Matrigel solution to each well of a 384-well biosensor microplate, or 100 μL to a Transwell insert at 4 °C.

4. Incubate the microplate or Transwell insert inside an incubator at 37 °C under air/5% CO_2 for 24 h to form a 3D gel on the biosensor surface, or on the Transwell insert.

3.3 Label-Free Single Cell 3D^2 Invasion Assay

This section describes the protocol to perform label-free single cell 3D^2 invasion assay. This assay starts with placing a single spheroid onto the top surface of Matrigel, followed by that individual cells spontaneously transfer from the spheroid to the top Matrigel surface, invade through the matrix, and eventually adhere onto the biosensor surface (Fig. 3). The RWG imager monitors in real time the cell adhesion events, which, in turn, are used as an indicator to determine the invasive potential of a cancer cell line. This assay effectively mimics the first critical step of cancer metastasis; that is, the local invasion of cancer cells through surrounding ECM, which involves adhesion, proteolytic remodeling of the ECM, and migration [47–50].

This assay enables multiparametric analysis of cell invasion and adhesion (Fig. 4), The initial adhesion point, designated as the coordinate of (0, 0), is first identified as the pixelated location at which the first positive DMR occurs. The lateral distance of any pixel relative to the initial adhesion point is then calculated. The maximal DMR at each pixel is extracted from its DMR signal. The adhesion time at each pixel is also extracted from its DMR as the time reaches a specific DMR amplitude (e.g., 500 pm for HT29, or 200 pm for the isogenic cell lines). The total adhesion events are obtained by counting the number of pixels that give rise to a DMR exceeding a predetermined threshold signal (e.g., 500 pm for HT29, or 200 pm for the isogenic cell lines). The total adhesion area is determined using ImageJ as the area that gave rise to a signal above the background. Correlation analysis can then be performed. For HCT116, PTEN deletion was found to increase the invasion distance (Fig. 5a vs. Fig. 5b), and the total adhesion events (Fig. 5c), suggesting that PTEN deletion increases the invasion

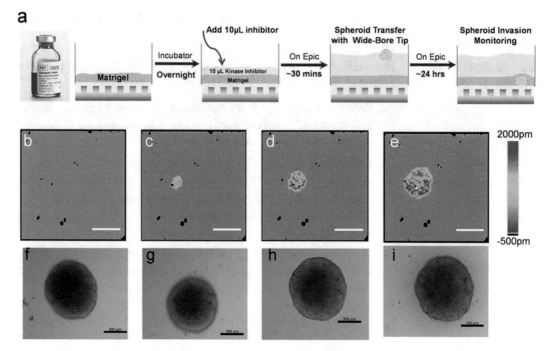

Fig. 3 Label-free single cell $3D^2$ invasion assay. (**a**) The assay consists of four critical steps: coating the biosensor surface with Matrigel; adding medium containing an inhibitor compound to the well; transferring a spheroid from a ULA round bottomed microplate and placing it onto the top Matrigel surface; and monitoring the invasion and adhesion in real time. (**b–e**) The time series DMR images of a biosensor before and after a single spheroid was placed onto the biosensor surface coated with 10 μL 0.1 mg/mL Matrigel: 0 min (**b**), 1 h. (**c**), 6 h (**d**), and 24 h (**e**). Spatial scale bar: 500 μm. Intensity scale bar: false colored (−500 to 2000 pm). (**f–i**) The time series light microscopic images of a biosensor before and after a single spheroid was placed onto Matrigel coated biosensor surface: 0 min (**f**), 1 h. (**g**), 6 h (**h**), and 24 h (**i**). Scale bar: 500 μm. Condition of spheroid formation: 40K per well seeding density, 4 days culture. (**b–i**) is adapted with permission from ref. 26. Copyright (2015) American Chemical Society

speed of the cells in spheroid through 3D matrix. Furthermore, the invasion and adhesion of both isogenic cell lines was found to be sensitive to PI3K inhibitors wortmannin and LY294002 (Fig. 5d).

1. Prepare all compound solutions in the complete medium.

2. Add 10 μL the complete medium in the absence and presence of a compound to the Matrigel coated biosensor plate.

3. Place the Matrigel coated biosensor microplates, RWG imager, cell media, pipette, pipette tips, and compound solutions all inside an incubator at least 2 h before the assay. This step is to ensure that temperature reaches equilibrium, so temperature mismatch, if occurring, will be minimal.

4. Place the Matrigel coated biosensor microplate onto the imager and wait for 30 min.

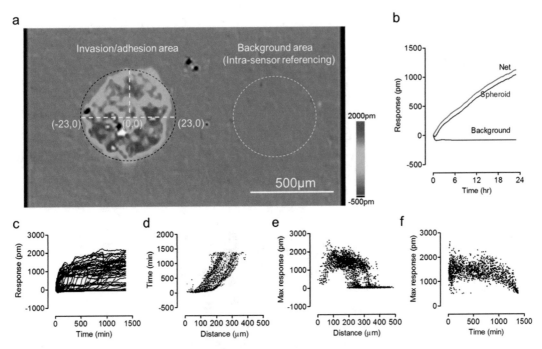

Fig. 4 Label-free single cell analysis for the invasion and adhesion of HT29 cell spheroids. (**a**) A DMR image taken 24 h after a spheroid was placed on the top Matrigel surface. Scale bar: 500 μm. (**b**) Real-time population averaged DMR on the Matrigel coated surface. (**c**) Pixelated real-time DMR signals for the line indicated in (**a**). (**d**) The correlation between the adhesion time and the lateral distance relative to the first adhesion point. (**e**) The correlation between the maximal response and the relative distance. (**f**) The correlation between the adhesion time and the maximal response. The biosensor was coated with 10 μL 0.1 mg/mL Matrigel. Condition of spheroid formation: 40K per well seeding density, 4 days culture. This figure is adapted (**b–j**) is adapted with permission from ref. 26. Copyright (2015) American Chemical Society

5. Monitor the baseline until all sensors reached a steady baseline. This step is to ensure that the baseline drifts <10 pm within 5 min.

6. Establish a baseline by recording resonance images for 5 min.

7. Pause the imager temporally for spheroid transfer. Critical to this step is to minimize the blackout period (<2 min) between the baseline and the first recording after spheroid placement.

8. Pick up a 30 μL media solution containing the single spheroid from a well of a ULA round bottomed microplate using a wide-bore tip. Critical to this step is to visually inspect the solution inside the wide-bore tip to ensure the spheroid pickup.

9. Gently place the single spheroid onto the top Matrigel surface in a well of a 384-well biosensor microplate. Critical to this step is to ensure that the pipette tip does not damage the coating, and the spheroid is placed within the biosensor (2 × 2 mm in the center of each well). In some cases, two tumor spheroids can be transferred into a single well to increase the probability of placing spheroids within the biosensor area.

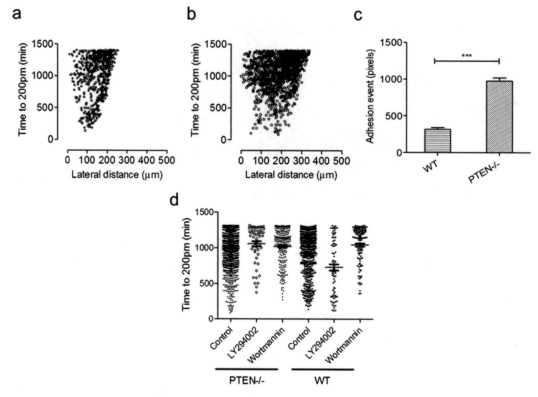

Fig. 5 PTEN deletion increases the invasion rate of CRC cells in spheroid through 3D Matrigel matrix. (**a**, **b**) Correlation analysis between the adhesion time and lateral distance for the parental (**a**) and PTEN−/− (**b**) cells. (**c**) The adhesion events versus cell types. (**d**) The adhesion time to reach 200 pm under different conditions. Coating: 0.2 mg/mL Matrigel. Data represent mean ± s.d. for **c**, **d** ($n = 3$). ***$p < 0.001$. This figure is adapted from ref. 27 with permission

10. Restart the imager and continuously record resonance images for ~24 h.

11. Collect light microscopic images at the top Martigel surface at the end of the assay (*see* **Note 4**).

12. Normalize the starting resonance wavelength at each pixel to zero, calculate the wavelength shifts, extract kinetic DMR signals at each pixel of a sensor, and save the data into Excel using Epic Processor Software.

13. Perform intra-sensor referencing to remove background signal from the adhesion signals of all pixelated locations (*see* **Note 5**).

14. Analyze the background corrected data to determine the initial adhesion point, the lateral distance relative to the initial adhesion point, the maximal DMR amplitude, the adhesion time, the total adhesion event, and the adhesion area using Excel (*see* **Note 6**).

15. Perform correlation analysis between two different parameters using Prism software.

3.4 Cell Monolayer Wound Healing Assay

This section describes the wound healing assay protocol to detect the migratory potential of CRC cells. This assay is to first culture cells to form a monolayer, then physically create a wounded area, and finally monitor the wound recovery over time using light microscopy (Fig. 6a). The migratory behavior of cells, one of the key hallmarks of a metastatic cancer cell, is reflected by the rate of recovery of the scratch region (Fig. 6b).

1. Harvest cells from T75 flask using $1 \times$ trypsin–EDTA.

2. Centrifuge the cells down, remove the supernatant, and resuspend the cell pellet in freshly prepared complete cell culture medium.

3. Add cells to a 6-well plate and grow cells to confluency at 37 °C with 5% CO_2.

4. Wash the confluent cells three times with the serum free medium, and starve the cells overnight.

5. Produce a scratch with a sterile 200 μL pipette tip in a marked region of the well.

6. Collect immediately light microscopic images of the marked scratch area.

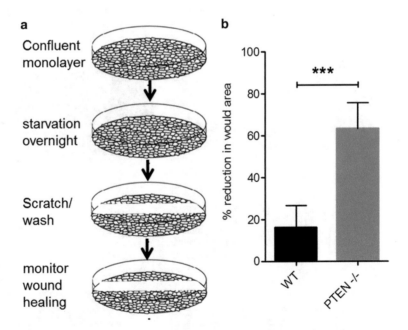

Fig. 6 2D cell wound recovery assay showing that PTEN deletion increases the migratory potential of HTC116 cells. (**a**) Critical steps of the assay. (**b**) The 2D migration rate of HCT116 versus HCT116-PTEN$-/-$ cells. Data represents mean $\pm$ s.d. ($n = 5$). ***$p < 0.001$. This figure is adapted from ref. 27 with permission

7. Culture the cells in serum free medium at 37 °C with 5% CO_2 for 48 h.

8. Collect light microscopic images of the marked scratch areas in the end.

9. Determine the percentage of reduction in wound size (i.e., scratch area) using ImageJ.

3.5 Transswell Invasion Assay of Individual Cancer Cells

This section describes the Transwell invasion assay protocol to detect the impact of PTEN deletion on the invasive potential of HCT116 cells. This assay measures the invasion of individual cells through Matrigel coated substrates having defined micropores (Fig. 7a). This assay mimics local invasion of tumor cells through the ECM or basement membrane. The invasiveness is quantified by counting the number of invaded cells and normalizing to those in uncoated Transwell inserts (Fig. 7b, c).

1. Harvest cells from T75 flask using 1× trypsin–EDTA.

2. Centrifuge the cells down, remove the supernatant, and resuspend the cell pellet in serum free media.

3. Add 100 μL cell suspension to the top side of the Matrigel coated Transwell inserts with a cell density of 2.5×10^4 cells/ mL. Use the uncoated Transwell inserts as the control.

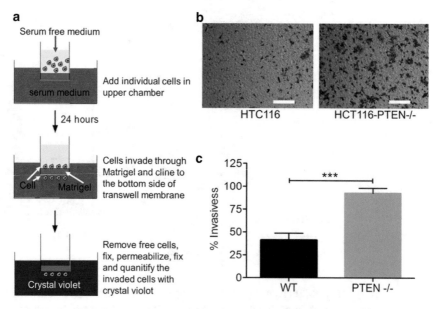

Fig. 7 Transwell invasion assay showing that PTEN deletion increases the invasiveness of the CRC cells. (**a**) Critical steps of the assay. (**b**) Representative light microscopic images of HCT116 versus HCT116-PTEN−/− cells adhered on the underside of Transwell inserts 24 h after invasion. (**c**) The invasion rate through Matrigel coated Transwell of HCT116 versus HCT116-PTEN−/− cells. Data represents mean ± s.d. ($n = 5$). ***$p < 0.001$. This figure is adapted from ref. 27 with permission

4. Place the inserts having cells to a 24-well plate, each well being filled with 750 μL of the cell culture media with 5% FBS as a chemoattractant, to form invasion chambers.

5. Incubate the invasion chambers for 24 h at 37 °C with 5% CO_2.

6. Remove the inserts and wash twice with 1× phosphate buffered saline.

7. Discard the non-invading cells using a cotton swab moistened with culture media.

8. Fix the adhered cells in 4% paraformaldehyde, permeablize in 100% methanol, and stain using crystal violet stain.

9. Manually count the number of invading cells on the underside of the Transwell using a light microscope.

10. Calculate the percentage invasiveness by normalizing the number of invading cells in the Matrigel coated inserts to that in the uncoated inserts.

4   Notes

1. HCT116 is a microsatellite-unstable and metastatic human CRC cell line and can grow in multilayers but is incapable of polarization or intestinal epithelial differentiation. Its isogenic PTEN−/− line is directly derived via homologous knockout of PTEN in HCT116 by deletion of exon 5 which encodes the active site of the protein [51]. The wild-type cell line does not bear any somatic PTEN mutations, but has an H1047R mutation of PI3KCA in exon 20 (kinase domain) [52].

2. Cells can undergo genetic drifting during passaging. Given the high sensitivity of DMR assays, the DMR of a cell line may be sensitive to its passages.

3. Optimal seeding density for spheroid formation is cell line dependent. This can be assessed based on the size and shape of cell clusters formed, as well as the morphology of spheroids after manually transferred to the biosensor plate using a wide-bore pipette tip.

4. Light microscopic images can be acquired at the end of the assay since RWG DMR imaging is noninvasive. DMR imaging has a short sensing depth (~200 nm) away from the biosensor surface, so it only examines cell adhesion on the biosensor surface (i.e., right below the Matrigel coating) (Fig. 3b–e). On the other hand, light microscopic imaging has much thicker imaging depth, so it can examine the spheroid and the pattern of cells after spontaneously transferring from the spheroid to the top Matrigel surface (Fig. 3f–i).

5. The intra-sensor referencing can be used to eliminate any artifacts associated with mismatch in temperature and bulk index during the spheroid placement step. RWG biosensor is known to be sensitive to temperature and bulk index mismatch [53]. The intra-sensor referencing is possible since the size of a spheroid is much smaller than the dimension of a biosensor. Here, a sensor is divided into a background zone (i.e., the cell free area), and a cell adhesion zone (Fig. 4a). Subtracting the background signal from the cell adhesion signal led to background-free responses for the cell adhesion zone (Fig. 4b), or all pixelated locations (exemplified in Fig. 4c).

6. (a) The initial adhesion point approximates closely to the initial contact point of a spheroid with the top Matigel surface, at which the cell has the shortest distance to invade through. The time for the initial adhesion to occur indicates how fast the first cell invades through the matrix and reaches the sensor surface. (b) The lateral distance relative to the initial adhesion point is not the travel distance of an invading cell. Despite this, given that all cells need invade the same 3D matrix, the lateral distance can be used as a useful parameter to investigate the invasion behavior of spheroidal cells. (c) The maximal DMR at each pixel is an indicator of cell adhesion degree. (d) The adhesion time mostly reflects the time required for a cell to invade through the matrix and arrive at the sensor surface, since the kinetics of cell adhesion is relatively rapid (Fig. 4c). (e) The total adhesion events and area are two indicators for total numbers of cells invading through the Matrigel.

References

1. Steeg PS, Theodorescu D (2008) Metastasis: a therapeutic target for cancer. Nat Clin Pract Oncol 5:206–219. doi:10.1038/ncponc1066

2. Desgrosellier JS, Cheresh DA (2010) Integrins in cancer: biological implications and therapeutic opportunities. Nat Rev Cancer 10:9–22. doi:10.1038/nrc2748

3. Eccles SA, Box C, Court W (2005) Cell migration/invasion assays and their application in cancer drug discovery. Biotechnol Annu Rev 11:391–421. doi:10.1016/S1387-2656(05) 11013-8

4. Albini A, Benelli R (2007) The chemoinvasion assay: a method to assess tumor and endothelial cell invasion and its modulation. Nat Protoc 2:504–511. doi:10.1038/nprot.2006.466

5. Kam Y, Guess C, Estrada L, Weidow B, Quaranta V (2008) A novel circular invasion assay mimics in vivo invasive behavior of cancer cell lines and distinguishes single-cell motility in vitro. BMC Cancer 8:198. doi:10.1186/1471-2407-8-198

6. Carragher N (2009) Cell migration and invasion assays as tools for drug discovery. Clin Exp Metastasis 26:381–397. doi:10.3390/pharmaceutics3010107

7. Kramer N, Walzl A, Unger C, Rosner M, Krupitza G, Hengstschläger M, Dolznig H (2013) In vitro cell migration and invasion assays. Mutat Res 752:10–24. doi:10.1016/j.mrrev.2012.08.001

8. Quail DF, Maciel TJ, Rogers K, Postovit LM (2012) A unique 3D in vitro cellular invasion assay. J Biomol Screen 17:1088–1095. doi:10.1177/1087057112449863

9. Chitcholtan K, Asselin E, Parent S, Sykes PH, Evans JJ (2013) Differences in growth properties of endometrial cancer in three dimensional (3D) culture and 2D cell monolayer. Exp Cell Res 319:75–87. doi:10.1016/j.yexcr.2012.09.012

10. Härmä V, Virtanen J, Mäkelä R, Happonen A, Mpindi JP, Knuuttila M, Kohonen P, Lötjönen J, Kallioniemi O, Nees M (2010) A

comprehensive panel of three-dimensional models for studies of prostate cancer growth, invasion and drug responses. PLoS One 5: e10431. doi:10.1371/journal.pone.0010431

11. Tanner, K. and Gottesman, M.M. (2015) Beyond 3D culture models of cancer. Sci Transl Med 7:283 ps9. doi:10.1126/scitranslmed. 3009367.

12. Justice BA, Badr NA, Felder RA (2009) 3D cell culture opens new dimensions in cell-based assays. Drug Discov Today 14(1-2):102–107. doi:10.1016/j.drudis.2008.11.006

13. Pickl M, Ries CH (2009) Comparison of 3D and 2D tumor models reveals enhanced HER2 activation in 3D associated with an increased response to trastuzumab. Oncogene 28:461–468. doi:10.1038/onc.2008.394

14. Luca, A.C., Mersch, S., Deenen, R., Schmidt, S., Messner, I., Schäfer, K.L., Baldus, S.E., Huckenbeck, W., Piekorz, R.P., Knoefel, W. T., Krieg, A., and Stoecklein, N.H. (2013) Impact of the 3D microenvironment on phenotype, gene expression, and EGFR inhibition of colorectal cancer cell lines. PLoS One 8(3): e59689. doi:10.1371/journal.pone.0059689.

15. Fisher KE, Pop A, Koh W, Anthis NJ, Saunders WB, Davis GE (2006) Tumor cell invasion of collagen matrices requires coordinate lipid agonist-induced G-protein and membrane-type matrix metalloproteinase-1-dependent signaling. Mol Cancer 5:69. doi:10.1186/1476-4598-5-69

16. Brekhman V, Neufeld G (2009) A novel asymmetric 3D in-vitro assay for the study of tumor cell invasion. BMC Cancer 9:415–427. doi:10. 1186/1471-2407-9-415

17. Pathak A, Kumar S (2011) Biophysical regulation of tumor cell invasion: moving beyond matrix stiffness. Integr Biol 3:267–278. doi:10.1039/c0ib00095g

18. Koch TM, Münster S, Bonakdar N, Butler JP, Fabry B (2012) 3D traction forces in cancer cell invasion. PLoS One 7:e33476. doi:10. 1371/journal.pone.0033476

19. Cheung KJ, Gabrielson E, Werb Z, Ewald AJ (2013) Collective invasion in breast cancer requires a conserved basal epithelial program. Cell 155:1639–1651. doi:10.1016/j.cell. 2013.11.029

20. Liu L, Duclos G, Sun B, Lee J, Wu A, Kam Y, Sontag ED, Stone HA, Sturm JC, Gatenby RA, Austin RH (2013) Minimization of thermodynamic costs in cancer cell invasion. Proc Natl Acad Sci U S A 110:1686–1691. doi:10.1073/pnas.1221147110

21. Kniazeva E, Putnam AJ (2009) Endothelial cell traction and ECM density influence both capillary morphogenesis and maintenance in 3D. Am J Physiol Cell Physiol 297: C179–C187. doi:10.1152/ajpcell.00018. 2009

22. Blacher S, Erpicum C, Lenoir B, Paupert J, Moraes G, Ormenese S, Bullinger E, Noel A (2014) Cell invasion in the spheroid sprouting assay: a spatial organization analysis adaptable to cell behaviour. PLoS One 9:e97019. doi:10. 1371/journal.pone.0097019

23. Ghosh S, Joshi MB, Ivanov D, Feder-Mengus C, Spagnoli GC, Martin I, Erne P, Resink TJ (2007) Use of multicellular tumor spheroids to dissect endothelial cell–tumor cell interactions: a role for T-cadherin in tumor angiogenesis. FEBS Lett 581:4523–4528. doi:10.1016/j. febslet.2007.08.038

24. Oxmann D, Held-Feindt J, Stark AM, Hattermann K, Yoneda T, Mentlein R (2008) Endoglin expression in metastatic breast cancer cells enhances their invasive phenotype. Oncogene 27:3567–3575. doi:10.1038/sj.onc.1211025

25. Sabeh F, Shimizu-Hirota R, Weiss SJ (2009) Protease-dependent versus-independent cancer cell invasion programs: three-dimensional amoeboid movement revisited. J Cell Biol 185:11–19. doi:10.1083/jcb.200807195

26. Febles NK, Ferrie AM, Fang Y (2014) Label-free single cell quantification of the invasion of spheroidal colon cancer cells through 3D Matrigel. Anal Chem 86:8842–8849. doi:10. 1021/ac502269v

27. Chandrasekaran S, Deng H, Fang Y (2015) PTEN deletion potentiates invasion of colorectal cancer spheroidal cells through 3D Matrigel. Integr Biol 7:324–334. doi:10.1039/c4ib00298a

28. Salmena L, Carracedo A, Pandolfi PP (2008) Tenets of PTEN tumor suppression. Cell 133:403–414. doi:10.1016/j.cell.2008.04. 013

29. Fang Y, Ferrie AM, Fontaine NH, Mauro J, Balakrishnan J (2006) Resonant waveguide grating biosensor for living cell sensing. Biophys J 91:1925–1940. doi:10.1529/biophysj. 105.077818

30. Fang Y (2007) Non-invasive optical biosensor for probing cell signaling. Sensors 7:2316–2329

31. Benton G, Kleinman HK, George J, Arnaoutova I (2011) Multiple uses of basement membrane-like matrix (BME/Matrigel) in vitro and in vivo with cancer cells. Int J Cancer 128:1751–1757. doi:10.1002/ijc.25781

32. Vinci M, Gowan S, Boxall F, Patterson L, Zimmermann M, Court W, Lomas C, Mendiola M, Hardisson D, Eccles SA (2012) Advances in

establishment and analysis of three-dimensional tumor spheroid-based functional assays for target validation and drug evaluation. BMC Biol 10:29. doi:10.1186/1741-7007-10-29

33. Ferrie AM, Deichmann OD, Wu Q, Fang Y (2012) High resolution resonant waveguide grating imager for cell cluster analysis under physiological condition. Appl Phys Lett 100:223701. doi:10.1063/1.4723691

34. Ferrie AM, Wu Q, Deichmann O, Fang Y (2014) High frequency resonant waveguide grating imager for assessing drug-induced cardiotoxicity. Appl Phys Lett 104:183702. doi:10.1063/1.4876095

35. Fang Y, Ferrie AM, Fontaine NH, Yuen PK (2005) Characteristics of dynamic mass redistribution of EGF receptor signaling in living cells measured with label free optical biosensors. Anal Chem 77:5720–5725. doi:10.1021/ac050887n

36. Fang Y (2011) Label-free biosensors for cell biology. Intl J Electrochem 2011:e460850. doi:10.4061/2011/460850

37. Schröder R, Janssen N, Schmidt J, Kebig A, Merten N, Hennen S, Müller A, Blättermann S, Mohr-Andrä M, Zahn S, Wenzel J, Smith NJ, Gomeza J, Drewke C, Milligan G, Mohr K, Kostenis E (2010) Deconvolution of complex G protein-coupled receptor signaling in live cells using dynamic mass redistribution measurements. Nat Biotechnol 28:943–949. doi:10.1038/nbt.1671

38. Verrier F, An S, Ferrie AM, Sun H, Kyoung M, Fang Y, Benkovic SJ (2011) GPCRs regulate the assembly of a multienzyme complex for purine biosynthesis. Nat Chem Biol 7:909–915

39. Ferrie AM, Wang C, Deng H, Fang Y (2013) Label-free optical biosensor with microfluidics identifies an intracellular signalling wave mediated through the β_2-adrerengic receptor. Integr Biol 5:1253–1261

40. Li G, Lai F, Fang Y (2012) Modulating cell-cell communication with a high-throughput label-free cell assay. J Lab Automation 17:6–15. doi:10.1177/2211068211424548

41. Pai S, Verrier F, Sun H, Hu H, Ferrie AM, Eshraghi A, Fang Y (2012) Dynamic mass redistribution assay decodes differentiation of a neural progenitor stem cell. J Biomol Screen 17:1180–1191. doi:10.1177/1087057112455059

42. Fang Y (2013) Troubleshooting and deconvoluting label-free cell phenotypic assays in drug discovery. J Pharmacol Toxicol Methods 67:69–81. doi:10.1016/j.vascn.2013.01.004

43. Fang Y (2014) Label-free drug discovery. Front Pharmacol 5:52. doi:10.3389/fphar.2014.00052

44. Fang Y (2014) Label-free cell phenotypic drug discovery. Comb Chem High Throughput Screen 17(7):566–578

45. Song MS, Salmena L, Pandolfi PP (2012) The functions and regulation of the PTEN tumour suppressor. Nat Rev Mol Cell Biol 13:283–296. doi:10.1038/nrm3330

46. Kasinskas RW, Venkatasubramanian R, Forbes NS (2014) Rapid uptake of glucose and lactate, and not hypoxia, induces apoptosis in three-dimensional tumor tissue culture. Integr Biol 6:399–410. doi:10.1039/c4ib00001c

47. Hanahan D, Weinberg RA (2011) Hallmarks of cancer: the next generation. Cell 144:646–674. doi:10.1016/j.cell.2011.02.013

48. Valastyan S, Weinberg RA (2011) Tumor metastasis: molecular insights and evolving paradigms. Cell 147:275–292. doi:10.1016/j.cell.2011.09.024

49. Stetler-Stevenson WG, Aznavoorian S, Liotta LA (1993) Tumor cell interactions with the extracellular matrix during invasion and metastasis. Annu Rev Cell Biol 9:541–573

50. Poincloux R, Lizárraga F, Chavrier P (2009) Matrix invasion by tumour cells: a focus on MT1-MMP trafficking to invadopodia. J Cell Sci 122:3015–3024. doi:10.1242/jcs.034561

51. Lee C, Kim JS, Waldman T (2004) PTEN gene targeting reveals a radiation-induced size checkpoint in human cancer cells. Cancer Res 64:6906–6914. doi:10.1158/0008-5472.CAN-04-1767

52. Samuels Y, Diaz LA Jr, Schmidt-Kittler O, Cummins JM, Delong L, Cheong I, Rago C, Huso DL, Lengauer C, Kinzler KW, Vogelstein B, Velculescu VE (2005) Mutant PI3K3CA promotes cell growth and invasion of human cancer cells. Cancer Cell 7:561–573. doi:10.1016/j.ccr.2005.05.014

53. Fang Y (2015) Label-free cell phenotypic profiling and screening: techniques, experimental design and data assessment. Methods Pharmacol Tox 53:233–252. doi:10.1007/978-1-4939-2617-6_2

Chapter 11

Label-Free Biosensors Based on Bimodal Waveguide (BiMW) Interferometers

Sonia Herranz*, Adrián Fernández Gavela*, and Laura M. Lechuga

Abstract

The bimodal waveguide (BiMW) sensor is a novel common path interferometric transducer based on the evanescent field detection principle, which in combination with a bio-recognition element allows the direct detection of biomolecular interactions in a label-free scheme. Due to its inherent high sensitivity it has great potential to become a powerful analytical tool for monitoring substances of interest in areas such as environmental control, medical diagnostics and food safety, among others. The BiMW sensor is fabricated using standard silicon-based technology allowing cost-effective production, and meeting the requirements of portability and disposability necessary for implementation in a point-of-care (POC) setting.

In this chapter we describe the design and fabrication of the BiMW transducer, as well as its application for bio-sensing purposes. We show as an example the biosensor capabilities two different applications: (1) the immunodetection of Irgarol 1051 biocide useful in the environmental field, and (2) the detection of human growth hormone as used in clinical diagnostics. The detection is performed in real time by monitoring changes in the intensity pattern of light exiting the BiMW transducer resulting from antigen–antibody interactions on the surface of the sensor.

Key words Bimodal waveguide interferometry, Evanescent field biosensor, Integrated optics, Immunoassay, Biofunctionalization, Biorecognition

1 Introduction

Most analytical methodologies used in fields such as environmental monitoring, medical diagnostics and food safety are based on time-consuming and expensive techniques that must be performed by specialized personnel in laboratory environments. This usually involves extensive time and effort due to the need for analysis and sampling, as well as interpretation of results. These can cause critical delays in clinical treatment decisions or removal of contaminated products from the market in order to avoid extended health

---

*These authors contributed equally to this work.

Avraham Rasooly and Ben Prickril (eds.), *Biosensors and Biodetection: Methods and Protocols Volume 1: Optical-Based Detectors*, Methods in Molecular Biology, vol. 1571, DOI 10.1007/978-1-4939-6848-0_11, © Springer Science+Business Media LLC 2017

impacts. Therefore it is imperative to use reliable diagnostic tools that ensure sensitive, rapid, and simple analysis, and to expedite the diagnostic process.

In this context, biosensors are beginning to overcome the main drawbacks of classical analytical techniques. Modern biosensors are often portable, easy-to-use, and highly sensitive, and can operate in real-time and in-situ. Most of these photonic biosensors employ working principles and components (as waveguides) from the field of Integrated Optics (IO) (for basic information about IO, readers are referred to references [1, 2]). The IO-based biosensors are normally fabricated with standard Si-technology which employs well-known fabrication processes and silicon (Si) or silicon nitride (Si_3N_4) as core waveguide materials (*see* **Note 1**). Within IO-based biosensors, photonic interferometric biosensors based on the evanescent wave principle are highly suitable for many applications due to their high sensitivity and great potential to be integrated in a POC device [3].

1.1 Interferometric Sensors Based on Integrated Optics

The working principle of interferometric biosensor relies in the creation of an interference pattern, generated by the superposition of two or more light waves. In a common interferometric device, the incoming light beam is split in two beams of equal intensity that travel through different optical paths (arms) and are recombined before arriving at a detector, which collects the interferometric signal, as Fig. 1 shows.

For biosensing applications, one of the arms is used as a reference while the other acts as sensing one. Interferometric biosensors are based on the evanescent wave detection principle, therefore any modification in the sensing arm, such as binding events or concentration variation, induce a change of the effective refractive index of the guided light wave producing an interference pattern at the output.

The output intensity, $I(t)$, of an interferometric sensor is related with the phase difference between the two arms, $\Delta\varphi(t)$ as follows:

$$I(t) \propto \cos(\Delta\varphi)$$

$$\Delta\varphi(t) = \frac{2\pi L}{\lambda} \Delta n_{eff}$$

where L is the length of the sensing arm, λ the wavelength of the light, and Δn_{eff} the difference between the effective refractive index of the two propagated modes.

The phase difference, in an experimental interferometric signal, is determined by the variation between the initial value, $\Delta\varphi_1$, and the final value, $\Delta\varphi_2$, given by the number of fringes, where a complete fringe represents a phase difference of 2π, as Fig. 1 shows.

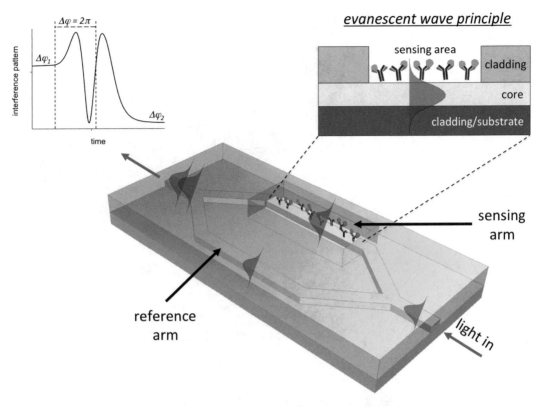

Fig. 1 Scheme of the working principle of an interferometric waveguide biosensor

Our work focuses on the development of IO nano-immunosensors based on BiMW transducers and their integration into a complete Lab-On-a-Chip (LOC) platform. BiMW is a novel interferometric transducer, which has been designed for sensitive, label-free, and direct detection of biomolecular interactions. In this IO interferometer, light is coupled in a straight single-mode rib waveguide, and after passing through a step-junction, two transversal modes with the same polarization are excited (*see* Fig. 2). Due to the presence of two propagated modes in a common path with different evanescent field profiles, any change in the refractive index causes an interference pattern at the device output. The intensity distribution at the output depends principally on the refractive index at the sensor surface, as well as on the dimensions of the BiMW structure (layers thicknesses and rib dimensions). The bimodal section of the BiMW includes a sensing window area free of cladding, in order to have access to the waveguide surface for sensing purpose (*see* Fig. 2), following the interferometric working principles explained previously. Therefore, by measuring the output light intensity the sensor response can be determined in a quantitative way [4].

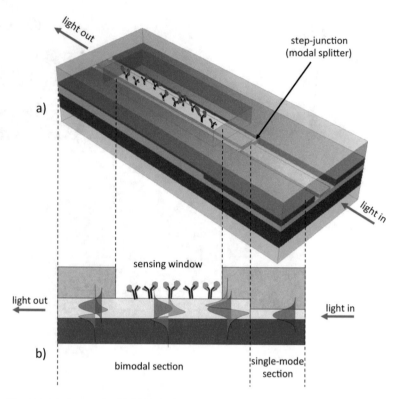

Fig. 2 (**a**) Scheme of a BiMW interferometric sensor. (**b**) Longitudinal side view of a BiMW showing the distribution of the electromagnetic field in several points

Interferometric sensors, and in particular BiMW interferometers, have proved to be one of the most sensitive devices for label-free analysis, reaching sensitivity to homogeneous changes in the refractive index of 10^{-7}–10^{-8} Refractive Index Unit (RIU), meaning that they have the potential to detect biomolecular interaction at pM levels. However, to use this technology for biosensing applications the transducer must be highly selective for instance, particularly with applications such as immunoassays (immunosensors). The limit of detection (LOD) achieved with an immunosensor is dependent not only on the transducer itself but also on the biorecognition element and the biofunctionalization protocol used to combine them.

Here we describe the design and fabrication process of BiMW interferometers, paying special attention to parameters affecting their performance (such as waveguide length or rib dimensions) in order to achieve the highest possible sensitivity. Additionally, we report the application of BiMW interferometry to biosensor development. Two applications in the fields of environmental monitoring and clinical diagnostics are described: the detection of Irgarol 1051 biocide and that of human growth hormone, respectively.

2 Materials

2.1 Equipment

1. Plasma cleaner reactor: Standard plasma system Femto, version A (40 kHz, 0–100 W) from Diener Electronic GmbH (Ebhausen, Germany).

2. Ultrasonic bath FB 15054, from Fisher Scientific (Madrid, Spain).

3. ABBE refractometer (Optic Ivymen System, Spain).

4. Clean Room facilities: the BiMWs are fabricated from a standard 4-in. (p-type) Si wafer with a thickness of 500 μm purchased from Siltronic Company. The fabrication of the BiMW is done at Clean Room facilities (class 100) using standard microelectronics optical photolithography, reactive ion etching, wet etching, and various deposition methods. Over the Si substrate a 2 μm thick layer of thermal oxide ($n_{bottom,clad} = 1.46$) is grown. Then, a 340 nm thick core layer of Si_3N_4 ($n_{core} = 2.00$) is deposited by Low Pressure Chemical Vapor Deposition (LPCVD). An absorbent layer of poly-crystalline Si ($n_{poly} = 4.06$) of 100 nm is deposited by LPCVD over a silicon dioxide (SiO_2) layer of 200 nm previously deposited. Finally, a SiO_2 ($n_{top,clad} = 1.46$) layer 1.5 μm thick is deposited by Plasma Enhanced Chemical Vapor Deposition (PECVD).

2.2 Setup Components

1. Fluid cell is made of polydimethylsiloxane (PDMS, Sylgard) fabricated using a methacrylate (PMMA) topographic master.

2. Syringe pump (NE300, New Era) for maintaining a constant flow during all the biochemical detection process.

3. An injection valve (V-451, Idex) allows the injection of different solutions without changing the flow rate.

4. Laser diode (ML101J27, Mitsubishi) with $\lambda_0 = 660$ nm and $P = 120$ mW, mounted on a compact temperature controlled laser diode mount (TCLDM9, Thorlabs), is used as light source. A Temperature controller (TED 200C, Thorlabs) and current controller (LDC220C, Thorlabs) are employed to stabilize the laser diode.

5. To couple the light into the waveguide a lenses system is used, composed by: collimated lens (C240TME-D, Thorlabs), polarization-dependent isolator (IO-3D-660-VLP, Thorlabs) and coupling objective 40× (Achro, Leica).

6. A four quadrants photodetector (S4349, Hamamatsu) is employed for collecting the light at the end of the device. The signals are amplified through standard benchtop instrumentation (PDA200C, Thorlabs).

7. A digital acquisition card (6251, National instruments) for reading the photodetector signal in real time. The signal acquisition is controlled by a LabVIEW-based application.

8. 3-axis precision micro-position stage (Nanomax-TS, Thorlabs) is used for laser-BiMW alignment.

9. A temperature sensor (AD590, Thorlabs) is used to achieve a temperature feedback circuit together with a thermo-electric cooler (TEC3-1.5, Thorlabs), operated through a bench top temperature controller (TED200C, Thorlabs) allowing a temperature resolution of 0.01 °C.

2.3 Reagents

1. Potassium phosphate monobasic (KH$_2$PO$_4$, P0662-500G), sodium phosphate dibasic (Na$_2$HPO$_4$, S0876-500G), sodium chloride (NaCl, S9625-1KG-D), potassium chloride (KCl, 60,128-250G-F), 2-(N-Morpholino)ethanesulfonic acid (MES, M3671-50G), sodium carbonate (Na$_2$CO$_3$, S7795-500G), sodium hydroxide (NaOH, S8045-500G), sodium dodecyl sulfate (SDS, 436,143-25G), hexadecyltrimethylammonium bromide (CTAB, H6269-100G), bovine serum albumin (BSA, A7906-10G), (3-aminopropyl)triethoxysilane (APTES, A3648-100ML), N-hydroxysulfosuccinimide sodium salt (sulfo-NHS, 56,485-1G), N-(3-dimethylaminopropyl)-N-ethylcarbodiimide hydrochloride (EDC, E1769-1G, *see* **Note 2**), N,N-dimethyl-formamide anhydrous (DMF, 227,056-100ML), toluene (244511-1 L), pyridine (270970-100ML), *p*-phenylene diisothiocyanate (PDITC, 258,555-5G), hydrofluoric acid (HF, 695,068-25ML-D), ammonium fluoride (NH$_4$F, 338,869-25G), and phosphoric acid (85 wt. % in H$_2$O, W290017-1KG-K) can be purchased from Sigma-Aldrich.

2. Acetone (161007.1211), ethanol absolute (161086.1211), methanol (161091.1211), nitric acid (65%, 473255.1611), and hydrochloric acid (HCl 37%, 471020.1611) can be purchased from Panreac.

3. Recombinant human Growth Hormone (r-hGH, 22 kDa) was provided by Dr. Parlow (National Hormone and Peptide Program (NHPP), National Institute of Diabetes and Digestive and Kidney Diseases (NIDDK), CA, USA). The monoclonal antibody (anti-hGH) was produced and characterized at the Department of Immunology and Oncology from the National Center of Biotechnology (CNB-CSIC, Madrid, Spain) [5].

4. Irgarol 1051, Irgarol 1051 derivative *4e* (hapten) [6], and anti-Irgarol 1051 serum (As87) were provided by Prof. Marco (Nanobiotechnology for Diagnostics Nb4D group, IQAC-CSIC, Barcelona, Spain).

5. Deionized water from a Milli-DI® Water Purification System, Merck Millipore, USA.

2.4 Buffer Composition

1. Buffered hydrofluoric acid (BHF): mixture of HF and NH_4F, 1:6 (v/v)

2. MES buffer: solution of 0.1 M MES and 0.5 M NaCl in water, pH adjusted to 5.5. Store at 4 °C.

3. Phosphate buffer saline (PBS): solution of 137 mM NaCl, 2.7 mM KCl, 8 mM Na_2HPO_4, and 2 mM KH_2PO_4 in water, pH adjusted to 7.4. Store at 4 °C. We recommend preparing a ten-fold concentration PBS (10× PBS, pH 7–7.5) as stock solution, which can be stored at room temperature (RT). Then, prepare a working solution of 1× PBS by dilution from the stock solution. Alternatively, PBS can be purchased from Sigma-Aldrich (P7059-1 L).

4. Carbonate buffer (CB): 0.1 M Na_2CO_3 at pH 9.

3 Methods

One of the first steps to develop a BiMW transducer of high sensitivity is to simulate a range of different core and cladding thicknesses, taking into account the rib dimensions and the tolerances at the clean room facilities where the fabrication will be done. The sensitivity of an interferometry-based transducer is directly related with the dimension of the interaction surface (*see* Subheading 1.1), thus higher sensitivity is obtained with a long sensing area. For that reason a compromise between the total length of the BiMW chip and the quantity of devices per wafer needs to be reached, for a cost-effective fabrication.

Once the optimum dimensions have been defined by modelling, lithographic masks are designed according the chosen dimensions and then the BiMW devices are fabricated at clean room facilities. The fabrication processes should be reproducible and must guarantee the same reliable performance for all the BiMW transducers.

To characterize the device, BiMW is placed in a custom-designed setup. The light from a laser diode is end-fire coupled using a lens system, as Fig. 3 shows. For light readout, a two-section photodetector is employed as the interference of the common path BiMW interferometer generates a variation of the output intensity distribution, but the total intensity should be constant. The photodetector must be placed directly at the BiMW output, and total intensity must be divided in the up and down quadrant of the photodetector (*see* Fig. 3). The sensor response, S_R, is determined through the equation:

$$S_R = \frac{I_{up} - I_{down}}{I_{up} + I_{down}}$$

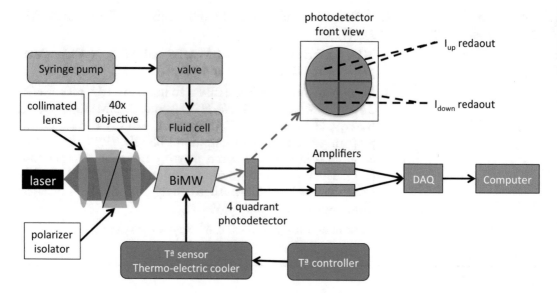

Fig. 3 Scheme of the experimental setup employed for the evaluation of the BiMW sensors

where I_{up} and I_{down} are the intensities acquired for the up and down quadrants of the photodetector, respectively.

For the application of the BiMW interferometer as a biosensor the surface must be modified with a recognition element or biological receptor which, when immobilized onto the surface, increases the transducer selectivity for the substance to be monitored. The environmental and clinical diagnostic BiMW interferometric biosensor applications will be illustrated.

Irgarol 1051 is a protective algaecide (biocide, s-triazine compound) used in antifouling paints to prevent attachment and growth of biofouling organisms in the aquatic environment. It leaks out from the paint and by its toxicity Irgarol 1051 prevents accumulation of organisms onto the painted surface (boat hulls or marine installations). Traces of Irgarol 1051 have been detected in the marine environment [7, 8], and due to its broad toxicity (photosystem-II inhibitor), Irgarol 1051 is potentially toxic toward valuable aquatic species (mainly small aquatic plants), and its presence in the environment can cause significant environmental problems. In fact, a number of studies have demonstrated its negative impact on marine organisms such as corals.

Human growth hormone (hGH) is produced by the pituitary gland and promotes growth, regulates the activity of vital organs and helps to preserve the health of the whole organism. Its deficiency commonly affects children and has several negative effects such as hypoglycemia in newborn infants or stunted growth in children. hGH deficiency is results from several genetic diseases such as Turner syndrome and Prader-Willi syndrome. Thus the hGH level in blood or urine can be used for the diagnosis of

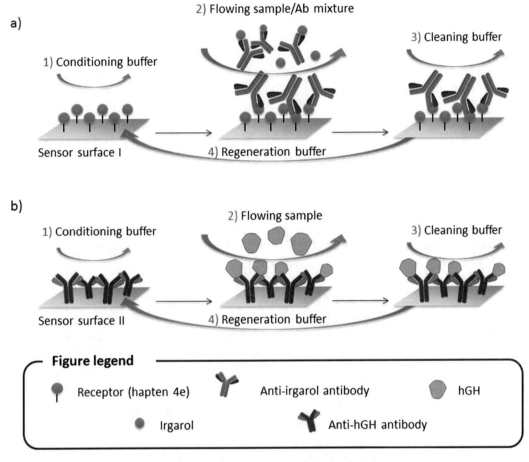

Fig. 4 Immunoassay formats: (**a**) Competitive immunoassay format selected for Irgarol 1051 detection (**b**) Direct immunoassay format selected for hGH detection. For sensor surface details, *see* Fig. 7

those disorders. Additionally, r-hGH has been used by many sportsmen in an attempt to improve athletic performance, and the control of hGH level in urine is usually employed as a doping test.

Both biosensor applications are based on antigen–antibody specific interactions. For the analysis of Irgarol 1051, a low molecular weight pollutant, a competition immunoassay has been selected (*see* Fig. 4a). The transducer surface is functionalized with an Irgarol 1051 derivative (hapten *4e*). A fixed amount of an anti-Irgarol antibody is mixed and allowed to interact with the sample and then, the mixture is injected to the flow cell where the sensor is placed. The free antibodies interact with the immobilized *4e* receptor and generate a binding response, which is inversely related to the concentration of Irgarol 1051 in the sample (*see* Fig. 5a). The detection of hGH is carried out by using a direct immunoassay (*see* Fig. 4b). In this case the transducer surface is functionalized with an anti-hGH antibody. The hGH present in the

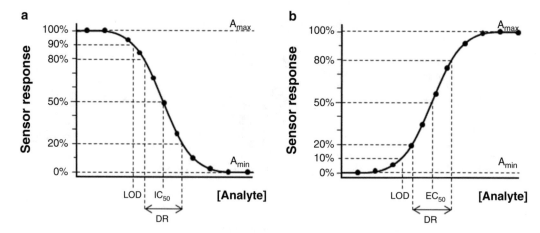

Fig. 5 Standard calibration curves showing the relationship between the analyte concentration ([Analyte], logarithmic scale) and the sensor response for: (**a**) competitive and (**b**) direct immunoassays

Table 1
Layer parameters employed for the evaluation of the single-mode and bimodal condition

Layer	Material	Thickness (nm)	Refractive index
Top cladding	Water/SiO$_2$	1500	1.33/1.46
Core	Si$_3$N$_4$	150–340	2.00
Bottom cladding	SiO$_2$	2000	1.46

injected sample interacts with the immobilized antibody and generates a binding response directly related to its concentration (*see* Fig. 5b).

3.1 BiMW Simulation and Fabrication

3.1.1 BiMW Simulation

(a) To design a BiMW we need to choose optimal materials, which ensure a cost-effective fabrication, a further integration onto a complete LOC platform, and a highly sensitive device. As the transducers are designed for biosensors applications, SiO$_2$/Si$_3$N$_4$/SiO$_2$ waveguides are selected [9]. Table 1 summarizes the parameters employed in the calculation, relative to the materials' refractive indices and their thicknesses.

(b) Once the material is defined, core dimensions are analyzed to ensure the best sensitivity of the device. For that, different considerations have to be taken into account [4] (*see* Fig. 6):

- Single-mode behavior (mode TE$_{00}$ or TM$_{00}$) in section A in Fig. 6b.

- Bimodal behavior (mode TE$_{00}$ and TE$_{10}$ or TM$_{00}$ and TM$_{10}$) in sections B, C, and D in Fig. 6b.

- Performances achievable in the clean room facilities.

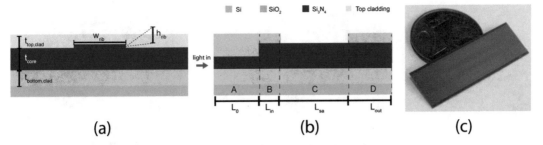

Fig. 6 (**a**) Front cross section of the employed rib-waveguide, indicating the main parameters described on the text. (**b**) Longitudinal side view of the BiMW device. Light is injected in the single-mode section (A). After an abrupt step junction, two modes propagate till the end of the device. Regions B and D are covered by silicon dioxide while C is exposed to the external medium (sensing area). (**c**) Picture of a chip with 20 BiMW sensors

Table 2
Optimal parameters for core dimensions in a high sensitivity BiMW

Section	t_{core} (nm)	h_{rib} (nm)	w_{rib} (nm)
Single-mode	150	1.5	3000
Bimodal	340	1.5	3000

In bimodal section, bulk and surface sensitivity must be calculated for different core thicknesses (t_{core}). Once t_{core} has been fixed, the thickness of the single-mode section is chosen to determine how the energy carried by the fundamental mode, propagating in section A in Fig. 6b, is distributed to the two modes propagating in section B in Fig. 6b.

Table 2 summarizes the optimal parameters for core dimensions in a BiMW for a $\lambda = 660$ nm and TE polarization (*see* Fig. 6a).

(c) Once the core dimensions are defined, the sensor design is completed by choosing the length of the different sections. It is required to take into account the following considerations:

- The need to adapt a microfluidic system to the BiMW for biosensing applications.

- To ensure high sensitivity while keeping a reasonable total length.

- Distribution of the BiMWs in 4-in. silicon wafer to optimize cost-effective fabrication.
  The following values were chosen: $L_0 = 3$ mm, $L_{in} = 4.5$ mm, $L_{sa} = 15$ mm, and $L_{out} = 8.5$ mm (*see* Fig. 6b).

(d) In a 4-in. silicon wafer, 240 BiMW are distributed in 12 different chips (1 cm wide × 3.1 cm long) with 20 sensors per chip (*see* Fig. 6c).

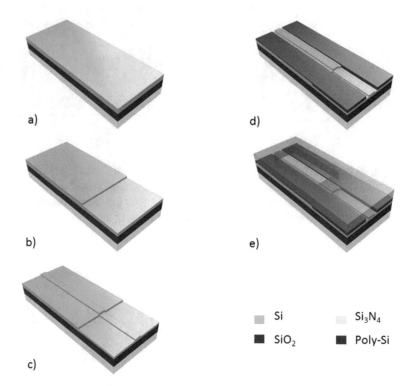

Fig. 7 Scheme of the fabrication process of integrated BiMW sensors. (**a**) A layer of SiO$_2$ and Si$_3$N$_4$ is deposited on a Si wafer. (**b**) Step-junction etched in Si$_3$N$_4$ layer for single-mode section definition. (**c**) Rib structure is defined. (**d**) Polycrystalline deposition over the Si$_3$N$_4$ and etching process to open desired window on this layer. (**e**) SiO$_2$ deposition and sensing area opening

3.1.2 BiMW Fabrication

Figure 7 summarizes the fabrication process. Below, each step of Fig. 7 is explained.

(a) The fabrication processing starts with a 4-in. (p-type) Si wafer with a thickness of 500 μm, over which a 2 μm thick layer of thermal oxide is grown. Then a 340 nm thick core layer of Si$_3$N$_4$ is deposited by LPCVD.

(b) The thickness of the Si$_3$N$_4$ is reduced 190 nm to obtain a 150 nm single-mode section using a wet etching process with 75% phosphoric acid at 160 °C. A hard mask constituted by a layer of SiO$_2$ deposited by PECVD and defined by photolithography is employed in this step to protect unexposed regions.

(c) Using a wet etching process with BHF and a standard photolithography process, the rib structure of the waveguide (3 μm in width and 1.5 nm in height) is defined. The low etching rate of the BHF for Si_3N_4 allows precise control of the process.

(d) A 100 nm layer of poly-crystalline Si is grown by LPCVD over 200 nm of SiO_2 deposited by PECVD, which is introduced to improve the adhesion to the underlying Si_3N_4 surface. A standard lithography process is used to define the structure. Dry etching (RIE) is employed to remove the poly-crystalline Si layer and 150 nm of silicon dioxide, to ensure vertical sidewalls. Then, a wet etching process is used to eliminate the remaining 50 nm of SiO_2. Poly-crystalline Si layer is employed as absorbing material to define lateral bands along the waveguide path.

(e) A SiO_2 layer 1.5 μm thick is deposited by PECVD as top cladding layer. In order to open the sensing area a standard lithography process is done, employing a wet etching process with BFH to remove the desired SiO_2.

3.2 Setup Configuration

(a) The BiMW chip is placed in an aluminum homemade holder as shown in Fig. 8. Into the holder a temperature stabilization is included, and the entire system is mounted on a 3-axis stage. In this way, it is possible to focus the light in the desired BiMW sensor optimally.

(b) As the total output intensity must be divided in the up and down quadrant of the photodetector, this one is joined in a 2-axis stage assembly to the aluminum chip holder, to ensure the right position of the photodetector at the BiMW

Fig. 8 Photograph of a BiMW experimental setup

output. As photodetector has four quadrants, two BiMW could be measure simultaneously, using two quadrants per sensor. This position is essential if a modulated signal is required [10].

(c) A digital acquisition card connected to a computer is employed to process the photodetector signal. A homemade Labview software acquires a continuous photodetector signal in real time.

(d) A diode laser is placed in a temperature controlled mount. To guarantee an optimal end-fire incoupling of the laser beam, a lenses system is employed. A lens is used to collimate the divergent laser beam of the diode. After this lens a polarization-dependent isolator is placed to avoid optical back reflections and to guarantee the required polarization. Finally, a 40x objective is employed to focus the collimated beam to the BiMW input for end-fire coupling.

(e) All the biological interactions must be performed in a liquid medium. For that reason, a PDMS fluid cell must be placed on top of the sensor chip. This fluid cell has channels of 11 μL volume, and an inlet and outlet for liquid flow. To avoid liquid leaks between the PDMS channels and the BiMW chip, and in order to properly align the fluid cell, four screws and a methacrylate sheet are employed to press, as Fig. 9 shows.

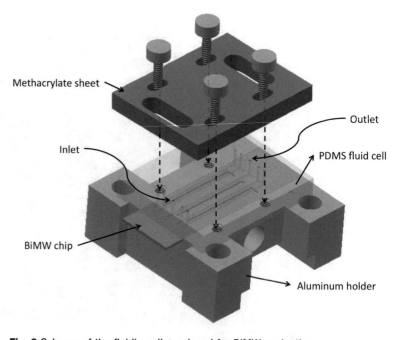

Fig. 9 Scheme of the fluidic cell employed for BiMW evaluation

(f) The flow delivery system is mounted by connecting a syringe pump with an injection valve and this one to the cell flow inlet by using Teflon tubes. The length of the tubes should be reduced as much as possible to minimize diffusion effects. The loop of the injection valve has a variable length, depending on the desired volume. Here, 150 μL and 250 μL loops have been used for Irgarol 1051 and hGH immunoassays, respectively. An extra tube is used from the cell outlet to the waste.

3.3 Sensing Characterization: Homogeneous Sensing

1. The determination of the BiMW biosensor bulk sensitivity allows evaluating the response of the transducer, independently of the biofunctionalization process. For that, milli-Q water is supplied to the sensor in a continuous rate (40 μL/min), and solutions providing small refractive index changes, such as different concentrations of HCl are injected over the sensor.

2. Previously to the injections, the refractive index of the solutions is checked with a commercial refractometer, and the difference of refractive index, Δn, between the continuous buffer running and the injected solutions is calculated.

3. Taking into account the sinusoidal relationship between the phase change, $\Delta\varphi$, and the output intensity, I, in an interferometric device [3], changes of the refractive indexes over the sensor area, induce output signal alterations through the evanescent field interaction.

4. Phase variation, $\Delta\varphi$, is plotted versus index variation, Δn, obtaining a calibration curve as Fig. 10 shows. The slope of the fitting curve, expressed in rad/RIU, represents the bulk sensitivity of the sensors. For BiMW, the sensitivity (S_{bulk}) is around $1700 \cdot 2\pi$ rad/RIU, for TE polarization.

5. The LOD, $\Delta n_{\min}$, is calculated through the equation:

$$\Delta n_{\min} = \frac{\Delta\varphi_{\min}}{S_{\text{bulk}}} = \frac{\Delta S_{\text{R,min}}}{S_{\text{bulk}}}\frac{\pi}{V} = \frac{3 \cdot \sigma_{S_{\text{R}}}}{S_{\text{bulk}}}\frac{\pi}{V}$$

where $S_{\text{R,min}}$ is the minimum sensor response, defined as three times the system noise, $\sigma_{S_{\text{R}}}$. V is the visibility of the output signal, which corresponds to π rad in the phase variation. LOD for BiMW sensors are in the range of 10^{-7}–10^{-8} RIU.

3.4 Surface Biofunctionalization

The most widely applied approaches to chemical modification of Si_3N_4 surfaces are based on the molecular grafting of the native oxide layer present onto this material by forming self-assembled monolayer (SAM) of alkyl-silanes. Among the wide variety of commercially available silanes, APTES ((3-aminopropyl)triethoxysilane) is one of the most employed for chemical modification of Si-based photonic devices. Surfaces modified with APTES present

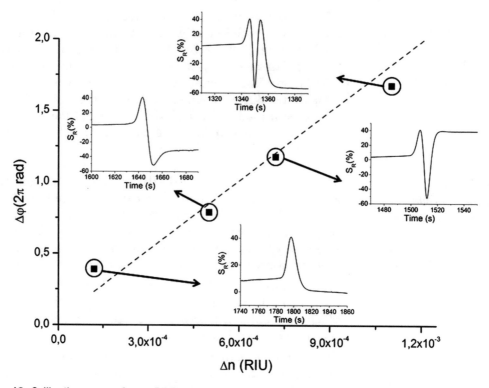

Fig. 10 Calibration curve for a BiMW sensor. Insets represent the temporal interferometric signals corresponding to the entrance of solutions with different refractive indexes

3-aminopropyl moieties, which allow the further covalent immobilization of bioreceptors presenting carboxyl groups in their structure. Moreover, by using different cross-linker molecules, the variety of bioreceptors that can be covalently attached is multiplied.

We describe the use of APTES to chemically modify the BiMW chips. APTES-modified chips are then functionalized by covalently attaching the hapten *4e*, resulting in a biosensor surface which can detect Irgarol 1051 by a competitive immunoassay (Surface I). In the second approach, APTES-modified chips are biofunctionalized by covalently attaching the anti-hGH antibody, using *p*-phenylene diisothiocyanate (PDITC) as a cross-linker molecule (Surface II). PDITC displays two isothiocyanate groups, which can couple the free amine groups of a bioreceptor. Additionally, PDITC has the potential to form well-organized assemblies driven by $\pi-\pi$ stacking, resulting in improved antifouling properties [11]. Surface II is used to detect hGH by a direct immunoassay.

In order to avoid contamination of the surface and to enhance silanization efficiency and reproducibility, the BiMW chips are thoroughly cleaned and oxidized forming a fresh prepared oxide layer with a high density of silanol groups (reactive hydroxyl groups). This procedure is carried out immediately prior to the

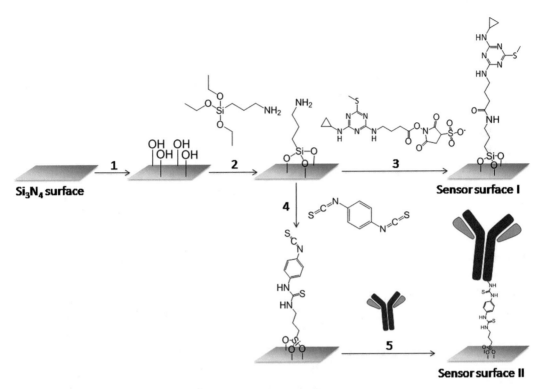

Fig. 11 Functionalization scheme/protocol: (*1*) cleaning and oxidation; (*2*) APTES silanization; (*3*) covalent attachment of hapten 4e; (*4*) activation of APTES-modified surface with PDITC, and (*5*) covalent attachment of anti-hGH antibody. Sensor surface I: 4e-functionalized surface, used for Irgarol 1051 immunodetection. Sensor surface II: anti-hGH antibody-functionalized surface, used for hGH immunodetection

surface biofunctionalization. Avoid storage of cleaned and oxidized chips, which are susceptible to contamination.

Figure 11 shows a scheme of the different steps of the functionalization protocol. Briefly, the BiMW chips are cleaned, oxidized (Fig. 11, step 1), and immediately silanized with APTES (Fig. 11, step 2). Then, sulfo-NHS-ester activated *4e* hapten is covalently attached (Fig. 11, step 3) getting sensor surface I. For preparing sensor surface II, APTES-modified chips are activated with PDITC (Fig. 11, step 4) and finally anti-hGH antibody is covalently bound (Fig. 11, step 5).

3.4.1 Cleaning Procedure

(a) Wipe the BiMW chip down using a cleanroom paper soaked in acetone.

(b) Sequentially rinse the chip with acetone, ethanol, and deionized water, and dry under nitrogen flow.

(c) Immerse the chip in 1% SDS aqueous solution and sonicate for 5 min.

(d) Rinse the chip with deionized water and dry again under nitrogen flow.

(e) Immerse the chip in a mixture of methanol and 37% fuming hydrochloric acid (1:1, v/v) and sonicate for 10 min (*see* **Note 3**).

(f) Finally, generously rinse the chip with deionized water and dry carefully under nitrogen flow.

3.4.2 Oxidation Procedure

(a) Treat the cleaned BiMW chip with oxygen plasma (100 W, 45 sccm) for 5 min (*see* **Note 4**).

(b) Immerse the chip in 15% HNO_3 at 75 °C for 25 min.

(c) Rinse generously the chip with deionized water, dry carefully under nitrogen flow and immediately immerse it in the silanization solution.

3.4.3 BiMW Interferometer Silanization Procedure Using APTES

(a) Clean and oxidize the BiMW chip according to the previous sections.

(b) Immediately immerse the BiMW chip in a solution of 1% APTES and incubate at RT for 1 h. Two different solvents have been evaluated: absolute ethanol and toluene.

(c) Amply rinse the chip with the silanization solvent and deionized water, sequentially.

(d) Dry the chip thoroughly under nitrogen flow and thermally treat it in an oven at 110 °C for 1 h.

3.4.4 Covalent Immobilization of Irgarol Derivative "4e"

(a) At this point, an APTES-modified BiMW (silanization solvent: toluene) chip is positioned on the experimental apparatus, and a PDMS multichannel gasket (flow-chamber) is pressed against it, forming the assay flow channels (*see* Fig. 9). The immobilization of the Irgarol derivative *4e* is carried out inside the measuring setup, using a *stopped-flow* incubation approach, i.e., by filling the flow channel with the reaction mixture and stopping the flow. After the appropriate incubation time, the reaction is stopped by restarting the flow and washing out the surface with deionized water.

(b) Prepare a solution mix of hapten *4e*, EDC, and sulfo-NHS in MES buffer. Concentrations must be optimized for each application; however, concentration typically ranges from 10 to 50 μg/mL of hapten, and 0.2–0.4 M and 0.05–0.1 M of EDC and sulfo-NHS, respectively. Pre-incubate this mixture for 30 min to 4 h at room temperature (amine-reactive sulfo-NHS ester hapten intermediate solution).

(c) Initiate a continuous flow (20 μL/min) of deionized water and wait until stabilization of the interferometric signal.

(d) Inject into the flow cell a volume of the NHS-activate *4e* solution large enough to fill the whole assay flow channel (25–50 µL). Once the channel is filled up, stop the flow and incubate from 4 h to overnight.

(e) Remove the NHS-activate *4e* solution by flushing deionized water through the flow channel for 5 min.

(f) BiMW chip is now functionalized and ready to be used for Irgarol 1051 immunodetection.

3.4.5 Covalent Immobilization of Anti-hGH Antibody

(a) Prior to the covalent attachment of the anti-hGH antibody, the APTES-modified chip (silanization solvent: ethanol) must be activated with PDITC, which acts as cross-linker. After PDITC activation the antibody can be immobilized through the amine groups present in its structure.

(b) Prepare a 20 mM PDITC solution (5 mL) in anhydrous DMF containing 1% pyridine.

(c) Immerse the APTES-modified chip in PDITC solution and incubate in the dark for 2 h.

(d) Wash off unbound reagents by sequentially rinsing the chip with acetone, ethanol and water, and dry it under nitrogen stream.

(e) Add 50–100 µL of a solution of anti-hGH antibody in carbonate buffer onto the PDITC-modified chip, cover it with a coverslip and incubate overnight, in the dark. Concentrations must be optimized for each application. For hGH immunoassay, a 50 µg/mL solution was used.

(f) Rinse the functionalized chip with PBS and carefully dry it under nitrogen stream.

(g) At this point, the functionalized BiMW chip is positioned on the experimental apparatus and the PDMS flow-chamber mounted. Before starting sample evaluation the sensor surface is blocked to avoid nonspecific adsorption and reduce background signal (*see* **Note 5**).

(h) Initiate a continuous flow of PBS and wait until the interferometric signal stabilizes.

(i) Inject into the flow cell a 2 mg/mL BSA solution in PBS until the flow channel is full. Then stop the flow and incubate for 1 h (*see* **Note 6**).

(j) The BiMW chip is now functionalized and ready to begin the sample evaluation. If not used immediately chips must be dried and stored at 4 °C. Before use, the chip is positioned again on the experimental apparatus and the sensor surface is conditioning by maintain a continuous flow of PBS for 1 h.

3.5 Detection by a Competitive Immunoassay

3.5.1 Selection of the Antibody Concentration

(a) Initiate a continuous flow of PBS and wait until stabilization of the interferometric signal. This buffer will be used to perform the Irgarol 1051 immunodetection assay.

(b) Prepare series of solutions (300 μL) with different dilution factors of the antibody (serum As87). For Irgarol 1051 detection, dilutions ranging from 1:10,000 to 1:500 were used.

(c) Inject into the flow cell the antiserum solutions with a regeneration step in between, and measure the phase shift (sensor response). Regeneration consists in disrupting the receptor–antibody interaction while maintaining the receptor onto the surface with minimal damage, by washing up the sensor surface with the so-called regeneration solution. Regeneration solution composition must be empirically optimized (*see* **Note 7**). For Irgarol 1051 immunoassay, a 50 mM NaOH solution was employed.

(d) Construct a plot of sensor response vs. antiserum concentration. Choose an antibody concentration value (fixed concentration to perform the inhibition assay) below the saturation limit, able to give high enough response intensity (phase shift), so that at inhibition, signals would be still clear. Take into account that by decreasing the antibody concentration, a better limit of detection can be achieved (*see* **Note 8**).

3.5.2 Detection of Irgarol 1051

(a) Initiate a continuous flow of PBS (20 μL/min) and wait until the interferometric signal is stabilized.

(b) Prepare a series of solutions of different concentrations of Irgarol 1051 in PBS (300 μL, standard solutions). Concentrations of the standard solutions to be measure depend on the limit of detection for each specific application. For Irgarol 1051, the standard solutions were in the ng/L to μg/L range.

(c) Mix the series of standard solutions with a volume of As87 serum so that the final dilution factor in solution equals that previously selected. Incubate the mixture for 10 min.

(d) Pump one antibody-Irgarol standard solution into the flow cell at 20 μL/min. This step produces a shift phase of the wave ($\Delta\varphi$) due to the interaction of the free antibody with the immobilized antigen.

(e) Wash off the unbound antibody by pumping PBS for a few minutes (20 μL/min).

(f) Regenerate the surface by injecting 50 mM NaOH (20 μL/min, 120 s), and wait until stabilization of the signal (surface conditioning). At this point, the surface is ready for the measurement of a second standard solution. In Fig. 12,

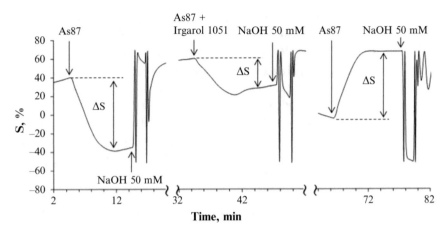

Fig. 12 Real monitoring of anti-Irgarol 1051 antibody (As87) detection. The presence of Irgarol 1051 (1 µg/L) in solution leads to a decrease of the sensor response. The sensor surface was regenerated with 50 mM NaOH

the sensor response for three consecutively samples (in PBS) is represented, in which the inhibition of the signal by the presence of Irgarol 1051 in the sample can be noticed.

(g) Construct a plot of sensor response vs. Irgarol 1051 concentration and fit the data to an appropriate curve-fitting model for obtaining the calibration curve (*see* **Note 9**). Limit of detection (LOD) should be in the desired range for the application. If not, try to optimize immobilization and immunoassay conditions. Irgarol 1051 levels typically range from 0.1–1.7 µg/L in marinas and 1–40 ng/L in coastal and estuarine waters [12], so a LOD at low ng–pg/L level is desirable.

(h) Mix the Irgarol samples (unknown concentrations) with an equal concentration of anti-Irgarol serum used with the standard solutions. Incubate the mixture for 10 min.

(i) Pump the antibody-sample mixture into the flow cell at 20 µL/min. After monitoring the antigen–antibody interaction, regenerate and condition the sensor surface in order to prepare it for the next analysis.

(j) The concentration of Irgarol 1051 in the measured samples can be estimated by interpolating the sensor response in the fitting curve.

3.6 Detection of hGH by a Direct Immunoassay

1. Initiate a continuous flow of PBS and wait until the signal is stabilized.

2. Prepare a series of solutions of different concentrations of hGH in PBS (400 µL, standard solutions). For hGH, the standard solutions were in the ng/L to µg/L range.

3. Inject into the flow cell the hGH solutions with a regeneration step in between (regeneration solution: 20 mM HCl), and measure the phase shift (sensor response).

4. Construct a plot of sensor response vs. hGH concentration and fit the data to an appropriate curve-fitting model for obtaining the calibration curve. The hormone hGH usually appears at ng/L levels in urine, so a LOD at low ng/L is desirable.

5. Inject into the flow cell the sample (unknown hGH concentration), wait until stabilization of the signal and measure the sensor response.

6. Estimate the concentration of hGH in the sample by interpolating the measured sensor response in the fitting curve.

3.7 Data Analysis

To quantify the concentration of the target analyte in the sample, the sensor response is interpolated into the calibration curve. In order to minimize the estimation error and obtain accurate concentration values it is extremely important to employ an appropriate curve-fitting model. The most frequently model used to analyse immunoassay data is the 4-parameter logistic (4-PL) model, which fit the data by using the equation:

$$y = A_{min} + \frac{(A_{max} - A_{min})}{1 + \left(\frac{x}{c}\right)^{b}}$$

where y is the sensor response, x is the concentration of the target analyte, A_{max} and A_{min} are the asymptotic ends, b is the slope at the inflection point of the curve and c is the concentration at the inflection point and represents the half maximal inhibitory concentration (IC_{50}) and the half maximal effective concentration (EC_{50}) in a competitive and a direct assay, respectively [13] (*see* **Note 10**). Typically, the detection limit (LOD) for a competitive and a direct assay is calculated as the analyte concentration for which the signal is inhibited or increase by 10% of sensor signal range ($A_{max} - A_{min}$), respectively, and the dynamic range (DR) of the biosensor is evaluated as the analyte concentrations that produced a sensor signal between 20 and 80% of ($A_{max} - A_{min}$) (*see* Fig. 4).

4   Notes

1. Polymer-based biosensors are increasingly popular, but the instability of the polymers, especially when they are in contact with a buffer solution, restricts their use. Progress in polymer material research could bring about advances in polymer-based IO biosensors in the near future.

2. EDC is hygroscopic and quickly oxidizes in air. A good way to handle EDC is to open the container under a dry inert

atmosphere such as nitrogen (e.g., inside a glovebox), separate it into aliquots, using containers which provide a tight seal, and then store aliquots at $-20\,°C$. EDC solution must be prepared immediately before use.

3. Concentrated hydrochloric acid is corrosive and releases acidic gas. Therefore, handling of hydrochloric acid under a fume hood for approved acids and the use of personal protective equipment is strongly recommended.

4. Oxygen plasma treatment can be replaced by UV/Ozone plasma treatment (UV/Ozone ProCleaner™, Biorforce Nanosciences) for 1 h. Both oxidative pretreatments render highly hydrophilic surfaces (water contact angle lower than $5°$). However, we recommend oxygen plasma rather than UV/Ozone plasma treatment, since it is less time-consuming and, in our experience, leads to surfaces with a slightly lower roughness.

5. Prevention of nonspecific adsorptions of sample matrix components is particularly important when working with label-free optical transducers in which an overvalued sensor response due to nonspecific adsorption of matrix components is difficult to discriminate and leads to false positive (direct immunoassay) or false negative (competitive immunoassay) results.

6. Alternately, adjust the flow rate to a value that keeps the injected sample volume in contact with the sensor surface during at least 1 h. Thus flow rates in the range of 2.0–2.5 and 3.5–4 µL/min are recommended for sample volumes of 150 and 250 µL, respectively.

7. The regeneration procedure depends on the antigen/antibody pair used. The ideal regeneration buffer should effectively disrupt the antigen-antibody interaction preserving the receptor activity. In practice, all regeneration buffers cause some damage on the bio-layer immobilized onto the transducer, limiting the number of cycles that the same sensor surface can be reused. Usually, either low pH (0.01–1 M HCl, 1–20% formic acid, 0.01–1 M phosphoric acid) or high pH (10–100 mM, 0.2 M Na_2CO_3) solutions, alone or in combination with chelating and surfactant reagents, are employed. Less but also employed are solutions of high ionic strength (1 M NaCl, 2–4 M $MgCl_2$) or chaotropic agents (8 M urea, 6 M guanidinium chloride), and nonpolar water-diluted solvents (20% acetonitrile, 10–100% ethanol). For high-affinity interactions, the used of a cocktail of different regeneration solutions may be necessary [14].

8. Antibody concentration has a strong effect on the sensitivity of the immunoassay and must be carefully optimized. In competitive immunoassays it is an advantage to use a low

concentration of the antibody since by decreasing antibody concentration the effect of the competitor (target analyte) is magnified, usually leading to a lower limit of detection (higher sensitivity). However, extremely low concentrations of antibody give a poor signal-to-noise ratio. Thus, a concentration leading to a response around 70% of that at the saturation limit is usually selected.

9. For fitting purposes, several data analysis and graphing software products such as SigmaPlot, OriginPro, or GraphPad Prism can be use.

10. 4-PL model gives rise to symmetric curves about the c point. If the experimental data rise to asymmetric curve, more complex functions, such as a five parameter logistic model, must be adopted [15, 16].

Acknowledgments

The nanoB2A is a consolidated research group (Grup de Recerca) of the Generalitat de Catalunya and has support from the Departament d'Universitats, Recerca i Societat de la Informació de la Generalitat de Catalunya (2014 SGR 624). ICN2 is the recipient of Grant SEV-2013-0295 from the "Severo Ochoa Centers of Excellence" Program of Spanish MINECO. The authors acknowledge to the European Union (BRAAVOO Grant Agreement No 614010). The authors thank Dr. Parlow (NIDDK, CA, USA), Dr. Rodríguez-Frade and Dr. M. Mellado (CNB-CSIC, Madrid, Spain), and Prof. Marco (IQAC-CSIC, Barcelona, Spain) for the supply of the inmunoreagents.

References

1. Hunsperguer RG (2009) Integrated optics: theory and technology. Springer, New York, NY

2. Nishihara H, Haruna M, Suhara T (1989) Optical integrated circuits, McGraw-Hill Optical and Electro-optical Engineering. Series, London

3. Kozma P, Kehl F, Ehrentreich-Förster E, Stamm C, Bier FF (2014) Integrated planar optical waveguide interferometer biosensors: A comparative review. Biosens Bioelectron 58:287–307

4. Zinoviev KE, González-Guerrero AB, Domínguez C, Lechuga LM (2011) Integrated bimodal waveguide interferometric biosensor for label-free analysis. J Light Technol 29:1926–1930

5. Mellado M, Rodriguez-Frade JM, Kremer L, Martinez-Alonso C (1996) Characterization of monoclonal antibodies specific for the human growth hormone 22 K and 20 K isoforms. J Clin Endocrinol Metab 81:1613–1618

6. Ballesteros, B., Barceló, D., Sanchez-Baeza, F., Camps, F. and Marco, M.P (1998) Influence of the Hapten Design on the Development of a Competitive ELISA for the Determination of the Antifouling Agent Irgarol 1051 at Trace Levels. Anal Chem 70, 4004–4014.

7. Ali, H.R, Arifin, M.M., Sheikh, M.A., Shazili, N.A.M. and Bachok, Z. (2013) Occurrence and distribution of antifouling biocide Irgarol-1051 in coastal waters of Peninsular Malaysia. Mar Pollut Bull 70, 253–257.

8. Martínez K, Ferrer I, Hernando MD, Fernández-Alba AR, Marcé RM, Borrull F, Barceló D (2001) Occurrence of Antifouling Biocides in

the Spanish Mediterranean Marine Environment. Environ Technol 22:543–552

9. Prieto F, Sepúlveda B, Calle A, Llobera A, Domínguez C, Abad A, Montoya A, Lechuga LM (2003) An integrated optical interferometric nanodevice based on silicon technology for biosensor applications. Nanotechnology 14:907–912

10. Dante S, Duval D, Fariña D, González-Guerrero AB, Lechuga LM (2015) Linear readout of integrated interferometric biosensors using a periodic wavelength modulation. Laser Photon Rev 9:248–255

11. Gandhiraman RP, Gubala V, Nam LCH, Volcke C, Doyle C, James B, Daniels S, Williams DE (2010) Deposition of chemically reactive and repellent sites on biosensor chips for reduced non-specific binding. Colloids Surfaces B 79:270–275

12. Peñalver A, Pocurull E, Borrull F, Marcé RM (1999) Solid-phase microextraction of the antifouling Irgarol 1051 and the fungicides dichlofluanid and 4-chloro-3-methylphenol in water samples. J Chromatogr A 839:253–260

13. Dunn J, Wild D (2013) Calibration curve fitting. In: Wild D (ed) The immunoassay handbook, 4th edn. Elsevier, Oxford, pp 323–336

14. Fisher MJE (2010) Amine coupling through EDC/NHS: a practical approach. In: Mol NJ, Fischer MJE (eds) Surface Plasmon Resonance Methods and Protocols, 1st edn. Humana Press, Springer New York, pp 55–73

15. Gottschalk PG, Dunn JR (2005) The five-parameter logistic: a characterization and comparison with the four-parameter logistic. Anal Bioanal 343:54–65

16. Giraldo J, Vivas NM, Vila E, Badia A (2002) Assessing the (a)symmetry of concentration-effect curves: empirical versus mechanistic models. Pharmacol Ther 95:21–45

Chapter 12

DNA-Directed Antibody Immobilization for Robust Protein Microarrays: Application to Single Particle Detection 'DNA-Directed Antibody Immobilization

Nese Lortlar Ünlü*, Fulya Ekiz Kanik*, Elif Seymour, John H. Connor, and M. Selim Ünlü

Abstract

Protein microarrays are emerging tools which have become very powerful in multiplexed detection technologies. A variety of proteins can be immobilized on a sensor chip allowing for multiplexed diagnostics. Therefore, various types of analyte in a small volume of sample can be detected simultaneously. Protein immobilization is a crucial step for creating a robust and sensitive protein microarray-based detection system. In order to achieve a successful protein immobilization and preserve the activity of the proteins after immobilization, DNA-directed immobilization is a promising technique. Here, we present the design and the use of DNA-directed immobilized (DDI) antibodies in fabrication of robust protein microarrays. We focus on application of protein microarrays for capturing and detecting nanoparticles such as intact viruses. Experimental results on Single-particle interferometric reflectance imaging sensor (SP-IRIS) are used to validate the advantages of the DDI method.

Key words Protein microarrays, DNA-directed antibody immobilization, Label free detection, Single particle detection, SP-IRIS

1 Introduction

Protein microarrays have become indispensable tools with a wide variety of applications including biomarker detection, protein-protein interaction analysis, and detection of small molecule targets as in drug discovery [1, 2]. They are also promising candidates for multiplexed clinical diagnostics and disease progression monitoring [3]. Here, we first describe the challenges of protein microarrays to motivate DNA-directed antibody immobilization (DDI) for production of robust protein microarrays. For characterization

*These authors contributed equally to this work.

Avraham Rasooly and Ben Prickril (eds.), *Biosensors and Biodetection: Methods and Protocols Volume 1: Optical-Based Detectors*, Methods in Molecular Biology, vol. 1571, DOI 10.1007/978-1-4939-6848-0_12,
© Springer Science+Business Media LLC 2017

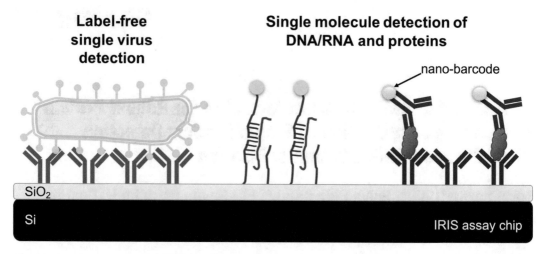

Fig. 1 Conceptual representation of IRIS detection platform for digital detection of individual viral particles and antigens labeled with gold nanoparticles

of the protein microarrays, we use Interferometric Reflectance Imaging Sensor (IRIS) technology developed in our laboratory. The IRIS platform has two distinct operation modalities [4]: (1) high-throughput label-free measurement of biomass accumulation; and (2) digital detection of single particles with high-magnification, also known as Single Particle Interferometric Reflectance Imaging Sensor (SP-IRIS). First modality is utilized for label-free characterization of probe immobilization and the latter is used for comparison of protein chips for single particle (virus) detection applications. Figure 1 shows the detection of single virus particles and single molecule detection of proteins and DNA/RNA, conceptually. With SP-IRIS, particles, which are too small to be detected by the standard microscopy, such as viruses, can be easily detected, quantified and differentiated according to their size and shape. Moreover, it becomes possible to detect individual molecules, such as proteins and DNA, via labeling with small metallic nanoparticles with high sensitivity [4].

The high-throughput protein microarrays are based on the technology developed for DNA chips that made profound impact in genomic analysis. Fabrication of DNA chips benefit from the ease of uniform immobilization of DNA probes on the sensor surface. DNA probes of different sequences can be chemically functionalized at a particular end (e.g. amine, thiol, or carboxyl functionalization) or have identical regions in the proximity of the solid sensor surface to assure uniformity. In contrast, proteins represent a great diversity of size and conformation resulting in significant variability of surface immobilization. Therefore, despite many advantages of protein microarrays in proteomics and multiplexed diagnostics applications, the technical difficulties associated with fabrication

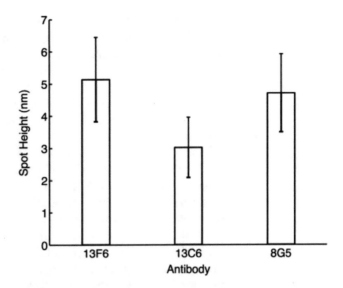

Fig. 2 Variation of antibody immobilization for different types of antibodies. In this particular example, we compare 13F6 (anti-EBOV glycoprotein antibody), 13C6 (anti-EBOV glycoprotein antibody) and 8G5 (anti-wild type VSV glycoprotein antibody) were used. Error bars show the variation among 60 different spots in ten different spotting runs

and reliability of the protein chips limit the potential of this promising technology. Perhaps the most fundamental challenge to overcome is related to the immobilization of proteins on the microarray surface in a repeatable manner. Furthermore, how the proteins are bound on the surface as well as the orientation of the protein probes significantly affects their capture efficiency. For example, on an amine-reactive sensor surface, the protein probes may have multiple covalent interactions that limit their accessibility to bind to a target molecule [5]. Moreover, challenges such as variability of spot properties and protein immobilization within and across chips influence the accuracy and robustness of protein microarrays. As illustrated in Fig. 2, when different types of antibodies are spotted, a significant difference in spot heights (as characterized by IRIS), thus a variation in antibody surface densities, is observed.

In this chapter, we describe DNA-directed antibody immobilization (DDI) for robust protein microarrays specifically for applications in single particle (or digital) detection - an exciting recent technological development that provides sensitivity beyond the reach of ensemble measurements [6–8] but brings along additional challenges. We have demonstrated an optical imaging technique termed Single-Particle Interferometric Reflectance Imaging Sensor (SP-IRIS)—a versatile platform that allows for a large range of nanoparticle detection including both natural nanoparticles (such as viruses) and synthetic nanoparticles in a highly-multiplexed microarray format [9]. The technology is based on interference of light from an optically transparent multi-layer dielectric structure. The interference of light reflected from the sensor surface is modified by the presence of particles producing a distinct signal that reveals the presence and size of the particle that is not otherwise visible under a conventional microscope. Size discrimination of the imaged virions in label-free virus detection allows for rejection of non-specifically bound particles to achieve a limit-of-detection competitive with the state-of-the-art laboratory technologies [10]. SP-IRIS has also shown promising results for detection of protein [11] and DNA molecules labeled with small gold nanoparticles—showing attomolar sensitivity and meeting the requirements for most in vitro tests.

Using our experimental results on digital detection of viruses via SP-IRIS, we illustrate the additional and more stringent requirements for robust protein microarrays. For a typical molecular detection assay, (for example when protein probes are used to capture protein biomarker targets), the signal is expected to vary linearly with the probe density. Therefore, assay variability due to probe immobilization differences can be accounted for and calibrated as described in [12]. In detection of intact viruses (or nanoparticles much larger than individual protein molecules), the influence of probe density is much more significant as illustrated by the experimental results in Fig. 3. The amount of virus that is captured on the antibody spots increases rapidly with the surface antibody density. Also, virus capture nearly vanishes when probe density drops below 5×10^9 Ab/mm^2 corresponding to less than 20% surface coverage.

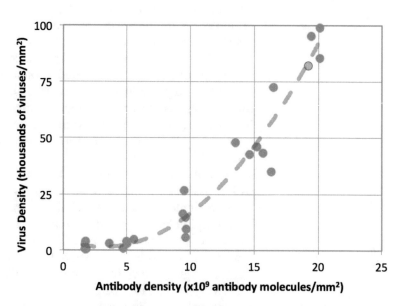

Fig. 3 The dependence of virus capture on the density of antibody immobilization on the sensor surface illustrating the strong variation

We speculate that multiple antibody-antigen binding interactions are required for the capture of viruses. A sparse coverage of immobilized antibodies on a solid surface does not allow for multiple interactions with the proteins on the relatively rigid surface of the target virus resulting in significant reduction of capture efficiency. One potential solution is to attach the antibodies to the sensor surface with flexible tethers thus allowing them to conform to the surface of the virus to make multiple bonds even for relatively sparse surface coverage. This is precisely the most significant motivation for using DNA-directed antibody immobilization in the context of single particle (such as an intact virus) detection.

Motivated by the limitations of direct immobilization of antibodies, various alternative methods have been explored for protein immobilization to promote antibody activity and improve assay sensitivity. DNA-directed immobilization (DDI), one of the recent alternative methods for protein immobilization, combines both the robustness of DNA microarrays with the diagnostic utility of proteins through the use of protein-DNA conjugates to functionalize a DNA surface for subsequent antigen capture [13–15]. Previous studies demonstrated that DDI enhanced molecular (antigen) capture efficiency, refined spot homogeneity and improved assay reproducibility contrasted to direct covalent immobilization of antibodies on the solid surface [16–18]. To validate the improved capture efficiency in the context of single particle detection, we have compared DDI to direct antibody immobilization [19]. A systematic study demonstrated that antibodies attached to the sensor surface with flexible tethers (DNA) provide additional conformational freedom, elevate the capture probes from the solid sensor surface and increases the target capture efficiency. Therefore, DDI not only provides a more robust protein microarray fabrication process directly benefiting from well-established DNA chips but also enhances sensitivity. Additional advantages of DNA-directed antibody immobilization include the ability to reprogram the sensor surface by using a different set of antibodies conjugated to the same DNA sequences and the resilience of DNA microarrays in elevated temperatures required for assembly of microfluidic cartridges [19]. In DDI, selected antibodies are covalently attached to specific DNA sequences which are complementary to surface probe ssDNA sequences. The surface probe sequences are immobilized onto the sensor surface and antibody-attached DNA sequences hybridize to the surface probes on the surface [19] as represented in Fig. 4.

Below, we describe the detailed method for DNA-directed immobilization of antibodies and application to single virus detection. We illustrate the advantages using experimental results on SP-IRIS platform.

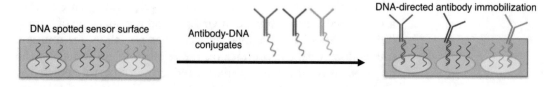

Fig. 4 A schematic representation of SP-IRIS substrate surface with ssDNA spots and the conversion of the DNA chip into a triplex antibody array through DDI of antibodies. Reprinted with permission from Seymour et al. Anal. Chem., 2015, 87 (20), pp. 10505–10,512. Copyright 2016 American Chemical Society

2 Materials

2.1 Single-Particle Interferometric Reflectance Imaging Sensor (SP-IRIS) (Validation Platform: Can Be Replaced by Another Single Particle Detection Method)

1. Scientific CCD camera. Retiga 4000R (Qimaging, Corp., Surrey, BC, Canada).

2. Sample illumination source. 530 nm LED source (Thorlabs Inc., Newton, NJ, USA).

3. Microscope objectives. 50× 0.8 NA Nikon objective for dry samples or a 40× 0.9 NA Nikon objective for samples in liquid (Nikon Inc., Melville, NY, USA) (*see* **Note 1**).

4. CRISP (The Continuous Reflection Interface Sampling and Positioning) autofocus system. MFC-2000 (Applied Scientific Instrumentation, Eugene, OR, USA).

5. 50:50 Beam splitter (Thorlabs Inc., Newton, NJ, USA).

6. Lenses (Thorlabs Inc., Newton, NJ, USA). Achromatic doublets for the visible spectrum. Catalogue numbers: f_1: AC254-030-A-ML, f_2: AC254-050-A-ML and f_3: AC254-200-A-ML.

7. Diffuser (Thorlabs Inc., Newton, NJ, USA). N-BK7 Ground Glass Diffuser, 220 Grit.

8. XY stage (Micronix USA, Santa Ana, CA, USA).

9. Mechanical components for optical setup. Cage system and lens holders (Thorlabs Inc., Newton, NJ, USA).

10. Computer. Data acquisition and processing.

2.2 Substrate Fabrication

1. Silicon (Si/SiO_2) chips with a patterned thermally grown silicon dioxide for single particle detection experiments (Silicon Valley Microelectronics, Santa Clara, CA, USA) (*see* **Note 2**).

2. Dimensions of the chips: 10 mm × 10 mm × 0.5 mm, with a 2.3 mm × 2.3 mm active center region for spotting DNA probes.

2.3 Surface Chemistry

1. Copoly(N,N-dimethylacrylamide (DMA)-acryloyloxysuccinimide (NAS)-3-(trimethoxysilyl)-propylmethacrylate (MAPS)) (Copoly(DMA-NAS-MAPS)) (MCP-2) (Lucidant Polymers Inc., Sunnyvale, CA, USA) (*see* **Note 3**).

2. Solution A2. Coat-On™ Coating buffer (Lucidant Polymers Inc., Sunnyvale, CA, USA).

3. Acetone and isopropanol.

4. Plastic petri dish.

5. Plasma asher.

6. Vacuum drying oven.

2.4 Protein Microarray Preparation

1. HPLC purified 5′-aminated ssDNA molecules (Integrated DNA Technologies, Inc., Coralville, IA, USA). A surface probe and a linker (for antibody conjugation) sequences.

2. Monoclonal antibody against Ebola virus (EBOV) glycoprotein as a model antibody (13F6) (Mapp Biopharmaceutical Inc., San Diego, CA, USA).

3. DNA spotting buffer ($2\times$): 300 mM sodium phosphate buffer, pH 8.5. 42.6 g of Na_2HPO_4 and 7.4 g of NaH_2PO_4 is dissolved in 1 L of filtered deionized water (18.2 MΩ resistivity, 0.2 μm-filtered) (Barnstead, NANOpure Diamond water purification system). The pH is adjusted to 8.5 with HCl. Stored at room temperature.

4. Tris–HCl buffer ($1\times$): 50 mM Tris–HCl with 150 mM NaCl, pH 8.5. 6.06 g of Tris is dissolved in 800 mL filtered deionized water. pH is adjusted to 8.5 with the appropriate volume of concentrated HCl. Final volume is brought to 1 L with deionized water. Stored at room temperature.

5. Blocking solution: 50 mM ethanolamine in Tris–HCl buffer. 310 μL of ethanolamine is added to 100 mL of Tris–HCl buffer (*see* **Note 4**).

6. Wash buffer (PBST): PBS buffer with 0.1% Tween-20. Stored at room temperature.

7. Sodium nitrate solution: 0.5 M sodium nitrate. 43 g of $NaNO_3$ is dissolved in 1 L of water.

8. Antibody-DNA conjugation kit (Thunder-Link Oligo Conjugation Kit, Innova Bioscience, Cambridge, UK).

9. Piezoelectric microarray spotter. Scienion S3 Fleaxarrayer (Berlin, Germany).

10. 24-well plate.

11. Multipurpose rotating shaker.

12. For comparison purpose only, antibodies are immobilized directly (*see* **Note 5**).

2.5 In-Liquid Virus Detection (Selected for Validation Experiments: Other Techniques Can Be Substituted)

1. Microfluidic cartridges. Multilayer polymer laminate devices: Disposable, custom-designed (ALine, Inc., Rancho Dominguez, CA, USA). From the bottom pf the costom-designed chip, Layer 1: acrylic, Layer 2: Si-PSA, Layer 3: polycarbonate, Layer 4: PET and Layer 5: PET.

Fig. 5 The images of an SP-IRIS chip and microfluidic cartridge for in-liquid SP-IRIS measurements (on the *left*). Cross section model of the microfluidic cartridge and objective which indicates the fluidic path (in *blue*), the sensor (in *gray*) and the chip and imaging window (in *yellow*) not to scale (on the *right*). Adapted with permission from Sherr et al. ACS Nano, 2016, 10 (2), pp. 2827–2833. Copyright 2016 American Chemical Society

2. Syringe pump. PHD 2000 (Harvard Apparatus, Holliston, MA, USA).

3. Silicon Tubing. FEP Nat 1/16 × 0.010 × 20 ft (Upchurch Scientific, IDEX Health and Science, Middleborough, MA, USA).

4. Adaptors, syringes, fitting.

5. Figure 5 shows an SP-IRIS chip, a microfluidic cartridge and a schematic representation of the imaging setup.

2.6 Data Processing

1. MATLAB (MathWorks, Natick, MA, USA).

2. Custom particle detection software developed in MATLAB.

3 Methods

3.1 DNA Sequence Design

1. DNA sequences are designed to minimize the hairpin, self-dimer and heterodimer structures to increase the hybridization efficiency and prevent cross hybridization (*see* **Note 6**).

2. Twenty base pair of DNA sequence used as the surface probe (40 mer) is complementary to the antibody conjugated sequence used in the antibody-conjugation.

3. Amine modification is introduced at the 5′-end of the surface probe DNA sequence in order to achieve covalent immobilization on the copoly(DMA-NAS-MAPS) polymer coated chip surface. Table 1 shows the DNA sequences used in this work for antibody immobilization.

4. Lyophilized oligonucleotides are dissolved in ultrapure water to have a final concentration of 100 µM. 100 µL aliquots

Table 1
Model surface probe DNA sequence and its corresponding sequence used in antibody-conjugation in this work

Antibody	DNA sequences conjugated to the antibodies	DNA sequences immobilized on the chip surface
Anti-EBOV GP (13F6)	B: 5'AAAAATACAGAGTTA GTCGCAGTGG3'	B': 5'ATCCGACCT TGACATCTCTACCACTGCGACTAACTCTGTA3'

Complementary sequences are *underlined*. This particular sequences are selected as an example to be used in DNA-directed immobilization

are made. The concentration of the DNA solution is determined spectrophotometrically by measuring the absorbance at 260 nm. The aliquots are stored at −20 °C. When needed, a single aliquot is thawed, diluted to desired concentration and used immediately.

3.2 Protein Microarray Fabrication

1. The silicon chips are functionalized with the polymer copoly (DMA-NAS-MAPS). Prior to polymer coating, the chips are cleaned by sonicating in acetone for 5 min, rinsing in isopropanol and deionized water. Then, the chips are dried by blowing nitrogen gas.

2. The chips are cleaned by plasma ashing with oxygen plasma at 300 sccm and 500 W for 10 min.

3. 1% (w/v) copoly(DMA-NAS-MAPS) is dissolved in a mixture of Solution A2 and water (1:1) by vigorous vortexing.

4. The clean chips are placed in a plastic petri dish and the polymer solution is poured onto the chips until they are fully covered with the solution (*see* **Note 7**).

5. The chips are incubated in the polymer solution for 30 min at room temperature while shaking.

6. After functionalization, the chips are rinsed and dipped in deionized water for 3 min with shaking. This is repeated for three times. Then, the chips are dried with nitrogen gas.

7. In a vacuum oven, the chips are baked for 15 min at 80 °C (*see* **Note 8**).

8. To prepare protein microarrays via DNA-directed immobilization technique, surface probe oligonucleotide (Sequence B') is spotted on the polymer-functionalized chips at an optimum concentration of 30 μM in sodium phosphate buffer (150 mM, pH 8.5). In order to determine the effect of probe density, a varying degree of immobilization for antibody-conjugate was created and virus capture capacity was determined by spotting DNA surface probe at different concentrations which are 3, 6,

12, 18, 24 and 30 µM. Each concentration is spotted in six replicates on the chip (*see* **Note 9**).

9. An automated piezoelectric arrayer is used to spot surface probe DNA. A microarray pattern is designed for six replicate spots for the probe DNA. The space between the spots is set as 288 µm.

10. The humidity and the temperature inside the spotter chamber are set to 67% and 20 °C, respectively (*see* **Note 10**).

11. After spotting, DNA-spotted chips are kept in the spotter chamber at 67% humidity overnight at room temperature.

12. Following the immobilization, the chips are incubated in blocking solution for 30 min to quench the excessive NHS groups left after immobilization.

13. Then, the chips are washed with wash buffer for 30 min while shaking.

14. Afterwards, the chips are rinsed with phosphate buffer saline and deionized water, and dried with nitrogen.

15. Antibody-DNA conjugates are synthesized using Thunder-Link oligo conjugation kit according to manufacturer's protocol. 40 µM 5′-aminated linker DNA (Sequence B) is reacted with 1 mg/mL antibody (13F6) (*see* **Note 11**).

16. Antibody immobilization is achieved by incubating the DNA-chips with the antibody-DNA conjugate solution (at 5 µg/ml in PBS with 1% BSA) for 1 h in a 24-well plate on a shaker.

17. After DNA-DNA hybridization and antibody immobilization, the chips are washed with PBS twice, 0.5 M sodium nitrate solution twice, and then, dipped in cold 0.1 M sodium nitrate solution. Finally, the chips are dried with nitrogen gas.

18. For comparison purpose only, antibodies are immobilized directly (*see* **Notes 12** and **13**).

3.3 Quality Control of the Modified Chips

1. Biomass quantification is performed for direct antibody-immobilized and DNA directed antibody-immobilized chips using IRIS (*see* **Note 14**).

2. To test the activity of DNA conjugate for binding to its target, biomass measurements of the conjugate hybridization and virus binding for both directly immobilized and DNA-tethered antibody are performed. The spots are imaged before antibody-conjugate hybridization, after hybridization (thus, antibody immobilization) and after virus (VSV) capture, shown in Fig. 6. After the immobilization and virus capture, the spot height change is identified with IRIS measurements concluding that although DNA-spots have much less probe density (⅓ of antibody-spot density), they capture almost the

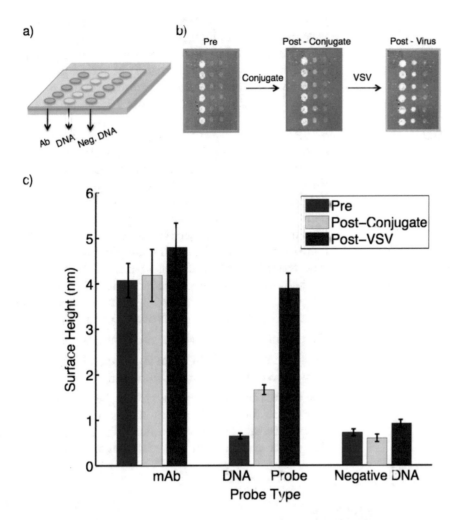

Fig. 6 IRIS average spot height measurements for direct-antibody and DNA-directed immobilized-antibody spots. One nanometer of surface thickness corresponds to 1.2 ng/mm² of antibody density. (**a**) Schematic representation of the IRIS chip spotted with three different probe types: antibody, probe DNA and a negative DNA sequence (not complementary to target antibody-conjugated DNA). (**b**) IRIS images of the chip at different stages of the experiment: before hybridization of DNA B-conjugate, after hybridization of the conjugate and after incubating with VSV virus. (**c**) IRIS surface height measurements for three different stages of the experiment

same amount of virus with the direct-antibody spots. This shows the significant improvement in the capture efficiency due to DDI for protein immobilization. The height change in the antibody spots shows that they are only able to capture a small amount of virus even though they are spotted at the highest surface density. This shows that even though the protein immobilization on the surface is achieved successfully, functionality of the protein for virus detection after immobilization is questionable. Moreover, it is also seen from the figure that variability in DNA-tethered protein immobilized spots is very small. Hence, DDI is found to be an optimum technique for obtaining robust and reproducible protein microarrays. These results also suggest that DDI technique can be easily adapted to immobilization of various proteins without a need for further optimizations (*see* **Note 15**).

3.4 SP-IRIS

1. SP-IRIS optical setup is based on a top-illumination microscope configuration. To create the optical path with the illuminating LED, 2-lens system is used and the LED is imaged onto the back aperture of the objective for Kohler illumination. LED light is first collimated with 30 mm-lens in the illumination path. Then, the collimated light gets focused onto the back focal aperture of the objective using a 50 mm-lens. Hence, the light beam is directed onto the sample stage which, then, gets reflected off the SP-IRIS sensor surface. With a beam splitter, the light beam is focused; therefore, it is directed down to the sample stage and the reflected light from the sensor surface is captured by the camera (*see* **Notes 16** and **17**).

2. Via the microscope objective, the light beam is focused.

3. After, the reflected light beam is collected by the objective and transmitted through the beam splitter.

4. The specular reflected light as well as the scatter light from the sample are focused via the tube lens onto the CCD camera where intensity image is recorded. The schematic representation of the system, light path and imaging principle are shown

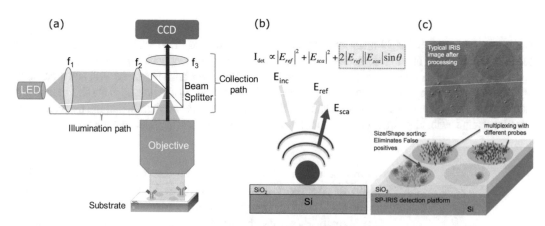

Fig. 7 (**a**) Schematic representation of the SP-IRIS setup. LED is used for illumination and bright field image is reflected on a CCD camera. (**b**) With the help of cross term (*highlighted*) in the equation, visibility of low-index nanoparticles is enabled with their enhanced intensity by order of magnitude compared to scattered intensity (middle term). (**c**) A representation of multiplexed SP-IRIS chip (*bottom*) and an actual image of detected particles on the chip after processing (*top*). Reprinted with permission from Avci et al. Sensors, 2015, 15, pp 17649–17665. 2016 Creative Commons (http:/creativecommons.org/licenses/by/4.0/)

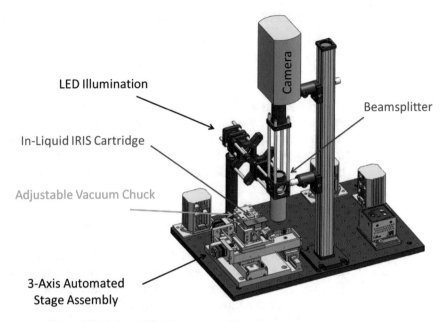

Fig. 8 Representative image of the SP-IRIS setup

in Fig. 7. Figure 8 shows a representative image of the setup and its configuration.

5. For imaging, the microfluidic device is placed on the sample holder (stage) and endpoint images are taken.

3.5 Data Processing

1. Each spot is imaged in SP-IRIS.

2. The images for each spot are analyzed for bound virus particles.

3. By subtracting pre-incubation particle counts from post-incubation counts and dividing the result by the analyzed spot area, the net particle densities are determined (*see* **Note 18**).

4. Six spots for each condition are used to average the particle densities.

5. The morphological features as well as non-diffraction limited peaks are removed from the image with appropriate filters.

6. The signal coming from the particle is based on the interference between the scattered field from the particle and the reference field arising from the reflection off the sensor. The signal from the particle is normalized relative to the nearby background intensity. Custom particle detection software is developed in MATLAB to find particle-originated local intensity maxima which fall within the point spread function of the optical system (*see* **Note 19**).

7. The interaction of the light with small particle is considered as an induced dipole where its strength is proportional with the

polarizability of the particle (α), as given in the following equation:

$$\alpha = 4\pi\varepsilon_0 r^3 \frac{\varepsilon_p - \varepsilon_m}{\varepsilon_p + 2\varepsilon_m} \qquad (1)$$

where r is radius of the particle, ε_p and ε_m are particle and surrounding medium permittivity, respectively [20]. In interferometric imaging techniques, in the intensity recorded at the detector, strong reference field and weak scattered fields coming from the particle are mixed and can be expressed with the following equation:

$$I \propto |E_{\text{ref}}|^2 + |E_{\text{sca}}|^2 + 2|E_{\text{ref}}||E_{\text{sca}}|\cos\theta \qquad (2)$$

In this equation, for weakly scattering particles, $|E_{\text{sca}}|^2$ is negligible. Hence, the signal is dominated by the cross term as seen in Fig. 7b. This signal appears as a dot in the image as seen in Fig. 7c. Based upon the contrast of the particle, the size is determined with the model. Since the virus particles scatter the light upon binding the surface, enhanced signal coming from the layered chip surface is detected on the CCD camera.

3.6 Single Particle Detection (In Liquid Virus Detection Measurements)

1. The flow cell is obtained by assembling the individual layers. The acrylic layer is placed as the base and then, biosensor chip is put in the middle of the cartridge. After placing the second layer, the third layer is placed on top providing imaging window like a polymer cover glass. Then, the fourth and fifth layers are placed and adaptors are attached to the fifth layer. FEP tubing is assembled to the adaptors. The syringe pump is connected via tubing to provide the flow. The flow is directed in the channels to the chip surface with the precisely designed microfluidic device as shown in Fig. 5.

2. The flow cell is placed on the sample stage of SP-IRIS underneath the objective for image acquisition. DNA-immobilized SP-IRIS chips are already placed in the microfluidic cartridges.

3. Acquisition software is opened and the camera is turned on.

4. Number of images to be averaged is determined and selected. Maximum exposure time before saturation of the camera is selected.

5. First, 13F6-DNA, B-conjugate is flowed at a concentration of 5 μg/mL in PBS with 1% BSA for 1 h to achieve protein immobilization on the chip.

6. To get pre-incubation particle counts in spots with different concentration of protein immobilized, SP-IRIS images are acquired.

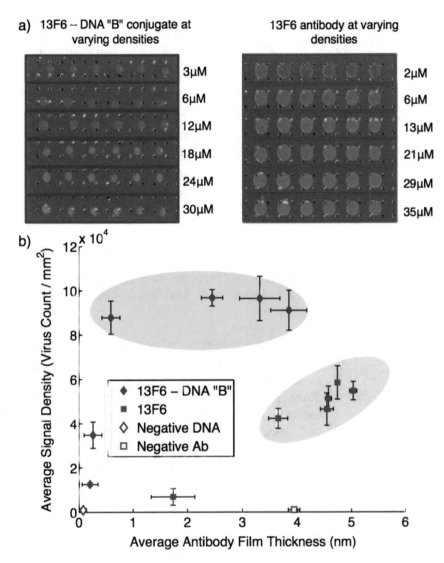

Fig. 9 (**a**) Left IRIS image showing different surface densities of immobilized antibody-DNA conjugate after hybridization of the B′ sequence with the 13F6-DNA "B" conjugate. *Right image* shows 13F6-antibody spots at different concentrations. (**b**) Effect of surface antibody density (1 nm = 1.2 ng/mm²) on virus capture efficiency for both direct antibody and DNA-conjugated antibody immobilization. Average capture virus densities (*n* = 6 spots) obtained from SP-IRIS images versus average optical thickness of the antibody spots obtained from IRIS images were plotted. The *ellipses* indicate the range of surface antibody heights where the optimal virus capture occurs. Reprinted with permission from Seymour et al. Anal. Chem., 2015, 87 (20), pp. 10505–10512. Copyright 2016 American Chemical Society

7. Then, the chips are incubated with 10^4 PFU/mL EBOV-VSV by flowing it over the chip for 30 min.

8. After incubation, 400 µL sodium phosphate buffer is flowed over the microchannel and SP-IRIS images are acquired (*see* **Note 20**).

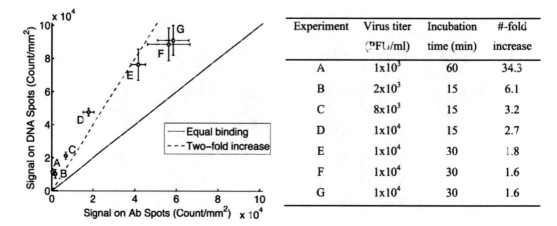

Experiment	Virus titer (PFU/ml)	Incubation time (min)	#-fold increase
A	1×10^3	60	34.3
B	2×10^3	15	6.1
C	8×10^3	15	3.2
D	1×10^4	15	2.7
E	1×10^4	30	1.8
F	1×10^4	30	1.6
G	1×10^4	30	1.6

Fig. 10 Comparing captured virus densities on direct-spotted and DNA-tethered antibodies from independently performed experiments. Each *letter* represents a data point next to it. Experimental conditions corresponding to A–G experiments are summarized in the table. Error bars show the variation between six spots for a given experiment. Reprinted with permission from Seymour et al. Anal. Chem., 2015, 87(20), pp 10505–10512. Copyright 2016 American Chemical Society

9. Finally, end point images of virus-captured chips are acquired.

10. Figure 9 shows the comparison of the virus capture performance of both direct antibody-immobilized and DNA-directed-antibody immobilized assays. IRIS is used to show the surface morphologies of antibody and DNA-tethered antibody spots at different concentrations. According to Fig. 9b, DNA-conjugated 13F6 performs optimally over a larger range of surface densities compared to the directly spotted antibody [19].

11. Figure 10 depicts the performance of the protein immobilization technique by varying the capture conditions. In general, since limited number of target is present at lower concentrations of the virus, a lower capture efficiency is obtained in both direct and DDI techniques. However, when the available targets increase in the solution, a drastic increase in DNA-tethered antibody microarray is observed; whereas, direct-antibody immobilized microarray does not show the same sensitivity. DDI makes the capture proteins highly available and accessible on the sensor surface. Hence, an improved binding capacity is achieved in shorter time.

4   Notes

1. Adjusting the NA offers a tradeoff between nanomaterial contrast and usable field of view (FOV).

2. Oxide thicknesses are chosen to maximize the phase angle term.

3. It can be synthesized as described in the reference [21]. One of the most affecting components of protein microarrays is microarray immobilization surface. For DDI technique, specifically a reactive-functional surface is needed in order to achieve the covalent attachment of the surface probe DNA sequences to the sensor surface. Therefore, MCP-2 is an excellent option for this purpose. It bears functional NHS groups which take part in covalent immobilization of the amino-modified oligonucleotides. Moreover, due to its unique 3D structure, it provides right confirmation for oligonucleotides by enhancing immobilization with its hydrophobic regions. Also, MCP-2 lowers non-specific bindings of proteins providing a perfect sensor surface for sensitive and selective detections [22].

4. NHS groups are active for further possible reactions. Therefore, it is important to block them after the completion of immobilization. Otherwise, it may cause further crosslinking; thus, a decrease in the efficiency in protein immobilization due to the less availability of the probes or proteins.

5. Antibody spots directly immobilized for comparison purpose only require antibody spotting buffer ($1\times$): $1\times$ PBS with a final concentration of 25 mM trehalose, pH 7.5. Ready-made $1\times$ PBS (Thermo Fisher, GIBCO PBS $1\times$). Trehalose stock solution is prepared at 0.25 M. Threhalose is added to the antibody solution (which is in PBS) just before spotting. After diluting antibody solution, trehalose is added to give a final concetration of 25 mM. 8.6 g of trehalose is dissolved in 10 mL of filtered deionized water to have 0.25 M trehalose stock solution.

6. $5'$-polyA sequence is added as a spacer to DNA sequence which is conjugated to the antibody.

7. It is important to use a plastic petri dish in the polymer coating process of the chips. Since a glass petri dish competes with the chip in functionalization, for higher coating efficiency, a plastic petri dish is used.

8. It is critical to store the functionalized chips in a desiccator after fabrication. The chips can be stored in the desiccator for up to 3 weeks under vacuum. It would be worthwhile to bake the chips at 80 °C for 15 min in a vacuum oven prior to use in order to have completely dried and NHS-active polymer coating on the chips for successive spotting.

9. Surface antibody density is an important criterion which drastically affects assay sensitivity. Surface density of the antibodies should be optimized for a given application to optimize the target capture.

10. During spotting of DNA, it is very important to keep the humidity at an optimum level in the spotter chamber. The optimal level we found is 67% for the humidity. After spotting, the chips are kept in the spotter chamber at 67% humidity overnight. However, for antibody spotting, optimum spotting is found as 57–59%. These changes in the humidity make huge variations in the spot morphology. If DNA is spotted at the same humidity with antibody, then, a smaller and non-uniform spot morphology is observed.

11. DNA concentration is optimized to obtain 1–2 DNA sequences per antibody. After conjugation, yield of reaction is determined by Bradford assay. The monoclonal antibody is conjugated to DNA sequence, B′, in 1:1.5 ratio. 13F6-DNA, B, is antibody-conjugates against Ebola-GP pseudotyped VSV.

12. Different antibodies have different probe characteristics. Therefore, in order to obtain a uniform spot morphology and surface density, for each type of protein, an optimization study is needed. This is not time and cost efficient. On the other hand, oligonucleotides only differ in sequences and length among each other; therefore, as long as the length of the sequence is kept constant, DNA immobilization characteristics generally stay the same. DNA spotting does not depend on the type or immobilization. Hence, a standard and parameter-independent protein immobilization can be achieved with DDI. Thereupon, this technique is preferred in this study.

13. Antibody spots directly immobilized for comparison purpose only. (a) Antibody solutions are prepared as 3 mg/mL in antibody spotting buffer. The humidity during spotting is kept between 57 and 59%. The spotted chips are kept in the spotter chamber at 67% humidity overnight. (b) To evaluate the effect of antibody immobilization density on the virus capture, anti-EBOV GP antibody (13F6) is spotted at varying concentrations on SP-IRIS substrates (5, 4, 3, 2 ,1 and 0.5 mg/mL).

14. IRIS has a different modality of operation. The important difference between IRIS and SP-IRIS is the magnification of the optical system. In IRIS, numerical aperture of the objective lens is different. Hence, low-magnification can be achieved; whereas, using SP-IRIS, high-magnification is achieved. This modality is based on spectral reflectivity. When the biomass accumulates on the sensor surface, the thickness on the sensor increases. This increase generates a quantifiable change in the spectral reflectivity depending on the optical path difference (OPD) between substrate surface and biomass accumulated substrate surface. With the help of low-magnification modality, multiplexed monitoring of biomolecule immobilization or

capture can be precisely acquired. Therefore, using IRIS, a larger field of view (FOV), approximately 1 in.^{2}, can be captured. Hence, IRIS can be used for quality control of the assay chips [20]. Variations in the probe immobilization can be detected easily.

15. Spot morphology is another important criterion for single particle detection. It directly affects probe density, capture efficiency; thus, the reproducibility in the experiments. Non-uniform spot morphology causes spot-to-spot and assay-to-assay changes. Capture efficiency is directly proportional with the surface probes. A destroyed morphology causes non-binding of the target particle; therefore, false negative results may be obtained in quantification of the signal [23]. These irregularities may be crystallization appearing as bright spots in the image, coffee rings in the outer region of the spot or aggregations causing non-uniform regions in the spot. The detector response should not depend on the parameters. However, when antibodies are used as the surface probe, a robust microarray design may not always be achieved due to the problems in consistency in the spot morphology. On the other hand, DNA spotting is always uniform. More robust protein microarrays can be achieved via DDI. DNA spot morphology stays almost identical in repetitive spotting which provides a uniformity in the sensor response among spot-to-spot and assay-to-assay measurements.

16. The significant parameters in the optical setup are the illumination wavelength, uniformity, NA, magnification and camera pixel size. These parameters affect the resolution and the accuracy of sample sizing. In addition, since sample sizing is determined using the contrast of the central peak to near-neighbor background pixels; improper and non-uniform illumination cause sizing obscurity. Kohler illumination is implemented in the optical system due to its superiority over critical illumination.

17. To achieve a uniform illumination, a diffuser is placed in front of the LED light source.

18. In order to minimize signal variations arising from the spot morphology, it is desirable to take a pre-incubation image to later subtract from the post-incubation image to calculate the net number of bound particles.

19. Other image processing software such as ImageJ can be used as an alternative to MATLAB for post-processing and quantification.

20. During the microarray preparation processes, especially in between washings, the chips should always be wetted. Extra care should be taken.

References

1. MacBeath G, Schreiber SL (2000) Printing proteins as microarrays for high-throughput function determination. Science 289:1760–1763
2. Sun H, Chen GYJ, Yao SQ (2013) Recent advances in microarray technologies for proteomics. Chem Biol 20:685–699
3. Hall DA, Tacek J, Snyder M (2007) Protein microarray technology. Mech Ageing Dev 128 (1):161–167
4. Avci O, Lortlar ÜN, Yalcin A et al (2015) Interferometric Reflectance imaging sensor (IRIS)-a platform technology for multiplexed diagnostics and digital detection. Sensors 15 (7):17649–17665
5. Schwenk JM, Lindberg J, Sundberg M et al (2007) Determination of binding specificities in highly multiplexed bead-based antibody assays for antibody proteomics. Mol Cell Proteomics 6:125–132
6. Cretich M, Daaboul GG, Sola L et al (2015) Digital detection of biomarkers assisted by nanoparticles: application to diagnostics. Trends Biotechnol 33(6):343–351
7. Yurt A, Daaboul GG, Connor JH et al (2012) Single nanoparticle detectors for biological applications. Nanoscale 4(3):715–726
8. Walt D (2013) Optical methods for single molecule detection and analysis. Anal Chem 85 (3):1258–1263
9. Daaboul GG, Lopez CA, Chinnala J et al (2014) Digital sensing and sizing of vesicular stomatitis virus pseudotypes in complex media: a model for ebola and marburg detection. ACS Nano 8(6):6047–6055
10. Monroe MR, Daaboul GG, Tuysuzoglu A et al (2013) Single nanoparticle detection for multiplexed protein diagnostics with attomolar sensitivity in serum and unprocessed whole blood. Anal Chem 85(7):3698–3706
11. Sevenler D, Lortlar ÜN, Ünlü MS (2015) Nanoparticle biosensing with interferometric reflectance imaging. In: Vestergaard MC, Kerman K, Hsing I-M, Tamiya E (eds) Nanobiosensors and nanobioanalyses. Springer, Tokyo, pp 81–95
12. Monroe MR, Reddington A, Collins AD et al (2011) Multiplexed method to calibrate and quantitate fluorescence signal for allergenspecific IgE. Anal Chem 83(24):9485–9491
13. Niemeyer CM, Boldt L, Ceyhan B et al (1999) DNA-directed immobilization: efficient, reversible, and site-selective surface binding of proteins by means of covalent DNA-streptavidin conjugates. Anal Biochem 268:54–63
14. Ladd J, Boozer C, Yu Q et al (2004) DNA-directed protein immobilization on mixed self-assembled monolayers via a streptavidin bridge. Langmuir 20:8090–8095
15. Schroeder H, Adler M, Gergk K et al (2009) User configurable microfluidic device for multiplexed immunoassays based on DNA-directed assembly. Anal Chem 81:1275–1279
16. Washburn AL, Gomez J, Bailey RC (2011) DNA-encoding to improve performance and allow parallel evaluation of the binding characteristics of multiple antibodies in a surface-bound immunoassay format. Anal Chem 83:3572–3580
17. Wacker R, Niemeyer CM (2004) DNA- μFIA-a readily configurable microarray-fluorescence immunoassay based on DNA-directed immobilization of proteins. Chem Bio Chem 5:453–459
18. Wacker R, Schroder H, Niemeyer CM (2004) Performance of antibody microarrays fabricated by either DNA-directed immobilization, direct spotting, or streptavidin-biotin attachment: a comparative study. Anal Biochem 330:281–287
19. Seymour E, Daaboul GG, Zhang X et al (2015) DNA-directed antibody immobilization for enhanced detection of single viral pathogens. Anal Chem 87(20):10505–10512
20. Avci O, Adato R, Yalcin OA et al (2016) Physical modeling of interference enhanced imaging and characterization of single nanoparticles. Opt Express 24(6):6094–6114
21. Pirri G, Damin F, Chiari M et al (2004) Characterization of the polymeric adsorbed coating for DNA microarray glass slides. Anal Chem 76 (4):1352–1358
22. Yalçin A, Damin F, Özkumur E et al (2009) Direct observation of conformation of a polymeric coating with implications in microarray applications. Anal Chem 81(2):625–630
23. Romanov V, Davido SN, Miles AR et al (2014) A critical comparison of protein microarray fabrication technologies. Analyst 139 (6):1303–1326

Chapter 13

Reflectometric Interference Spectroscopy

Guenther Proll, Goran Markovic, Peter Fechner, Florian Proell, and Guenter Gauglitz

Abstract

Reflectometry is classified in comparison to the commercialized refractometric surface plasmon resonance. The advantages of direct optical detection depend on a sophisticated surface chemistry resulting negligible nonspecific binding and high loading with recognition sites at the biopolymer sensitive layer of the transducer. Elaborate details on instrumental realization and surface chemistry are discussed for optimum application of reflectometric interference spectroscopy (RIfS). A standard protocol for a binding inhibition assay is given. It overcomes principal problems of any direct optical detection technique.

Key words Label-free optical biosensor, Reflectometric interference spectroscopy (RIfS)

1 Introduction

The methods available for direct monitoring of biomolecular interaction can be divided into methods measuring changes in the refractive index of the interaction layer, and methods measuring changes in reflectometry at the layer [1]. Regarding the first method, BiaCore [2] has opened the market by introducing surface plasmon resonance (SPR) [3] as a very promising tool in biomolecular interaction analysis (BIA). In contrast to refractometry, reflectometry concentrates on the measurement of changes in optical thickness. Reflectometry has been introduced many decades ago as ellipsometry using polarized light. Interference at the interfaces of the layer causes a change in the relative amount of amplitude of the two polarized radiation beams and in phase. This interferometric method has been applied to a very simple analytical method, called reflectometric interference spectroscopy (RIfS) to be used as an example of a very robust, simple optical detection principle in chemical and biochemical sensing [4]. This label-free optical detection method for surface interactions is based on white light interference at transparent thin layers. At each interface of thin layers of

Avraham Rasooly and Ben Prickril (eds.), *Biosensors and Biodetection: Methods and Protocols Volume 1: Optical-Based Detectors*, Methods in Molecular Biology, vol. 1571, DOI 10.1007/978-1-4939-6848-0_13,
© Springer Science+Business Media LLC 2017

different materials with negligible absorption, radiation is partially reflected and transmitted. If the optical path length through these layers is less than the coherence wavelength, the different partial beams interfere, and form an interference pattern depending on the wavelength, the optical thickness which is given by the product of the refractive index of the layer and its physical thickness, the incident angle, and the refractive index of the surrounding medium [5, 6]. In case of perpendicular incidence, a non-absorbing layer, and low reflectances, the reflectance R is given by:

$$R = R_1 + R_2 + 2\sqrt{R_1 R_2} \cos\left(4\pi nd/\lambda\right) \qquad (1)$$

where R_1 and R_2 denote the Fresnel reflectance at the two interfaces, d is the physical thickness of the film, n its refractive index, and λ the wavelength of incident light (*see* **Note 1**). A typical interference pattern showing the modulation of reflectance with $\cos(1/\lambda)$ is given in Fig. 1.

The optical thickness $n \cdot d$ can be determined from the position of an extremum with a given order value m by

$$n \times d = \frac{m\lambda}{2}. \qquad (2)$$

RIfS uses the change in the optical properties in or at the top layer of a given layer system as detection principle (Fig. 1). The binding of an analyte molecule or particle to the sensor surface causes a shift of the interference pattern in the wavelength domain. To evaluate the binding signal, the locus of an extremum is tracked over time; thus, the change of the interference spectrum results in a time-resolved binding curve representing the binding of the analyte molecule to the sensor surface.

A major advantage of RIfS is its resistance to changes in temperature [7]. Refractometric methods such as SPR and ellipsometry, on the other hand, are very sensitive to temperature variations due to the high impact of temperature on the refractive index. Thus, temperature changes during a measurement cause negative effects with these methods, and quick changes of temperature between measurements are technically challenging. Since the

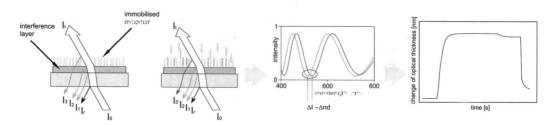

Fig. 1 Scheme of the RIfS detection principle. The *left part* shows the superimposition of the reflected light beams and the change in optical thickness during a binding event on the sensor surface. The *right part* shows the corresponding change of the characteristic interference spectrum and the resulting binding curve

refractive index n is dependent on the density given from the Clausius Mossotti equation, and the density is dependent on the thermal expansion, the refractive index decreases with increasing temperature. Due to the thermal expansion of the biopolymer-layer, the physical thickness d increases with increasing temperature. These two contrary temperature-dependent effects result in a rather low influence of temperature on the optical thickness, the product of the refractive index and the physical thickness.

2   Materials

Common chemicals of analytical grade are purchased from Sigma or Merck. Milli-Q water is deionized water with a conductivity of 18.2 MΩ cm^{-1}.

2.1   Transducer

The RIfS standard transducer consists of a glass substrate (D 263 glass, Schott AG, Germany) (~1 mm thick) coated with a 10 nm layer of a material with high refractive index (usually Ta_2O_5 or Nb_2O_5), and a top layer of SiO_2 (330 nm). As a reference use a transducer without SiO_2 layer. Due to the technical specifications of the components used for setting up a RIfS measurement system and the investigated sample types (e.g., liquids or gases), other transducers might be more beneficial. An ideal RIfS transducer is designed to create the two reflected light beams of interest at similar intensities. Therefore, it is recommended to use ray tracing programs (e.g., FilmWizard by SCI Scientific Computing International or WVASE by J. A. Woollam) to simulate the expected reflectivities for the visible wavelength range for a given RIfS setup, taking into account the expected optical properties of the investigated sample types.

2.2   Surface Chemistry

1. GOPTS (3-glycidyloxypropyltrimethoxysilane) purum: Toxic. Store at room temperature, keep under Argon, sensitive to humidity (Fluka).

2. AMD (aminodextran): MW 100 kD, level of amination 50%. Store at 4 °C under dry conditions (Innovent e.V. Technologieentwicklung Jena, Germany).

3. Dicarboxypoly- and diaminopoly(ethylene glycol) (PEG): MW 2000 Da. Store at −20 °C under dry conditions (Rapp Polymere, Tuebingen, Germany).

4. NHS (N-hydroxysuccinimide) purum: Store under room temperature (Fluka).

5. DIC (N,N'-diisopropylcarbodiimide) purum: Toxic. Store at room temperature, keep under argon, sensitive to humidity (Fluka).

6. EMCS (6-maleimidohexanoic acid *N*-succinimidyl ester) purum: Store at −20 °C under dry conditions (Sigma).

7. TBTU (2-(1H-benzotriazol- 1-yl)-1,1,3,3-tetramethyluronium tetrafluoroborate).

8. HOBT (1-hydroxybenzotriazole) Hydrate purum: Sore at −20 °C at dry conditions (Fluka).

9. DIPEA (*N*,*N*-diisopropylethylamine): Highly inflammable. Store at room temperature (Sigma).

2.3 Glass Substrates

1. BK 7 glass, $n = 151$, Schott, Mainz.

2. WG 345 glass, $n = 1699$, Schott, Mainz.

3. Interference transducer: D 263, 10 nm Ta_2O_5, 500 nm SiO_2, Schott, Mainz.

4. Goethe glass: multi-layer system: 1 m D 23, 45 nm, Ta_2O_5, 20 nm SiO_2, Schott, Mainz (Cut in small squares (10 × 10 mm)).

2.4 Parallel Setup

1. White light source 100 W/12 V, Osram, Munich.

2. Lenses, mirror, and positioning optics, Spindler&Hoyer, Göttingen

3. Polymer light guides (PMAA, $n = 1490$, coupling element 1 × 2 (50:50), 1 mm, fiber diameter with SMA 905 fiber connectors, Microparts, Dortmund.

4. Optical 4-×-1 multiplexer DiCon VX 500-C, Laser Components, Olching.

5. MMS diode row spectrometer with Liliput-PC, ZEISS, Jena.

6. 19″ industrial standard housing by RS Components, Walldorf-Mörfelden.

7. Commonly Microsoft Windows XP, Vista, Windows 7, Windows 8, Windows 8.1, Windows 10 operating systems.

2.5 Single Setup

1. Modified simultaneous spectrometer SPECKOL 1100, Zeiss, Jena

2. Commonly Microsoft Windows XP, Vista, Windows 7, Windows 8, Windows 8.1, Windows 10 operating systems.

2.6 Advanced Single Setup (See Fig. 4)

1. Light source providing light of a single color (White light source 100 W/12 V, Osram, Munich with Bandpassfilter 568 nm).

2. Phodiode based detection unit (Lightwave-Multimeter 2832-C with Si-photodiodes of the series 818 from Newport).

2.7 Liquid Handling

1. Ten position valve VICI, Valco Europa, Schenkon, Switzerland.

2. HPCL 3-way valve VICI, Valco Europa, Schenkon, Switzerland.

3. Six position valve, Bischoff, Leonberg.

4. Peristaltic pump Reglo-Digital MS2/8-160 ISM 832, Ismatec, Wertheim.

5. Peristaltic pump MS Fixo, Ismatec, Wertheim.

6. High grade steel capillaries, screws and fittings from Rheodyne, USA.

2.8 Software

1. Measure^{CR} for capturing spectras and controlling the systems. Depending on the selected spectrometer, any other software control can also be used. As most commercial available spectrometers are supplied with ready to use software or plug-ins for the widely used Lab-View solution (National Instruments Corporation), it is recommended to use one of these solutions.

2. IFZ^{CR} for evaluating the interferograms. Any other software solution capable for performing the same simple mathematical calculations like Matlab (The MathWorks, Inc.) can be used.

3. MS-Excel (Microsoft) and Origin (MicroCal Inc.) or any equivalent statistical software packages (open source software is available) can be used for further processing of measured data.

4. If an advanced single color setup for reflectometry measurements is used, no spectras have to be analyzed. Therefore, software only needs to control hardware components. All evaluation work can be done with software packages like Matlab or Origin.

3 Methods

3.1 Setup for RIfS

In the standard laboratory setup for RIfS (Fig. 2), the white light is guided via an optical fiber (1 mm PMMA fiber) to the back of the transducer mounted in a microfluidic flow cell, which in turn is attached to a liquid handling system. The reflected light is gathered in the same waveguide.

The fiber optics used is bifurcated (50:50 ratio), with one tail leading to the light source and the other to the UV–Vis spectrometer. Possible light sources are halogen lamps (e.g., 10 W halogen lamp with fiber in-coupling optics consisting of front surface spherical mirror, collimating lens, and an infrared absorption filter) or LEDs. For the spectral detection of the reflected light, diode array spectrometers are used normally. A gap (approx. 100 μm) between transducer chip and the fiber output is filled with glycerol (80%) for refractive index matching. Samples are handled by a flow system (Fig. 3) (e.g., two peristaltic pumps, inject-load valve, and 6-way valve).

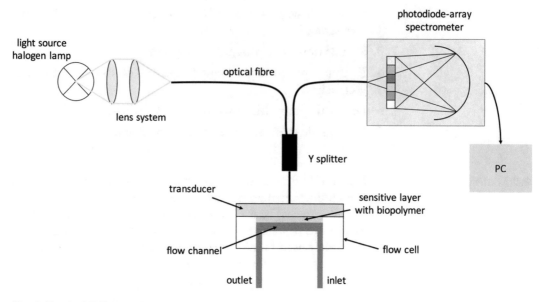

Fig. 2 Classical RIfS setup

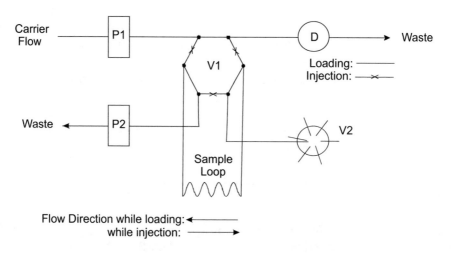

Fig. 3 Schematic drawing of the FIA system for RIfS

Moreover, this flow system can be equipped with an autosampler. The various samples are injected by the autosampler into a sample loop. From there, the sample is driven in continuous flow, passing the prepared transducer.

The raw spectra are corrected for the dark current of the spectrometer by subtraction (if necessary), and normalized to the reflectance spectrum from a glass chip without the SiO_2- and bio-layer. The position of an interference extremum at approx. 550 nm is determined by a parabolic fit to one halfwave of the interference spectrum. Optical thickness is calculated according to Eq. 2 (*see* **Note 2**).

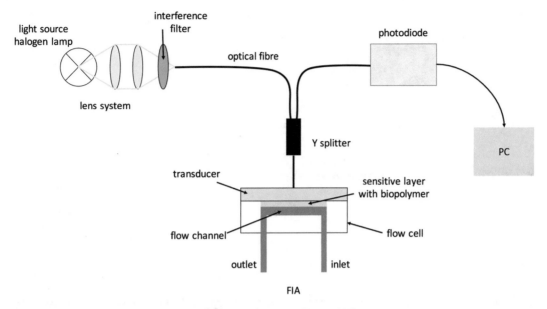

Fig. 4 Advanced setup for single-color RIfS detection according to [15]

The costs of the setup can be reduced by sequentially illuminating the transducer with different suitable wavelengths coming from LEDs or a white light source with appropriate filters, detecting the intensity without spectral resolution using a photodiode, and reconstructing part of the interference pattern by fitting a parabola through the interpolation points. The advanced development is a single-color (wavelength) RIfS approach, which uses a color filtered light source (Fig. 4) or even a LED, monitoring the change in intensity looking at a distinct wavelength rather than the shift of an extremum (*see* Fig. 1). This allows a very simple instrumentation setup and the use of a second wavelength as reference.

This detection principle can be used to realize a parallel screening system which allows optical online detection of specific biomolecular interaction in 96- or 384-well microplate formats (*see* Fig. 5) Reflectometric interference spectroscopy (RIfS):parallel set-up. Therefore, the whole area of the plate bottom consisting of a RIfS transducer is illuminated by a halogen light source combined with a filter wheel, which allows the subsequent passage of monochromatic light of seven different wavelengths. This is not only to reduce the costs, but is necessary because a CCD camera is used as detector which is able to detect the intensity of the whole area of interest in a single shot.

Singe-color RIfS also has an advantage in a parallelized setup. A commercial imaging device was developed by Biametrics GmbH that can measure up to 20,000 events per square centimeter on the transducer under flow-through conditions while obtaining information on thermodynamics and kinetics in parallel.

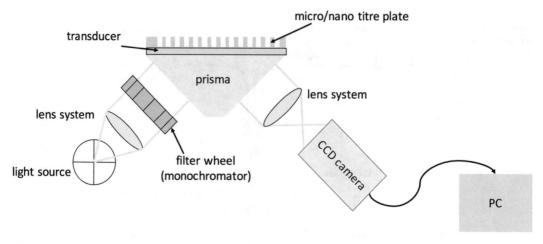

Fig. 5 Parallel RIfS setup

3.2 Surface Chemistry

All surface chemistry steps are applied to the coated side of the RIfS transducer (*see* **Note 3**) if not stated otherwise.

3.2.1 Silanization

(a) Transducer chips are cleaned with freshly prepared Piranha solution (mixture of 30% hydrogen peroxide and concentrated sulfuric acid at a ratio of 2:3; caution: hot and aggressive!) for 30 min in an ultrasonic bath to clean the chip and to generate silanol groups on the transducer surface. As an alternative, this activation step can be carried out using oxygen plasma.

(b) After rinsing with double distilled water and drying in a nitrogen stream, the surface is immediately activated by GOPTS at a surface concentration of approx. $10 \ \mu l/cm^2$ for 1 h in dryness (by assembling two slides face-to-face placed in a tray with ground joint) followed by cleaning with dry acetone and drying in a nitrogen stream (*see* **Notes 4** and **5**).

3.2.2 Biopolymer

When investigating interactions of biomolecules on surfaces, non-specific binding to the sensor must be minimized. Therefore, the glass slide is coated with a shielding layer which prevents nonspecific binding and additionally provides a large number of functional groups for the immobilization of the binding partner. The standard shielding chemicals used are dextrans which form an 3D-hydrogel loaded with a large number of binding sites, and polyethylene glycols (PEG) forming a two-dimensional brush-like monolayer with less binding sites but with a more defined surface. Amino and carboxy groups are normally functional groups for the immobilization process since the well-established peptide chemistry methods can be applied. Other shielding chemicals can be used depending on the application or the surface chemistry needed.

(a) Aminodextran.

The coupling of aminodextran as an aqueous solution (1:3) (approx. 15 μL/cm^2) to the silanized surface is carried out by incubating over night (min 16 h) in a water-saturated atmosphere (see **Note 6**). After thoroughly rinsing with double distilled water and drying, the prepared chips are stable for several months.

(b) Polyethylenglycol.

The coupling of diamino- or dicarboxypoly(ethylene glycol) (PEG) as a 1 mM solution in dichloromethane (DCM) (approx. 10 μL/cm^2) to the silanized surface is achieved by incubating overnight at 70 °C, followed by thorough rinsing with Milli-Q water and drying in a nitrogen stream.

(c) Change of functional groups.

For some applications (e.g., immobilization of DNA oligomers with an amino-linker) the follow-up functionalization via peptide chemistry works better with diaminopoly(ethylene glycol) treated with 5 M glutaric anhydride (GA) in N,N-dimethylformamide (DMF) (10 μL/cm^2) for 6 h to generate carboxylic groups on the surface. The same protocol can be used to modify AMD-surfaces with carboxylic groups.

(d) One step chemistry.

Depending on the expected properties of the biopolymer layer, complex silanes offering chemically beneficial side chains like 11-aminoundecyltrimethoxysilane can be used in a one step procedure combining silanization with biopolymer coating.

3.2.3 Immobilization of Amino Terminated Ligands

(a) The covalent coupling of amino-terminated molecules/ligands to biopolymers with carboxylic groups is done by standard peptide chemistry via activated esters (see **Note 7**). Therefore, the previously modified transducers are activated with a solution of NHS and DIC (1 M NHS, 1.2 M DIC in DMF, 10 μL/cm^2) for 2–4 h as sandwich pairs in a DMF saturated atmosphere (see **Note 5**). The transducers are rinsed first with DMF and then with dry Aceton, and dried in a nitrogen stream (see **Notes 4** and **5**).

(b) To achieve a high density of covalently bound molecules, it is necessary to use an excess of reagents. Therefore, the ligand is applied as a DMF solution (10 μL/cm^2) (if your ligand is insoluble in DMF, use another aprotic solvent of appropriate polarity) at a concentration of 1–3 μM to the activated surface for 4 h as sandwich pairs in a DMF saturated atmosphere (see **Note 6**). Clean the transducers by rinsing them with DMF and Milli-Q water and drying them in a nitrogen stream (see **Note 4**).

3.2.4 Immobilization of Carboxy Terminated Ligands

(a) For activation of the carboxy-terminated molecules, the ligand is dissolved in DMF at a concentration of 1–3 μM together with 1.5 M DIC. After quick mixing, this solution is immediately applied to a transducer modified with a biopolymer providing amino groups at a concentration of 10 μL/cm^2. Place a second transducer on top to form a sandwich pair (*see* **Note 6**).

(b) The reaction is finished after 4 h in a DMF saturated atmosphere (*see* **Note 6**). Clean the transducers by rinsing them with DMF and Milli-Q water and drying them in a nitrogen stream (*see* **Note 4**).

3.2.5 Immobilization of Thiol Terminated Ligands

(a) Dissolve 1 mg EMCS per 10 μL DMF and apply this solution to a transducer modified with a biopolymer offering amino groups at a surface concentration of 10 μL/cm^2 for 6–12 h in a DMF saturated atmosphere (*see* **Note 6**). After rinsing with DMF (*see* **Note 4**), dissolve the ligand at a concentration of 1–3 μM in DMF, and apply this solution to the transducers at a surface concentration of 10 μL/cm^2 for 6–12 h as sandwich pairs in a DMF saturated atmosphere (*see* **Note 6**).

(b) Clean the transducers by rinsing them with DMF and Milli-Q water and drying them in a nitrogen stream (*see* **Note 4**).

3.2.6 Immobilization of Biotin

(a) Biotin is immobilized by TBTU activation: D-biotin (1 mg, 4 mmol), TBTU (1.4 mg, 4.4 mmol), and DIPEA (4 mL, 23.3 mmol) are mixed with DMF (50 mL) until the active ester is formed and the reaction mixture appears homogeneous. This solution is dripped on to a transducer at a surface concentration of 10 μL cm^{-2} pretreated with a biopolymer providing amino groups. Two of such transducers are put together to form a sandwich.

(b) After a reaction time of 4 h in a saturated DMF atmosphere (*see* **Note 6**), the sandwich is separated and both transducers are rinsed with water and dried in a nitrogen stream.

3.3 Measurement Protocol for Standard Binding Inhibition Assay for the Determination of the Affinity of an Antigen–Antibody Interaction

Binding inhibition assay to determine the affinity constant of antigen-antibody interaction: The transducer is modified according to protocols given before either with a carboxy or an amino terminated antigen. Then fixed amounts of antibody are pre-incubated with fixed volumes of differently concentrated antigens (*see* **Note 8**) to reach binding equilibrium before injection of the mixture with the FIA system (*see* **Notes 9–12**) on the transducer.

Antigens with affinity to the antibody in the pre-incubated solution inhibit the binding of the antibody to the surface immobilized antigen. In the case of diffusion (mass transport) limited binding, the diffusion rate obeys first Fick's law. This results in a constant binding rate of the antibody to the surface and a linear binding curve. The slope of the binding curve is determined by the concentration of free antibody binding sites in solution. The slope of the binding curves decreases with increasing concentration of antigen in solution and reaches zero for high antigen concentrations. Ovalbumin (final concentration of 200 µg/ml) should be added to all antibody solutions to avoid loss of antibody by non-specific binding in the fluidic system. The model function describing the concentration of antibodies with free binding sites is fitted to the titration curve using a Marquart–Levenberg nonlinear least-square algorithm (software ORIGIN from Microcal, Northampton/USA).

3.4 Data Evaluation

3.4.1 Determination of the Concentration of Active Antibody

Measuring the concentration of active antibody in a sample working at mass-transport limited conditions is essential. Accordingly, the rate of diffusion to the surface is much slower than the binding reaction to the surface. The reaction kinetics can therefore be neglected. Mass-transport limited conditions can be achieved by a high binding capacity of the surface and a low antibody concentration in solution. Under mass-transport limited conditions according to first Fick's law, the resulting binding curve is linear, and its slope is proportional to the concentration of functional antibody. If all samples contain the same amount of protein, determined by UV spectroscopy or Bradford assay, it is possible to calculate the active antibody concentration in a.u. by the different slopes of the samples, where the sample with the highest slope is assigned the a.u. 1 per µg used protein.

3.4.2 Determination of Affinity and Kinetic Constants

To determine the affinity and the kinetics of an antibody (*see* **Note 13**), mass-transport to the surface must be much faster than the rate of binding to the surface. If this is the case, the mass-transport can be neglected. This can be achieved by a high concentration of bulk antibody and a low surface coverage of hapten derivative.

The time dependence of the surface coverage can then be described by a pseudo first-order binding reaction as following:

$$\frac{d\Gamma(t)}{dt} = k_a \cdot c_{Ab} \cdot (\Gamma_{max} - \Gamma(t)) - k_d\Gamma(t) \qquad (3)$$

with c_{Ab}: antibody concentration
   $\Gamma(t)$: surface coverage at time t.
   Γ_{max}: maximal surface coverage.
   k_a: association rate constant.
   k_d: dissociation rate constant.

Solving this differential equation, the surface coverage which is equal to the resulting signal curve is given by:

$$\Gamma(t) = \Gamma_{Eq}\left(1 - e^{-k_{obs}t}\right) \qquad (4)$$

with.

$$\Gamma_{Eq} = \Gamma_{max}\frac{K_{aff} \times c_{Ab}}{1 + K_{aff} \times c_{Ab}}, \qquad (5)$$

the equilibrium coverage and.

$$k_{obs} = k_a \times c_{Ab} + k_d, \qquad (6)$$

the observed rate constant.

The affinity constant K_{aff} is given by the fraction of the association rate constant, k_a, and the dissociation rate constant, k_d.

To obtain accurate values for Γ_{Eq} and k_{obs}, it is necessary that only the kinetics-controlled part of the signal curve is approximated, and that the surface and bulk are nearly in equilibrium at the end of the measurement. Now it is possible to determine k_a as the slope of the straight line of the plot with k_{obs} as ordinate and active antibody concentration as abscissa. In order to determine K_{aff} Eq. 5 can be rewritten as

$$\frac{1}{\Gamma_{Eq}} = \frac{1}{\Gamma_{max}} + \frac{1}{K_{aff} \times \Gamma_{max}} \times \frac{1}{c_{Ab}}. \qquad (7)$$

With this formula, it is possible to obtain the value for Γ_{max} with a linear fit of the $\frac{1}{\Gamma_{Eq}} \Big/ \frac{1}{c_{Ab}}$ plot. Inserting this value in Eq. 3, a nonlinear least square fit results in a value for K_{aff}. This method gives nearly the same values as a Scatchard plot $\left(\frac{\Gamma_{Eq}}{c_{Ab}} \Big/ \Gamma_{Eq}\right)$.

4   Notes

1. Very thick biolayers (more than approx. 100 nm) lead to measurements out of the linear correlation between change of optical thickness and shift of the interference spectrum.

2. Correct side of the RIfS transducers: mark the un-coated side with a diamond pen to avoid mistakes during the surface chemistry steps. The coated side of the transducers appears colored because of the light interference.

3. In the case a transducer shows a grey shadow after rinsing and drying repeat this cleaning procedure. If this does not work, start the surface chemistry from the beginning (otherwise the modification will not be homogeneous and will show high nonspecific binding).

4. All surface chemistry protocols which produce highly reactive groups are sensitive to humidity (deactivation). This can be

avoided by immediately applying the next step and working under dry ambient conditions.

5. Use a tray with ground joint. Place the transducer as sandwich-pairs (by assembling two slides face-to-face) in the tray on a solid support and add the same solvent which is used in the reaction for a solvent-saturated atmosphere.

6. In the case of limited ligands it is possible to use commercially available piezo-based microdosing devices for printing a ligand solution in Milli-Q water.

7. Usually phospahete buffered saline (PBS, 150 mmol NaCl and 10 mmol Di-potassiumhydrogenphosphate in Milli-Q water at pH 7.4) is used for antigen-antibody interaction analysis. In general all buffers can be used for RIfS measurements. Only restriction: do not use buffers with pH >8.5 because of destruction of the surface modification.

8. Testing of the modified transducer for nonspecific binding: use ovalbumine (OVA) or bovine serum albumin (BSA) in excess (e.g., 1 mg/mL) directly before the measurement.

9. Air disturbs the measurements: Use degassed buffers and avoid negative pressure (sucking) through the flow cell. In addition, it is helpful to keep the buffer under a slight positive argon pressure.

10. If possible use the same buffers during a complete measurement to avoid artifacts because of changes in the refractive index.

11. After the interaction process, the transducer surface can be regenerated. To remove antibodies from their antigens, a solution of 0.5% SDS (sodium dodecyl sulfate) at pH 1.9 is applied via the FIA system. In the case of hybridization experiments, a solution of either 0.25% SDS at pH 2.5 or a solution containing 6 M guanidinium hydrochloride and 6 M urea at pH 2 can be used.

12. RIfS is not restricted for biosensing; this technique can be used also for chemical sensing. Instead of biopolymers one can modify the transducer with for example nano-porous polymer layers, rubber like polymers or molecular imprinted polymers (MIPs). The determination of kinetics is also possible.

13. Nowadays, even more advanced devices based on the one-color approach are commercial available. These devices use the revised reflectometry read out called "1-lambda RIDe" (1-lambda Reflectometric Interference Detection). Currently, for example an imaging based version for the label-free readout of microarrays in the standard microarray format (b-screen®) is available from Biametrics GmbH. This technology enables the label-free detection of any microarray experiment like protein

or peptide microarrays, resulting in multiplexed kinetic experiments of up to 20,000 interactions in a single measurement.

14. Label-free detection like RIfS can be used for a great variety of applications ranging from standard kinetic analysis [8] to fragment-based screening [9]. New fields of application include the detection of small molecules by molecular imprinted molecules [10] or diagnostic applications [11–14].

References

1. Gauglitz G (2010) Direct optical detection in bioanalysis: an update. Anal Bioanal Chem 398 (6):2363–2372

2. www.biacore.com

3. Homola J, Yee SS, Gauglitz G (1999) Surface plasmon resonance sensors: review. Sens Actuators B 54:3–15

4. Schmitt HM, Brecht A, Piehler J, Gauglitz G (1997) An integrated system for optical biomolecular interaction analysis. Biosens Bioelectron 12:219–233

5. Brecht A, Gauglitz G, Nahm W (1992) Interferometric measurements used in chemical and biochemical sensors. Analysis 20:135–140

6. Brecht A, Gauglitz G, Kraus G, Nahm W (1993) Chemical and biochemical sensors based on interferometry at thin layers. Sens Actuators 11B:21–27

7. Proell F, Moehrle B, Kumpf M, Gauglitz G (2005) Label-free characterisation of oligonucleotide hybridisation using reflectometric interference spectroscopy. Anal Bioanal Chem 382(2):1889–1894

8. Fechner P, Pröll F, Albrecht C, Gauglitz G (2011) Kinetic analysis of the estrogen receptor alpha using RIfS. Anal Bioanal Chem 400 (3):729–735

9. Pröll F, Fechner P, Proll G (2009) Direct optical detection in fragment-based screening. Anal Bioanal Chem 393(6–7):1557–1562

10. Kolarov F, Niedergall K, Bach M, Tovar GEM, Gauglitz G (2012) Optical sensors with molecularly imprinted nanospheres: a promising approach for robust and label free detection of small molecules. Anal Bioanal Chem 402(10):3245–3252

11. Ewald M, Le Blanc AF, Gauglitz G, Proll G (2013) A robust sensor platform for label-free detection of anti-Salmonella antibodies using undiluted animal sera. Anal Bioanal Chem 405(20):6461–6469. doi:10.1007/s00216-013-7040-9

12. Krieg AK, Gauglitz G (2014) An optical sensor for the detection of human pancreatic lipase. Sens Actuators B 203:663–669. doi:10.1016/j.snb.2014.07.036

13. Ewald M, Fechner P, Gauglitz G (2015) A multi-analyte biosensor for the simultaneous label-free detection of pathogens and biomarkers in point-of-need animal testing. Anal Bioanal Chem 407(14):4005–4013. doi:10.1007/s00216-015-8562-0

14. Rau S, Hilbig U, Gauglitz G (2014) Label-free optical biosensor for detection and quantification of the non-steroidal anti-inflammatory drug diclofenac in milk without any sample pretreatment. Anal Bioanal Chem 406 (14):3377–3386. doi:10.1007/s00216-014-7755-2

15. Frank R, Möhrle B, Fröhlich D, Gauglitz G (2005) A lable-free detection method of biochemical interactions with low-cost plastic and other transperent transducers. Proc SPIE Int Soc Opt Eng 5826:551–560

Chapter 14

Hypermulticolor Detector for Quantum-Antibody Based Concurrent Detection of Intracellular Markers for HIV Diagnosis

Annie Agnes Suganya Samson and Joon Myong Song

Abstract

Antiretroviral treatment can reduce the death rate of human immunodeficiency virus (HIV) infection, and its effectiveness is maximized at the early stage of HIV infection. The present protocol demonstrates an early stage high-content HIV diagnosis based on multicolor concurrent monitoring of CD4, CD8, and CD3 coreceptors and F-actin cytoskeleton using quantum dot (Qdot)–antibody conjugates at the single cell level. Artificial HIV infection of peripheral blood mononuclear cells (PBMCs) can be achieved by treating PBMCs with gp120. Using the present methodology, we can determine the CD4–CD8 ratios of normal PBMCs and artificial HIV-infected PBMCs. In addition, this protocol enables monitoring of structural changes of actin filament alignments in PBMCs bound to gp120 proteins using the multicolor single cell imaging system. Overall, this approach presents a new model for accurate early stage HIV diagnosis. Simultaneously the approach provides information on actin cytoskeleton and subtypes of PBMCs as well as their CD4–CD8 ratios.

Key words High-content imaging, Quantum dots, HIV diagnosis, PBMCs and actin filament

1 Introduction

Current HIV diagnosis is performed to detect HIV in blood, saliva, and urine, and is divided into antibody or antigen tests according to the target to be detected. The antibody test detects antibodies arising from HIVs using ELISA or Western blot [1, 2]. However, after exposure of a human to HIV, HIV antibodies in thebody are not generally detected with blood tests by the time after infection and before seroconversion ("window period") during which markers of infection (*HIV*-specific antigen and antibodies) are still absent or too scarce to be detectable. After the window period the amount of HIV antibody reaches a high enough level to be detected. This can often lead to false negative errors. HIV genetic or p24 antigen tests are capable of complementing errors induced

Avraham Rasooly and Ben Prickril (eds.), *Biosensors and Biodetection: Methods and Protocols Volume 1: Optical-Based Detectors*, Methods in Molecular Biology, vol. 1571, DOI 10.1007/978-1-4939-6848-0_14,
© Springer Science+Business Media LLC 2017

by an HIV antibody test. An HIV genetic test amplifies DNA or RNA sequences inherent to HIV using polymerase chain reaction (PCR), real-time PCR, or reverse transcriptase PCR [3]. Although HIV genetic tests are very specific and accurate, these tests are expensive and difficult to administer and interpret compared to HIV antibody tests [4]. P24 antigen is a protein secreted from HIV that invades host cells, and its presence verifies that HIV has invaded host cells. Although the p24 antigen test works before HIV antibodies are produced in the period immediately after HIV infection, the detection sensitivity of this test is very low. Gp120 is a envelope glycoprotein exposed on the surface of HIV. The gp120 of HIV binds selectively to CD4 (cluster of differentiation 4) expressed on the surface of immune cells such as T helper cells, monocytes, and macrophages. The binding of HIV gp120 to CD4 causes a change in the conformation of gp120 that permits HIV to bind to a coreceptor expressed in the host cell, which is known as CCR5 or CXRX4. A structural change of HIV gp41 (glycoprotein 41) is then promotes fusion between HIV and the host cell. As a result of this fusion, actin filament rearrangement occurs within the host cell. At the early stage of HIV infection when the gp120 of HIV binds to the CD4 coreceptor of a T cell, the degree of conjugation between the CD4 coreceptor and the CD4 antibody decreases remarkably. This leads to a progressive reduction in the ratio of CD4–CD8. A ratio less than 1 is suggestive of advanced immunosuppression. Another characteristic arising from HIV infection is that morphological change of T cells is caused by actin filament rearrangement. The present protocol uses quantum dot (Qdot)–antibody nanoprobes for quantitative concurrent monitoring of actincytoskeleton, CD3, CD4, and CD8 at the single peripheral blood mononuclear cell (PBMC) level. The aim of this procedure is to develop an early stage high-content HIV diagnosis based on quantitative determination of the CD4–CD8 ratio and subtype of PBMCs, in addition to the monitoring of T cell morphological change.

Our hypermulticolor single cell imaging system utilizes slightly defocused PBMC images in the bright field mode. The slightly defocused cellular images will provide uniform intensity distribution at the single PBMC level. The PBMC images in the defocused mode should overlap those obtained by fluorescence intensity at a particular single PBMC level [5]. A threshold value is fluorescent intensity at a particular single PBMC region which has the lowest fluorescence intensity among the PBMCs, but greater intensity than that of the background signal. The threshold value should be applied to all overlapped PBMC images to select single PBMCs larger than the threshold value. This operation enables the quantification of total cells expressing CD coreceptors or actin filament structural changes for HIV diagnosis. This quantitative approach provides a statistical basis for measuring the CD4–CD8 ratio and determining the subtype of PBMCs.

2   Material

1. Human blood (Seoul National University Hospital).

2. HIV envelope glycoprotein gp120 (HIV-1 IIIB gp 120; Immuno Diagnostics, Inc., Woburn, MA, USA).

3. HISTOPAQUE-1077 (Sigma, Poole, Dorset, UK).

4. Acetone (prechilled).

5. Wash buffer (WB; 1× PBS, 1% FCS (fetal calf serum), 0.01% sodium azide).

6. Cold 1× PBS (phosphate buffer saline).

7. 3.7% formaldehyde.

8. 1% BSA (bovine serum albumin).

9. Mouse-antihuman CD3-Qdot605, mouse-antihuman CD8-Qdot565, mouse-antihuman CD4-Qdot655.

10. Alexa Fluor 488 phalloidin (200 mM).

11. ITK inhibitor (Interleukin-2-inducible T-cell Kinase) (Interleukin-2-inducible T-cell Kinase) (BMS509744; AdooQ Bioscience Irvine, CA, USA).

12. Hemocytometer.

13. Centrifuge.

14. Fluorescence microscope.

15. Optical components:

    - An Ar ion laser—350 nm (Melles Griot Laser Group, Carlsbad, CA, USA).
    - An interference filter (U-MGFPHQ, Olympus, Tokyo, Japan).
    - 20×, 40×, and 60× Objective lens (Olympus).
    - Acousto-optic tunable filter (AOTF; TEAF10-0.45-0.7-S, Brimrose).
    - Charge-coupled device (CCD) camera.
    - A long-pass filter.
    - Computer.
    - Software for Image analysis—MetaMorph (Version 7.1.3.0, Molecular Devices, Sunnyvale, CA, USA).

2.1   Special Equipment

2.1.1   Acousto-Optical Transmission Filter (AOTF) Microscope

Acousto-optical transmission filter (AOTF) microscope consists of different components, which includes a C-mount lens, an acousto-optic tunable filter (AOTF), a charge coupled device (CCD) camera, an Ar ion laser—350 nm, an interference filter, a fluorescence microscope, and a computer system equipped with image acquisition and analysis software (MetaMorph). In this assembled imaging system, AOTF will function as a tunable emission filter. A frequency of light gets diffracted at the Bragg angle that is built into the AOTF. The diffracted light that is the fluorescence emission at a particular wavelength will be transmitted and collected in a CCD camera. Fluorescence images at varying wavelength could be obtained using AOTF. The spectral range varies from 400 to 1000 nm wavelength. The AOTF microscope can select a single wavelength which will not overlap with other marker emission wavelengths. A single wavelength can be selected for observation of HIV markers. Maximum emission wavelengths of the QDs (quantum dots) used in this protocol occur at 565, 605, and 655 nm. No significant overlap occurs with emission wavelengths of QDs. Moreover, while using AOTF allowing transmission at a single wavelength, QD605 is barely detected when detection of QD655 is executed. Therefore, AOTF will provide much smaller spectral interference compared to bandpass filters when high-content HIV diagnosis is performed based on simultaneous monitoring of different markers. Simultaneous monitoring of different intracellular event (e.g., activation or deactivation of protein, receptor activity) in a single cell using fluorescent probes is termed as high-content based detection. This enhances the accuracy of early stage HIV diagnosis. As a probe for HIV markers, conventional dye antibody conjugate can be replaced with QD–antibody conjugate. Thus QD-antibody conjugate along with AOTF is advantageous for more accurate early stage HIV diagnosis. QDs having narrow emission wavelength ranges are shown to be much more suitable to high-content HIV diagnosis than conventional fluorescent dyes.

3 Methods

3.1 Isolation of PBMCs from Human Blood

Peripheral blood mononuclear cells (PBMCs) include lymphocytes (T cells, B cells, and NK cells), monocytes, and dendritic cells. PBMCs are considered as the population of immune cells. HIV infection is associated with hyperactivation of the immune system. The reproduction of HIV in T cells is closely correlated to the proliferation of the cells. In general, bulk populations of T cells (PBMC) are infected with HIV, which is connected with the increased T cell proliferation leading to greater viral replication. T cells in PBMC are extremely varied both functionally and phenotypically. This segment describes an artificial method to establish the HIV-infected PBMCs (peripheral blood mononuclear cells).

This segment describes an artificial method to establish the HIV-infected PBMC (peripheral blood mononuclear cells) model.

1. Add 5 mL of HISTOPAQUE-1077 to a 15 mL conical centrifuge tube.

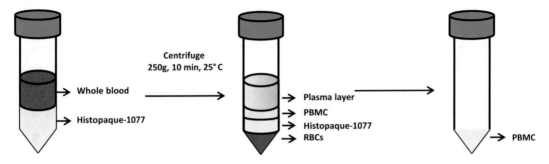

Fig. 1 Separation of PBMCs from blood sample

2. Add 5 mL human blood (*Note*: Blood sample should be carefully layered over the HISTOPAQUE-1077 to prevent mixing).

3. Centrifuge at $400 \times g$ for 30 min at 25 °C.

4. After centrifugation, three separate layers will be observed.

5. First, remove the upper transparent plasma layer of the three separate layers using a pipette.

6. Next, the opaque interface layer containing PBMCs should be gently transferred with a pipet into a clean 15 mL centrifuge tube (Fig. 1).

7. Fill the 15 mL centrifuge tube containing PBMCs with 1× PBS solution, mix carefully and then centrifuge at $250 \times g$ for 10 min at 25 °C.

8. After centrifugation, carefully discard the supernatant and resuspend the white PBMCs with 15 mL of cold 1× PBS solution. Centrifuge at $250 \times g$ for 10 min at 4 °C.

9. After centrifugation, discard the supernatant and resuspend the PBMC pellet in 6.25 mL of 1× PBS solution.

10. Perform cell counting using hemocytometer.

3.2 Treatment of PBMCs with gp120

Artificial HIV infection of peripheral blood mononuclear cells (PBMCs) is accomplished by treatment of PBMCs with gp120:

1. Centrifuge the PBMCs solution at $250 \times g$ for 10 min at 4 °C.

2. Discard the supernatant and wash the PBMC pellet twice with wash buffer.

3. For the treatment of PBMCs with gp120, take PBMCs (1×10^6 cells) in 30 μL of WB and add 15 μL gp120 (10 μg/mL).

4. Incubate the mixture for 6 h at 4 °C.

5. Finally, after the reaction with gp120 wash the PBMCs thrice with 5 mL WB (*Note*: For washing, centrifugation at $250 \times g$ for 10 min at 4 °C).

3.3 Treatment of PBMCs with Qdot–Antibody Conjugates

Monitoring of the CD4–CD8 ratio can greatly improve HIV diagnosis accuracy in early stages of infection, and it is important to accurately determine this ratio. This segment describes a staining protocol using quantum dot (Qdot)–antibody nanoprobes for quantitative concurrent monitoring of actincytoskeleton, CD3, CD4, and CD8 in normal and gp120 treated PBMCs at the level of single cells.

1. Fix the PBMC or gp120-treated (1×10^6) cells using 3.7% formaldehyde for 10 min and then wash twice with 5 mL $1\times$ PBS solution (centrifugation at $250 \times g$ for 10 min at 4 °C).

2. Place the fixed PBMCs in prechilled acetone for 3 min at −20 °C and then wash twice with $1\times$ PBS. (centrifugation at $250 \times g$ for 10 min at 4 °C).

3. Immerse the PBMCs pellet in 200 μL of 1% BSA aqueous solution.

4. Add 1 μL of each Qdot–antibody (Mouse-antihuman CD3-Qdot605, mouse-antihuman CD8-Qdot565, and mouse-antihuman CD4-Qdot655) stock solutions (1:200 dilutions) into 200 μL of 1% BSA aqueous solution containing 1×10^6 normal or gp120-treated PBMCs.

5. Incubate the solution for 1 h at room temperature.

6. After 1 h, wash the Qdot stained cells twice with $1\times$ PBS.

7. Finally, incubate the cells with Alexa Fluor 488 phalloidin in $1\times$ PBS for 20 minto detect F-actin filament in PBMCs.

3.4 Detection of CD3, CD4, CD8, and F-Actin filament in PBMCs

This section describes a system to simultaneously monitor fluorescence emission of different wavelengths. PBMCs stained with Alexa Fluor 488 phalloidin, antihuman CD3-Qdot605, antihuman CD8-Qdot565, and antihuman CD4-Qdot655 can be concurrently observed using this Hypermulticolor single cell imaging system. This system provides a scanning rate of 60 wavelengths (nm)/min, which will enable 60 different cellular images perminute. First, PBMCs on the sample stage are irradiated by UV using a microscope objective lens to excite all the Qdot–antibody conjugates and Alexa Fluor 488 phalloidin. Fluorescent emission from the PBMCs are collected using an identical lens, transmitted through an acousto-optical transmission filter (AOTF), and images detected as a function of wavelength using a charge-coupled device (CCD) (Fig. 2).

1. Alexa Fluor 488 phalloidin detects F-actin filament structural changes of PBMCs at 518 nm. The three quantum dot tagged antibody conjugates (antihuman CD3-Qdot605, antihuman CD8-Qdot565, and antihuman CD4-Qdot655) are bound to CD3, CD4, and CD8 coreceptors expressed in the cellular membrane of the PBMCs.

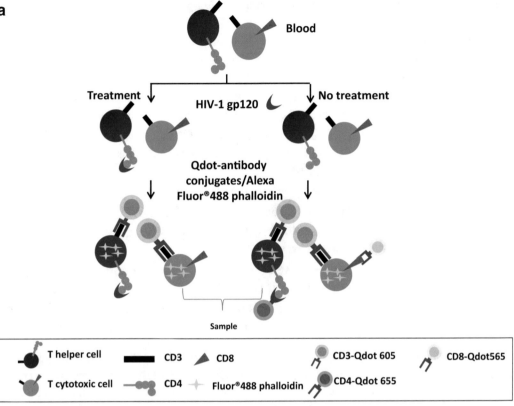

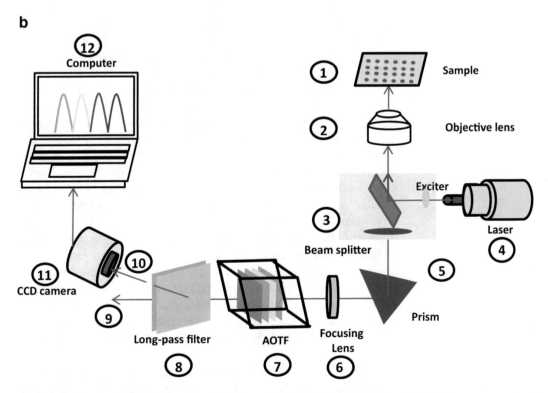

Fig. 2 (**a**) Multicolor single cell imaging. (**b**) Hypermulticolor single cell imaging detector. The components of the imaging system are as follows: (*1*) sample stage, (*2*) objective lens, (*3*) beam splitter, (*4*) laser, (*5*) prism, (*6*) focusing lens, (*7*) AOTF, (*8*) filter, (*9*) undiffracted light, (*10*) diffracted light, (*11*) CCD camera, and (*12*) computer

3.5 Treatment of gp120-Treated PBMC with ITK Inhibitor

The following steps illustrate an experimental procedure to verify the morphological change in T cells caused by actin filament rearrangement; characteristic changes arise during HIV infection.

1. Preincubate the PBMCs (1×10^6 cells) with 10 μM ITK inhibitor for 1 h before the gp120 treatment.

2. After 1 h, for the reaction of PBMCs with gp120; take PBMCs (1×10^6 cells) in 30 μL of WB and react with 15 μL gp120 (10 μg/mL).

3. Incubate the mixture for 6 h at 4 °C.

4. Wash the PBMCs thrice with 5 mL WB.

5. Fix the PBMCs cells in 3.7% formaldehyde for 10 min and then wash twice with 1× PBS solution.

6. Place the fixed PBMCs in acetone for 3 min at −20 °C and then wash twice with 1× PBS.

7. Incubate the cells with Alexa Fluor 488 phalloidin in 1× PBS for 20 min. Detect F-actin filament in PBMCs at 518 nm.

3.6 Hypermulticolor Detector

This protocol is based on quantitative multivariate imaging (QMI) cytometry combined with fluorescent semiconductor nanocrystals, which can be used for various quantitative cellular image-based analyses. This standardized protocol can be adapted to various cell types. Furthermore, the developed assay can overcome the difficulties inherent in flow cytometric assays. It is more miniaturized, cost effective, and user friendly to operate. A commercial flow cytometer has ~8 detectors; which limits the multicolor/multispectral analysis to detect eight target molecules. But in case of QMI cytometer provides a wide range of filtering wavelengths at ~3.75-nm intervals, making the system more compatible for high-throughput detection. Hence, this system enables simultaneous monitoring of several cellular molecules at single cellular level. So, this imaging system is described as Hypermulticolor detector system. By using a selective wavelength tuning, Hypermulticolor detector can minimize the interference of unwanted auto fluorescence of cells during cellular imaging analysis. Therefore, this protocol allows prompt, quantitative data to visualize and quantify molecular targets, while simultaneously highlighting there importance in disease diagnosis. Certainly, this hypermulticolor imaging system can be used for concurrent monitoring of various intracellular biomolecules by rapid multicolor/multispectral analysis, which cannot be achieved by filter-based conventional imaging systems (*see* **Note 1**). Subsequently, utilizing these concepts may have a significant role in the early diagnosis of HIV based on targeted molecules by using this developed protocol.

The Hypermulticolor system is based on uniform threshold intensity distribution (TID). It consists of an AOTF, C-mount lens, CCD camera, fluorescence microscope, and commercially available software for image acquisition and analysis. The technique used for data acquisition and quantitative analysis, involves following steps: (1) Coordinating the exposure time of CCD camera with AOTF scanning/filtering wavelength to obtain desired image plane at desired wavelength parameters. (2) Acquiring a focused image at an identical X and Y axis position. (3) Applying background correction and threshold adjustment. Certainly, quantitative analysis is possible by selecting a region around the threshold objects regardless of their cellular morphology. Image analysis can be carried out using MetaMorph software.

3.7 Imaging System

As mentioned before, the imaging system consists of various components (Fig. 3). A mercury arc lamp acts as an illumination source for cells (sample) in a fluorescence microscope and the laser beam is purified by an interference filter, allowing it to pass through the excitation filter and finally, reflected by a dichroic filter. The filtered laser beam is focused through a microscope objective lens ($20\times$, $40\times$, $60\times$) and then onto the sample platform. The quantum dot tagged probes in the cells (sample) are induced to emit fluorescence by the filtered laser beam. The fluorescent emission is collected by the same objective lens, passes through dichroic filter and comes out of the side port ($\phi = 25$ mm) of the microscope. In order to reduce the fluorescence beam diameter, the C-mount lens is attached to the side port of the fluorescence microscope. Then, the entire beam passes through the AOTF window (10×10 mm). The nano crystal in the AOTF splits the fluorescence beam into two separate beams, the diffracted light (at particular wavelength) and

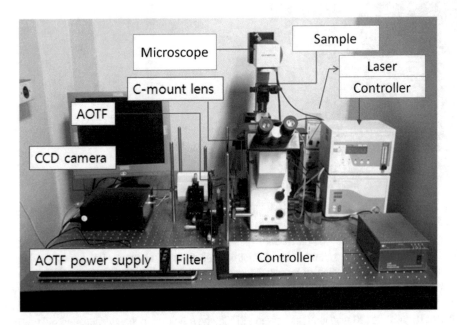

Fig. 3 Photo of the imaging system

undiffracted light. Thus, the fluorescence image of the cells can be detected at a particular wavelength and the fluorescence images are captured by a CCD camera. A long-pass filter is placed in front of the CCD camera to eliminate any laser scattering. The exposure time of the CCD camera is 1 s and the spectral resolution of the AOTF is adjustable up to 0.1 nm. Data analyses of fluorescence cell images are performed automatically using commercially available software (MetaMorph, Version 7.1.3.0, Molecular Devices).

3.8 Data Analysis   Quantitative data analysis can be done using different parameters like optical density, average gray value, Z position, angle, distance, area, width, image plane, elapsed time, stage label, wavelength, region label, intensity S/N, and threshold area. Single cell imaging cytometry can be used for simultaneous monitoring of cellular events, that is, expression of different cellular markers. Figure 4a

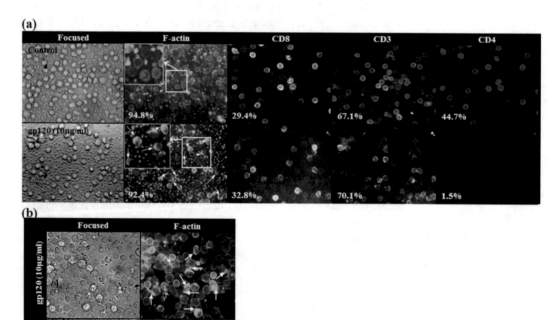

Fig. 4 (a) Concurrent monitoring of CD3, CD4, CD8 coreceptors and F-actin in PBMCs. The *top figures* were obtained from normal PBMCs while the *bottom figures* were obtained from gp120-treated PBMCs. F-actin images of normal PBMCs were compared with those of gp120-treated PBMCs. The gp120-treated PBMCs indicated with *arrows* showed ununiform distribution of F-actin filament. On the other hand normal PBMCs showed uniform distribution of F-actin filament. (**b**) Inhibitory effect of ITK on cytoskeleton structure of PBMCs treated with gp120 glycoprotein. *Top* cellular images are PBMCs treated with gp120 glycoprotein (10 μg/mL) while cellular images at the *bottom* are PBMCs treated with gp120 glycoprotein (10 μg/mL) and then ITK inhibitor (10 μM). Gp120 glycoprotein-treated PBMCs show uniform actin cytoskeleton by the ITK inhibitor treatment

represents the actin filament rearrangement of PBMCs by gp120-induced signaling transduction. The figure data clearly confirms that normal (control) PBMCs has uniform actin cytoskeleton than that of gp120 treated PBMCs. This is due to actin rearrangement that occurs when HIV-1 gp120 binds to PBMCs. Hence, a morphological change occurs in PBMCs. In fact, the rearrangement of actin cytoskeleton is a common happening that occurs in the HIV infected host cells and it is also considered as a good indicator to verify HIV infection. Qdot nanoprobes based high-content HIV diagnosis system make possible and efficient for the detection of different cellular markers like CD3, CD4, and CD8, simultaneously in both control and gp-120 treated PBMCs. The results demonstrate that there is significant reductions in the expression of CD4 in gp-120 treated PBMCs compared to control PBMCs. Moreover, ~0.89- and ~0.95-fold difference of CD8 and CD3 respectively, was observed in control and gp-120 treated PBMCs. Figure 4b shows inhibitory effect of ITK on the rearrangement of actin cytoskeleton induced by gp120 glycoprotein. ITK is required for effective HIV transcription, hence leading to increases in virus-like particle formation. Earlier studies have demonstrated that ITK affects gp120 protein-induced rearrangement of actin cytoskeleton that is needed for HIV to enter a host cell. Thus, ITK inhibitor has been studied to inhibit HIV entry into host cells and HIV replication. A nonuniform actin cytoskeleton was not observed in PBMCs treated with ITK inhibitor [6]. These data verifies that simultaneous monitoring of CD3, CD4, and CD8 coreceptors can clearly prove that actin rearrangement occurs as a result of HIV infection. To conclude, concurrent quantitative monitoring of actin cytoskeleton, CD3, CD4, and CD8 at the single cell level enables early stage HIV diagnosis without false negative error.

4   Note

1. Simultaneous monitoring of different intracellular event (e.g., activation or deactivation of protein, receptor activity) in a single cell using fluorescent probes is termed as high-content based detection. High-content based HIV diagnostic method is highly sensitive and more compatible for simultaneous monitoring of morphological changes and quantitative detection of targeted receptors in a single cell. This method is more sensitive than flow cytometry. Following are the benefits of incorporating this protocol in early stage HIV detection. It reveals a new model based on multicolor concurrent monitoring of CD4, CD8, and CD3 coreceptors and F-actin cytoskeleton using quantum dot (Qdot)–antibody conjugates at the single cell level. Due to brightness and photostability of Qdot, they provide relatively easy operation based on the use of a microscope.

This enables simultaneous monitoring of T cellular morphological changes as well as CD4–CD8 ratio, and greatly contributes to the accuracy improvement of early stage HIV diagnoses. False negative errors in the early stages of HIV infection is greatly reduced by the use of hypermulticolor single cell imaging cytometry. Hence, Hypermulticolor imaging system provides new opportunities to efficiently diagnose early stage HIV infection at single cell cytometric level.

References

1. Chun TW, Engel D, Berrey MM, Shea T, Corey L, Fauci AS (1998) Early establishment of a pool of latently infected, resting CD4(+) T cells during primary HIV-1 infection. Proc Natl Acad Sci U S A 95(15):8869–8873

2. Kahn JO, Walker BD (1998) Acute human immunodeficiency virus type 1 infection. N Engl J Med 339(1):33–39. doi:10.1056/NEJM199807023390107

3. Schvachsa N, Turk G, Burgard M, Dilernia D, Carobene M, Pippo M, Gomez-Carrillo M, Rouzioux C, Salomon H (2007) Examination of real-time PCR for HIV-1 RNA and DNA quantitation in patients infected with HIV-1 BF intersubtype recombinant variants. J Virol Methods 140(1-2):222–227. doi:10.1016/j.jviromet.2006.11.012

4. Tcherepanova I, Harris J, Starr A, Cleveland J, Ketteringham H, Calderhead D, Horvatinovich J, Healey D, Nicolette CA (2008) Multiplex RT-PCR amplification of HIV genes to create a completely autologous DC-based immunotherapy for the treatment of HIV infection. PLoS One 3(1):e1489. doi:10.1371/journal.pone.0001489

5. Kim MJ, Lee SC, Pal S, Han E, Song JM (2011) High-content screening of drug-induced cardiotoxicity using quantitative single cell imaging cytometry on microfluidic device. Lab Chip 11 (1):104–114. doi:10.1039/c0lc00110d

6. Tak YK, Song JM (2013) Early stage high-content HIV diagnosis based on concurrent monitoring of actin cytoskeleton, CD3, CD4, and CD8. Anal Chem 85(9):4273–4278. doi:10.1021/ac303727e

Chapter 15

Low-Cost Charged-Coupled Device (CCD) Based Detectors for Shiga Toxins Activity Analysis

Reuven Rasooly, Ben Prickril, Hugh A. Bruck, and Avraham Rasooly

Abstract

To improve food safety there is a need to develop simple, low-cost sensitive devices for detection of food-borne pathogens and their toxins. We describe a simple, low-cost webcam-based detector which can be used for various optical detection modalities, including fluorescence, chemiluminescence, densitometry, and colorimetric assays. The portable battery-operated CCD-based detection system consists of four modules: (1) a webcam to measure and record light emission, (2) a sample plate to perform assays, (3) a light emitting diode (LED) for illumination, and (4) a portable computer to acquire and analyze images. To demonstrate the technology, we used a cell based assay for fluorescence detection of the activity of the food borne Shiga toxin type 2 (Stx2), differentiating between biologically active toxin and inactive toxin which is not a risk. The assay is based on Shiga toxin inhibition of cell protein synthesis measured through inhibition of the green fluorescent protein (GFP). In this assay, GFP emits light at 509 nm when excited with a blue LED equipped with a filter at 486 nm. The emitted light is then detected with a green filter at 535 nm. Toxin activity is measured through a reduction in the 509 nm emission. In this system the level of detection (LOD) for Stx2 was 0.1 pg/ml, similar to the LOD of commercial fluorometers. These results demonstrate the utility and potential of low cost detectors for toxin activity. This approach could be readily adapted to the detection of other food-borne toxins

Key words LED, CCD, Microbial pathogens, Toxin activity, Shiga toxin, GFP, Webcam, Fluorescence, Fluorometer

1 Introduction

In recent years, charged-coupled devices (CCDs) [1–9] or complementary metal–oxide–semiconductors (CMOSs) [10–13] are increasingly utilized as optical detectors because of their low cost, small size, sensitivity, and low power consumption. Their ability to image large surfaces makes them ideal for sample detection because many samples can be imaged and analyzed simultaneously. The cost of technical and scientific-grade CCD and CMOS imagers for biodetection is typically high. While webcams, digital cameras, digital imagers for amateur astronomy, and other consumer

Avraham Rasooly and Ben Prickril (eds.), *Biosensors and Biodetection: Methods and Protocols Volume 1: Optical-Based Detectors*, Methods in Molecular Biology, vol. 1571, DOI 10.1007/978-1-4939-6848-0_15,
© Springer Science+Business Media LLC 2017

electronics incorporating imaging components are more affordable, there is a need for low-cost alternatives providing sufficient
sensitivity and imaging quality for biodetection at significantly
lower cost. This is especially important for point-of-care (POC)
diagnostics for global health where it is desirable to enhance the
quality of health care delivery for underserved populations at very
low cost.

**1.1 Considerations
for Optical Detector
Selection**

In both CCD and CMOS devices, photons striking the semicon-
ducting material generate electron–hole pairs, and the resulting
electric charge is processed into electronic signals. However, the
reading of these charges in each device is different. In CCDs, the
charge from each pixel is sequentially transported across the chip to
an analog-to-digital (A/D) converter, which converts the charge
into a digital value. The use of a single converter for all pixels
increases the uniformity of the output—an important factor in
image quality. However, blooming is a problem in CCD sensors
when a part of the image is over-exposed, causing leakage of electrical charge to adjacent pixels whose signals become corrupted.
The low noise associated with a CCD sensor enables longer exposure times for low-level light detection, provided the CCD has a
low thermal noise level relative to the incoming photonic signal.
This typically necessitates the use of a cooling system in order to
minimize temperature and the subsequent thermal noise for CCD
or CMOS sensors.

In most CMOS devices each pixel has its own charge-to-volt-
age conversion and the sensor may include transistors at each pixel,
noise correction, and digitization circuits to convert the output
signal to digital so that the chip outputs digital bits. When light
hits the sensor, noise may result from some photons hitting the
transistor instead of the pixel. By placing the transistors under the
pixels (i.e., back-side illuminated CMOS sensor) this type of noise
can be reduced. This configuration increases the chip complexity,
reducing the area available for light capture. Moreover, because
each pixel is doing its own conversion, uniformity is reduced,
especially at higher temperatures. Because the output of a CMOS
is massively parallel compared to a CCD, it is possible to achieve
higher image acquisition speeds and more efficient use of energy.
CMOS sensors are also immune to the blooming effect.

CCDs utilize global shutters in which the entire imager is reset
before integration in order to remove any residual signal in the
sensor pixels; the light collection starts and ends at exactly the same
time for all pixels with no motion blur. CMOS devices utilize a
rolling shutter in which light is not collected simultaneously by all
of the pixels. Thus the time that light collection starts and ends is
slightly different for each row; the top row of the imager is the first
one to start and finish collecting light. The row-by-row process of

light collection in a rolling shutter camera can result in motion blur and switch noise effect, particularly at high gain settings.

To summarize, CMOS sensors have cheaper manufacturing costs, greater energy efficiency, no blooming and faster data-throughput speeds, while CCDs have lower dark currents, larger areas available for light capture in low light level applications, and more pixel uniformity. Given the rapid improvement in sensor technologies such as complex pixel with correlated double sampling (CDS), many CMOS devices overcome some of the technology limitations and gradually blur the boundary between the capabilities of the two competing imaging technologies. Therefore, both CCD and CMOS technologies are now being broadly used for imaging in a variety of fields, with the exception that low-cost devices still primarily use CMOS sensors due to their substantially lower manufacturing costs.

1.2 Optics for Digital Detectors

The optics for digital detectors are relatively simple, consisting of a lens and filters. A good lens for biodetection has a focal length of 4–12 mm at f1.2. Such a wide aperture, wide-angle lens is capable of imaging a wide field-of-view (FOV) from a short distance, providing a more compact device configuration. However, to achieve this wide FOV at a short distance, the portion of the image captured along the edges of the lens ends up being curved, resulting in barrel distortion.

In addition to fixed focal length lenses, zoom lenses have variable focal lengths enabling them to image both wide and narrow FOVs. Their main disadvantage is they are slower, depending on the focal length used (usually operated at f/2.8–f/5.6), than comparable fixed focal length lenses (e.g., f/1.2), resulting in longer exposure times. Extension tubes are needed for greater magnification to achieve macro (i.e., close-up) imaging. The extension tube contains no optical elements and is placed between the sensor and the lens. It moves the lens farther from the sensor image plane and closer to the imaged surface. Alternatively, reversing the lens so small objects close to the image sensor can be magnified and projected onto the image sensor [14] can achieve a macro effect similar to extension tubes for imaging a very small detection surface at high magnification. There is a trade-off in using extension tubes: the farther the lens is from the sensor, the greater the loss of light. Furthermore, extension tubes may not be the only optical element affecting light intensity. For fluorometry, emission and excitation filters are also needed that further reduce the optical signal. Therefore, the choice of lens, extension tube, and filters must all be compatible in order to image with enough sensitivity over smaller surface areas on a detection plate.

1.3 Light Sources for Optical Detection

Light sources are critical for optical detection, especially for fluorescence-based detection. Traditional optical detectors utilize high intensity, high power and bulky bench-top excitation sources including tungsten, mercury, or xenon lamps. These light sources are usually expensive and nondurable. They also require high voltage, which further limits the portability of the device. Lasers [1, 7, 15, 16] have been used for illumination sources since they are low cost, highly efficient, small, simple to operate, durable and consume relatively little power. These characteristics make lasers ideal excitation light sources for portable detectors. However, lasers are inherently spot light sources that can only be used effectively with photodiodes or photomultiplier detectors. When using lasers as spot light sources to illuminate larger surface areas, mechanical translation stages can be used for raster scanning, which complicates the operation of the imaging system and slows down the imaging. Alternatively, a line generator and evanescence light can be used with a laser [1, 7, 15, 16] to expand the light source and cover larger surface areas compatible with the spatial imaging of a CCD or CMOS sensor. However, this may introduce light distribution and uniformity problems, and still complicates the design.

1.3.1 Electroluminescence (EL) Illumination

Electroluminescence (EL) illuminators are illuminated surfaces providing spatial illumination, which is inherently more compatible with CCD and CMOS optical detectors [17, 18]. EL illuminators are semiconductor surfaces that emit light in response to an alternating electric current. In EL, electrons that are excited into a higher state leave "holes" and when dropping back to lower-energy ground state, the excited electrons release their energy as photons upon recombining with their "holes". The spectrum of the light emitted by EL depends on the electroluminescent materials. Although EL material is widely used for signs and instrument dial illumination (indigo watches, clock radio, personal organizers, cockpit instrumentation, etc.), they are not commonly used for fluorescence excitation and biodetection. Nevertheless, they have great potential for biodetection applications given that the features of EL include spatial illumination (a great advantage for lab-on-a-chip (LOC) applications), a range of wavelengths, low cost, high efficiency, small footprint, simple operation, durability, and low power consumption. However, EL generally emits fewer photons (i.e., less light) than other light sources used for fluorescence excitation such as tungsten, mercury, xenon lamps, LEDs, or lasers. Therefore they require longer exposure times. Another limitation is that they can require high voltage (e.g., 100 V) to enhance photon emission.

1.3.2 Light-Emitting Diode (LED) Illumination

Light-emitting diodes (LEDs) have been used as illumination sources in various optical detectors such as portable real-time PCR detection [19] and fluorometers [18, 20, 21]. LEDs are semiconductor devices that emit light when an electric current passes through them. Similar to ELs, LEDs operate by electroluminescence, in which the emission of photons is caused by electronic excitation of a material. In the LED, the two semiconductor materials are of different composition, one negative, or n-type, and the other positive, or p-type, forming a p–n junction. Under an electric field, current flows across the p–n junction. The free electrons moving across a diode can fall into empty holes from the p-type layer. This involves a drop from the conduction band to a lower orbital, so the electrons release energy in the form of photons. The LED light is visible when the diodes are characterized by a wider gap between the conduction band and the lower electron orbit (i.e., wide band-gap). The size of this gap (depending on the semiconductor materials) determines the frequency of the photon (i.e., the wavelength the light). Unlike EL, LED intensity is high and suitable for fluorophore excitation in fluorescence detection. However, several LEDs are required to illuminate a surface, and unlike EL, which provides uniform illumination, the illumination of a panel of LEDs is not uniform. However, LEDs have great potential use for optical sensor illumination [20] because they are available in a wide variety of wavelengths. Furthermore, some LEDs can emit light over a range of wavelengths, enabling multi-wavelength analysis. They are low cost, highly efficient, small, simple to operate, durable and consume relatively little power. Their main advantage over ELs is they emit far more light than ELs and require low voltage DC for operation. Their main disadvantage is, unlike ELs, they are spot illuminators like lasers. Therefore, surface illumination with LEDs requires optical systems (mirrors, diffusers, lenses, etc.) that may complicate the design and introduce light distribution and uniformity problems similar to lasers.

1.3.3 Screens for Illumination of Portable Devices

The screen of a portable device can also be used for effective illumination. The screen enables high intensity uniform illumination in red, blue and green. Screens for a portable devices such as a smartphones can be organic LED (OLED) or liquid crystal display (LCD). The major difference between the two is the use of a backlight for an LCD display, where the pixels contain liquid crystals that regulate the wavelengths that pass through the pixel resulting in the desired color. One advantage of OLEDs over LCDs is the individual electrical control of each pixel in an OLED reduces power consumption, since even a "black" screen for an LCD requires electrical power to control the orientation of the liquid crystals to prevent any light from passing through, while the OLED is naturally black without any electrical power. However, the use of a single backlight for all pixels in an LCD results in greater uniformity.

1.4 Shiga Toxin Detection

In order to demonstrate the use of CCD-based biodetection we will discuss a specific system we have previously developed for the detection of Shiga toxin activity [22]. Sensitive methods for detection of Shiga toxin are an important component of the effort to prevent food poisoning. However, current detection technologies cannot distinguish between biologically active and inactive toxins. Current detection approaches rely mainly on immunological assays using expensive microplate readers or spectrophotometers. However, these approaches cannot **differentiate between** a biologically **active toxin and an inactive toxin not posing a risk.** In our previous work we developed a cell based assay [22] for fluorescence detection of the activity of the food borne Shiga toxin type 2 (Stx2) which can **differentiate between** biologically **active and inactive toxins.** The assay is based on Shiga toxin inhibition of protein synthesis measured by green fluorescent protein (GFP), so toxin activity can be measured as a function of GFP light emission. When excited by a blue LED equipped with a filter at 486 nm, a light emission peak occurs at 509 nm which can be detected with a green filter at 535 nm.

For this assay, mammalian Vero cells were transduced with an adenovirus coding GFP (Fig. 2a). The cells were treated with various amounts of Shiga toxin, and after incubation the cell fluorescence was detected with the CCD detector. The plate containing the cells was illuminated by the LED illuminator using the blue LED with a blue excitation filter. The detection plate was imaged with a green emission filter and the signals of the wells were quantified by the CCD camera operating in a video mode. The image was analyzed and quantified by the open-source imaging software ImageJ. By using this open source software it is possible to calculate the average value of each pixel for a series of frames.

2 Materials

2.1 CCD-Based Detector

1. Point Grey Research *Chameleon* monochrome camera equipped with a C-mount CCTV lens was used as the photodetector used to measure GFP (emission peak at 509 nm).

2. Pentax 12 mm f1.2 lens. Item # C61215KP (Spytown, Utopia, NY).

3. Alternative lens: Tamron manual zoom CCTV 4–12 mm, f1.2 lens. Item # 12VM412ASIR (Spytown, Utopia, NY).

4. Green emission filter HQ535/50 m (Chroma Technology Corp Rockingham, VT).

5. Blue excitation filter D486/20x (Chroma Technology Corp, Rockingham, VT).

6. LED illumination box containing red, green, blue, and white LEDs custom built by Luminousfilm (Shreveport, Louisiana, www.luminousfilm.com/led.htm).

2.2 Computer Control and Data Analysis

1. PC (laptop or desktop) with USB port.
2. Image analysis software ImageJ open source NIH software (http://rsb.info.nih.gov/ij/download.html).
3. Excel (Microsoft, Redmond, WA) and SigmaPlot (SigmaPlot, Ashburn, VA) data analysis software.

2.3 Biological and Cell Culture Regents

1. Stx2 was obtained from Toxin Technology (Sarasota, FL).
2. Vero cells: African Green Monkey adult kidney cells (ATCC CCL-81) were obtained from American Type Culture Collection (Manassas, VA).
3. HEK293 cells (ATCC CRL-1573) were obtained from American Type Culture Collection (Manassas, VA).
4. Dulbecco's Modified Eagle's Medium (DMEM) from Gibco (Carlsbad, CA).
5. Fetal bovine serum (FBS) (Hyclone, Waltham, MA).
6. Fetal calf serum (FCS) (Hyclone, Waltham, MA).
7. Penicillin and streptomycin from Gibco (Carlsbad, CA).
8. Trypsin (Gibco).
9. 96-well plates (Greiner 655,090 obtained from sigma).

2.4 Plate Assay Material

1. Black 3.2 mm acrylic (Piedmont Plastics, Beltsville, MD).
2. Clear 0.5 mm polycarbonate film (Piedmont Plastics, Beltsville, MD).
3. 3 M 9770 Adhesive transfer Tape (Piedmont Plastics, Beltsville, MD).
4. Epilog Legend CO2 65 W cutter (Epilog, Golden, CO).

3 Methods

3.1 Cell Culture

Vero cells were maintained at 37 °C in a CO_2 incubator in Dulbecco's Modified Eagle's Medium (DMEM) containing 10% fetal bovine serum and 100 units/ml of both penicillin and streptomycin. Cells were trypsinized when ready to harvest.

3.2 Preparation of High-Titer Viral Stocks

Since most of the virus remains associated with infected cells until very late in the infection process, high-titer viral stocks were prepared by concentrating the infected cells. Tissue culture flasks (175 mm) were seeded with HEK293 cells in DMEM containing 10% FCS and 100 units/ml penicillin/streptomycin. When cells reached 90% confluency, they were infected with the virus at a multiplicity of infection (MOI) of 10. When the cytopathic effect was nearly completed (after 48–72 h) and most of the cells were rounded but not yet detached, the cells were harvested (they were easily dislodged by tapping). Cells were pelleted by centrifugation at $800 \times g$ for 5 min at 4 °C. Both cell pellet and supernatant were collected. Since most progeny viruses remain cell-associated, infected cells were disrupted by freeze–thaw cycles followed by lysis as described below. To every cell pellet collected from 18 flasks (175 mm), 18 ml of PBS/1 mM $MgCl_2$/0.1% NP-40/1 mM $CaCl_2$, were added for hypotonic lysis followed by three rounds of freeze–thawing. Crude lysates were then pelleted to remove cellular debris by centrifugation at $9500 \times g$ for 20 min at 4 °C. The supernatants that contained the crude viral lysates were carefully removed. The viral supernatants were loaded onto CsCl step gradients made by layering three densities of CsCl (1.25, 1.33, and 1.45 g/ml) and centrifuged at $50,000 \times g$ for 2 h in a Beckman SW41 rotor at 14 °C. The lower band containing the intact packaged virus was removed, and a second step density gradient of 1.33 g/ml CsCl was centrifuged for 16 h at $48,000 \times g$ at 14 °C. The lower band containing the packaged virus was collected and dialyzed against 10% glycerol in saline solution. The stock viral titer, determined by plaque assays, was 10^{11} plaque-forming units/ml.

3.3 Plaque Assays for Purification and Titration of the Adenovirus

Plaque assays depend on the ability of the adenovirus to propagate in HEK293 cells. Six 35 mm tissue culture plates were seeded with HEK293 cells. The cells were incubated at 37 °C in a CO_2 incubator until the cells were 90% confluent. Serial dilutions (10^{-8}–10^{-13} of the adenovirus stock) were made in DMEM supplemented with 2% FBS. The diluted virus was added to the cells. After 2 h, the medium was removed and replaced with $1\times$ Modified Eagle's Medium and 1% sea-plaque agarose. The agar overlay was added to keep the virus localized after the cells had lysed. After 5 days, plaques were visible, and counted for titer determination after 7 days.

3.4 Cell Transduction

The cells were transduced with adenovirus coding for GFP as described in previous work [23] to produce label free cells that express the GFP gene used for the assay. Vero cells were plated on black wells 96-well plates at 1×10^4 cells in 100 µl of medium per well. Cells were incubated overnight to allow time for cells to attach to the plate and then the cells were transduced with Ad-GFP multiplicity of infection (MOI) of 100.

3.5 Shiga Toxin Activity Assay

Transduced cells were treated in triplicate with various amounts of Shiga toxin for 48 h. The toxin activity is measured through a decrease in fluorescence because the biologically active toxin inhibits protein synthesis. The controls in these experiments are cells without toxin. After incubation with toxin, the cell fluorescence was imaged with a green emission filter, and the signals of the wells were quantified by the CCD camera operating in a still single frame mode. Various concentrations of the toxins were employed and the effect of toxin on the cellular fluorescence was measured.

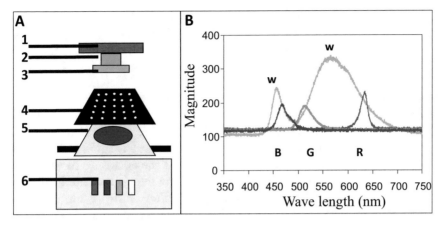

Fig. 1 Fluorescence detector. (**A**) The main system elements: CCD detector, LED multi-wavelength illuminator and cell toxicity assay. The CCD detector components: (*1*) a camera, (*2*) lens, (*3*) emission filter mounted on the end of the lens, (*4*) assay plate, (*5*) excitation filter, and (*6*) multi-wavelength LED. (**B**) The spectra (measured by a spectrometer) for the (*W*) *white*, (*B*) *blue*, (g) *green*, and (*R*) *red* LEDs

3.6 CCD Imager

The basic configuration of the CCD detector platform is shown schematically in Fig. 1a along with the spectra of the LED illuminator. The system consists of three main elements (a) the CCD detector module, (b) the assay chip module, and (c) the illumination/excitation module (*see* **Note 1**). All of these elements are contained within a black acrylic box designed for portability, reconfigurability, prevention of light loss and internal light reflection, and to minimize extraneous signals from ambient light.

3.7 Illumination Modules

The multi-wavelength spatial LED illuminator module [20] shown in Fig. 1a6 comprises a custom built multi-wavelength LED illumination box with two different types of LED: (1) white, which generates across the full optical spectrum, and (2) RGB, which generates at the red, green, and blue wavelengths. The LED illuminator contains four RGB LED strips and four white LED strips, both 18 LEDs per foot. Each LED emission color is controlled individually with a switch in the front panel. The top of the box consists of a diffusion panel (milky white plastic panel), which assures uniformity of the light and hence even illumination of the sample chip. The dimensions of the diffusion panel are 8.5 × 5.5 cm. The LEDs are powered by a 3 V DC battery. It is important to measure the illumination uniformity across the measured surface (*see* **Note 5**).

3.8 Optical System

The optical system includes a Pentax 12 mm f1.2 CCD lens. The fixed focal length enables the use of a wider aperture. The C mount fixed focal length lens was connected to the imager. The f1.2 lens permits operation at very low light levels. To minimize green light emission, which increases the noise and limits the detection sensitivity, a green emission filter HQ535/50 m was used (*see* **Note 2**). Similarly, a blue excitation filter (D486/20×) was used to block the blue light emission to the CCD and to enable the measurement of the 523 nm light from the excited GFP (*see* **Note 6**).

3.9 Assembling the CCD Detector

The assembly of the CCD-based fluorometer based on LED illumination was described in previous work [17, 18, 20]. The CCD imager (Fig. 1a1) is placed inside the light-sealed black acrylic enclosure (*see* **Note 4**). The camera is mounted in the enclosure with holes drilled in the top of the enclosure to enable air circulation for cooling the CCD. The lens is mounted onto the camera body (Fig. 1a2) and a green filter (EmF) is mounted on the lens (Fig. 1a3). The LOC or detection plate (Fig. 1a4), excitation filter (ExF) (Fig. 1a5), and the LED (Fig. 1a6) are interchangeable (*see* **Notes 1–7**)

3.10 Light Characteristics of the Multi-Wavelength LED

Unlike the monochromatic light emitted from lasers, the multi-wavelength LED illumination box provides illumination in the blue, green, red, and white ranges (Fig. 1b), covering a spectrum of 450–650 nm (red 610–650 nm, green 492–550 nm, and blue 450–495 nm) as described in our previous work [20]. The only region of the spectrum not represented by the RGB LEDs strip is the range between 550 and 610 nm. The white LED provide broad illumination spectra (440–680 nm with lower intensity in the range 472–510 nm). The multi-wavelength LED enables illumination in narrower spectra by using only one of the R, G, B, or white LEDs, or combinations of them. Within each LED spectrum, interchangeable filters can be used to narrow the excitation band (*see* **Note 6**).

3.11 Image Capturing

Images were captured using a camera incorporating a CCD sensor with high quantum efficiency and response linearity as well as the ability to generate high-quality, low noise images. The camera used in this case employed a monochrome sensor, and fluorescent emission was detected through a green filter. The images were analyzed and the data quantified with ImageJ [24, 25]. In PBS we observed no photobleaching when the cells were exposed for less than 5 min. The longest exposure time employed in our experiments was 20 s. The average sensor background signal was subtracted from sample images. This background signal was recorded by capturing images with exposure and gain settings identical to those used to image samples, but with a lens cap in place.

3.12 Image and Data Analysis

The exposure time for the signal will depend on the type of illumination and the light emitted, which is determined empirically (*see* **Note 8**). The spot intensity can be analyzed using any standard image processing software. For our system, we chose the freeware ImageJ developed at NIH (*see* **Note 8**), which enables 2D image analysis of the intensity of each spot in a row or column (Fig. 2b), as

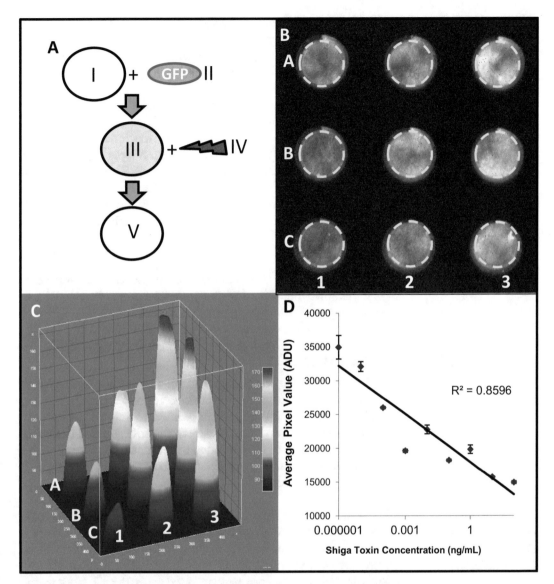

Fig. 2 Fluorescence detection of cell toxicity assay: (**a**) cell toxicity assay (*I*) Vero cells transduced with adenovirus with GFP (*II*) the resultant cells that express the GFP (*III*) emit light at 509 nm. A toxin effect on Vero cells (*IV*) will reduce light emission (*V*) after the addition of the toxin (*VI*) measured by CCD detector shown in (**b**). (**b**) Vero-GFP cells response to low concentrations of Shiga toxin. Vero-GFP cells were treated with various amounts of Shiga toxin. The signals were detected by the CCD operating in a still single frame mode, the amount of toxin used *3A* control no toxin, *3B* 0.01 pg/mL, *3C* 0.1 pg/mL, *2A* 1 pg/mL, *2B* 10 pg/mL, *2C* 100 pg/mL, *1A* 1 ng/mL, *1B* 10 ng/mL, *1C* 100 ng/mL. The corresponding ImageJ 3D image is shown in (**c**). Average signal brightness (ADU) was plotted against the various Shiga toxin concentrations (**d**)

well as 3D (Fig. 2c) spatial analysis (2D of the spatial position of the spots and the third dimension is the intensity of the spots), to provide visual representation of the image as shown in Fig. 2c. For spot analysis, the intensity value of every spot was exported to an Excel spreadsheet and to the scientific data analysis and graphing software Sigmaplot, which was used to plot the data. Several analysis methods were performed, including subtracting baseline noise level and calculating the signal-to-oise ratio (S/N).

3.13 Assay Plate Fabrication

A 9-well plate (Fig. 1a6) was fabricated as described in our previous work [21, 26, 27], using black acrylic with one side coated by 3 M 9770 black adhesive transfer tape laser machined to have the needed array of wells (*see* **Note 10** and **11**). A layer of thin polycarbonate sheet was attached to the adhesive transfer tape to form the bottom of the sample wells. To minimize light reflection, the polycarbonate material used for fabrication did not show detectable autofluorescence, and all the materials used to fabricate the device did not appear to inhibit enzymatic reactions. The assay plate fabrication includes:

1. *Micromachining*: The fluidics was made of 3.2 mm black acrylic, micromachined with a computer controlled 65 W Epilog Legend CO_2 laser system (*see* **Notes 9** and **10**). The well holes are 2 mm, which allows analysis of 20 μl samples. Smaller holes make the loading of the sample less reproducible (*see* **Notes 10** and **11**).

2. *Bonding:* The polycarbonate bottom was bonded to the acrylic with double sided pressure-sensitive adhesive transfer tape (*see* **Notes 12** and **13**) which was bonded to the micromachined layer so there is no contact between the adhesive and the fluids. Assembled devices were bonded together and processed with a laminating machine to eliminate air bubbles and improve uniformity of the bonding.

3.14 Detection of Active SHIGA Toxin Using CCD Based Optical Detector

Transduced cells were treated with various amounts of Shiga toxin in three replicates. As previously mentioned, the toxin activity is measured through decreases in fluorescence using controls without toxin. The cells were cultured for 48 h and after incubation with toxin the cell fluorescence was detected with the CCD detector. The plate containing the cells was illuminated by the LED illuminator using the blue LED with a blue excitation filter. The detection plate was imaged with a green emission filter, and the signals of the wells were quantified by the CCD camera operating in a still single frame mode (Fig. 2b). Various concentrations of the toxins were employed and the effect of toxin on the cellular fluorescence was measured as shown in Fig. 2b. The amount of toxin used was well 3A: control no toxin, 3B: 0.01 pg/mL, 3C: 0.1 pg/mL, 2A: 1 pg/mL, 2B: 10 pg/mL, 2C: 100 pg/mL, 1A: 1 ng/mL, 1B: 10 ng/mL, 1C: 100 ng/mL. The corresponding ImageJ 3D image is shown in Fig. 2c. Average signal brightness measured by the A/D unit (ADU) was plotted against the various Shiga toxin concentrations (Fig. 2d). The reduction in cell fluorescence signal is proportional to the amount of toxin due to toxin inhibition of protein synthesis, with the lowest signal at 100 ng/mL as shown in Fig. 2b well 1C. In other wells with lower levels of toxin (1A: 1 ng/mL, 1B: 10 ng/mL), the level of fluorescence increases slightly.

The image was analyzed and quantified using ImageJ software. By this method it is possible to calculate the average value of each pixel for a series of frames. Figure 2c illustrates the decreasing fluorescence signal with increasing amount of the toxin. The average signal intensities (brightness) displayed as Average Pixel Value (ADU count) of three replicates was plotted against the various Shiga toxin concentrations (Fig. 2c) which shows decreased fluorescence as the amount of toxin increased, suggesting that toxin level can be measured as a function of decreased fluorescence as shown in Fig. 2c. Another toxin (Clostridium difficile toxin B) was used with this assay with no effect on fluorescence, which demonstrates the specificity of the assay. A plot of the ADU versus Shiga toxin concentration can be seen in Fig. 2d, where there is a nearly linear variation of the fluorescence signal with toxin concentration.

3.15 Detection Limit of Biologically Active Stx2 Using CCD Measurements

To measure lower concentrations of toxin, three replicates of the transduced Vero-GFP cells were cultured for 24 h and treated with various amounts of Shiga toxin (100 ng to 0.01 pg/mL). The signals of the wells were detected by the CCD camera. The image was quantified with ImageJ. The limit of detection (LOD) was calculated as the mean pixel value in three control samples (cells with no toxin) minus three times the standard deviation of these samples. The plotted LOD is indicated as a horizontal dashed line (Fig. 3). Concentrations of toxin which yielded a mean signal out of three replicates that was under this limit were considered to be detected. Unlike the usual LOD calculation, the standard deviation was subtracted and not added because the toxin reduces signal. Thus, the LOD was determined to be 0.1 pg/mL, which is similar to a commercial fluorimeter.

3.16 Factors Contributing to System Sensitivity

There are several factors contributing to improve sensitivity of mobile low cost optical detection:

1. *Detector*: While camera phone lenses are not interchangeable, many webcams permit lenses to be changed (we used an $f/1.2$ lens to maximize the amount of light transmitted to the sensor. Using cooled CCS/CMOS devices to reduce thermal noise and improve SNR. In general, monochrome sensors are capable of higher detail and sensitivity than similar color sensors. Monochrome sensors capture all incoming light at each pixel—regardless of color. Each pixel therefore receives up to $3\times$ more light.

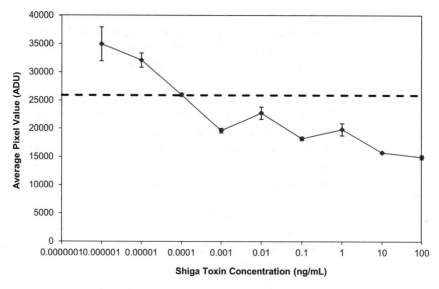

Fig. 3 Limit of Shiga toxin detection (LOD). Vero-GFP cells were treated with various amounts of Shiga toxin (100 ng to 0.01 pg/mL). The signals of the wells were detected by the CCD camera. The image was quantitated with ImageJ The level of detection (LOD) was calculated as the mean pixel value in three control samples (cells with no toxin) minus three times the standard deviation of these samples. The LOD is indicated as a *horizontal dashed line*. Concentrations of toxin which yielded a mean signal out of three replicates that was under this limit were considered to be detected

2. *Illumination source*: Increasing the power of the excitation source in fluorescent detection increases florescence emission. Increasing the intensity of the LED illumination (i.e., the use of more LEDs or using more powerful LEDs) increases the fluorescent signal. Using color LEDs enables illumination in the range of florescence excitation which may reduce noise and improve SNR (Fig. 1). The use of low cost lasers equipped with line generator or removing the laser lens may increase light intensity and provide narrow wavelength illumination.

3. *Exposure time*: For single frame imaging, cameras and some webcams allow for long exposure times (>1 s). Longer exposures can be used to detect low signal intensities; however, longer exposure will also increase image noise in high noise detectors.

4. *Video imaging*: The use of video imaging mode combined with the image stacking [24] computational approach improves SNR.

5. *Filters*: The quality of filters is critical. Using high quality narrow band filters at the emission/excitation wavelengths reduces noise and improves detection.

6. *Assays*: For immunoassays it is possible to increase primary antibody immobilization by increasing surface area for antibody binding using high surface area nanoparticles for

antibody binding such as gold nanoparticles [28] or carbon nanotubes [29, 30].

7. *Plate material*: For reducing light noise, the use of a light absorbing polycarbonate material combined with the compact design of the device enclosure minimizes light losses and cross talk.

4 Notes

1. Make sure the lens, filters, and sample plate are vertically centered and aligned.

2. Make sure the arrows on the coated filters are facing the CCD.

3. The filters should be leveled

4. The filters should be always in the same position.

5. The imager enclosure must be light sealed without any light leaks and use light reflecting material. A long exposure (e.g., 3 min) can be used to detect light leaks.

6. To measure light uniformity, it is important to measure the light with a long enough exposure time (e.g., 60 s) to measure the CCD uniformity when all of the wells are loaded with the same GFP fluorescence sample. If the lighting is not uniform, correction values for each well can be calculated and use for measurement composition.

7. To measure the effectiveness of the filters, it is recommended to perform two long exposures (e.g., 3 min) without the assay plate, one with the LED on and one with the LED off. Ideally, the two measurements should be very similar (the blue filters pass only blue light which is blocked by the green filters). The difference between measurements may suggest that the blue filters do not block all green light and/or the green filters do not block all blue light.

8. For data analysis, use the ImageJ high contrast visualization, which will not affect the values but will enable easy selection of the spots

9. A fully open aperture (e.g., f1.2) will enable shorter exposure time, but because of the limited depth of field, focusing will be more limited and the image less sharp. The laser power and speed for cutting polymers has to be determining empirically. It is recommended to use the minimum laser power to reduce overheating or burning the material.

10. The "holes" for the wells on the assay plate are 2 mm in diameter, which allows analysis of 20 μL samples. Smaller holes make the loading of the sample less reproducible because

the fluid meniscus within the wells causes light diffraction, which complicates quantification.

11. For strong bonding, remove 1 cm of the adhesive tape cover, align the tape with acrylic surface, and attached the exposed tape to the acrylic surface. With a ruler, press the tape to the surface and slowly move the ruler across the tape with little pressure in order to prevent air bubbles. When bonding the tape, it should be aligned with the acrylic. For assembling of the assay plate, remove the protective cover from the other side of the double side adhesive (taped to the acrylic) and remove the protective cover from the polycarbonate. Then, align the two pieces, apply pressure, and run through the assay plate and attach the polycarbonate to the acrylic.

12. Controlling optical noise: Fluorescence emission and scattered excitation light can propagate through the chip, causing cross talk between adjacent channels. This can become a major source of optical noise in the system [18, 31, 32] by increasing background noise, thus reducing the sensitivity of the measurements. To limit the effect of fluorescence background, PC, and not Mylar, which is the commonly used material for lamination based fabrication, was used as the main fabrication material due to its lower fluorescence background [18]. Using black material decreases the noise.

References

1. Ligler FS, Taitt CR, Shriver-Lake LC, Sapsford KE, Shubin Y, Golden JP (2003) Array biosensor for detection of toxins. Anal Bioanal Chem 377:469–477

2. Roda A, Manetta AC, Portanti O et al (2003) A rapid and sensitive 384-well microtitre format chemiluminescent enzyme immunoassay for 19-nortestosterone. Luminescence 18:72–78

3. Svitel J, Surugiu I, Dzgoev A, Ramanathan K, Danielsson B (2001) Functionalized surfaces for optical biosensors: applications to in vitro pesticide residual analysis. J Mater Sci Mater Med 12:1075–1078

4. Liu Y, Danielsson B (2007) Rapid high throughput assay for fluorimetric detection of doxorubicin—application of nucleic acid-dye bioprobe. Anal Chim Acta 587:47–51

5. Burkert K, Neumann T, Wang J, Jonas U, Knoll W, Ottleben H (2007) Automated preparation method for colloidal crystal arrays of monodisperse and binary colloid mixtures by contact printing with a pintool plotter. Langmuir 23:3478–3484

6. Tohda K, Gratzl M (2006) Micro-miniature autonomous optical sensor array for monitoring ions and metabolites 2: color responses to pH, K+ and glucose. Anal Sci 22:937–941

7. Feldstein MJ, Golden JP, Rowe CA, Maccraith BD, Ligler FS (1999) Array biosensor: optical and fluidics systems. Biomed Microdevices 1:139–153

8. Sohn YS, Goodey A, Anslyn EV, McDevitt JT, Shear JB, Neikirk DP (2005) A microbead array chemical sensor using capillary-based sample introduction: toward the development of an "electronic tongue". Biosens Bioelectron 21:303–312

9. Knecht BG, Strasser A, Dietrich R, Martlbauer E, Niessner R, Weller MG (2004) Automated microarray system for the simultaneous detection of antibiotics in milk. Anal Chem 76:646–654

10. Balsam J, Bruck HA, Rasooly A (2015) Smartphone-based fluorescence detector for mHealth. Methods Mol Biol 1256:231–245

11. Balsam J, Rasooly R, Bruck HA, Rasooly A (2014) Thousand-fold fluorescent signal amplification for mHealth diagnostics. Biosens Bioelectron 51:1–7

12. Balsam J, Bruck HA, Rasooly A (2014) Webcam-based flow cytometer using wide-field imaging for low cell number detection at high throughput. Analyst 139:4322–4329

13. Balsam J, Bruck HA, Rasooly A (2013) Orthographic projection capillary array fluorescent sensor for mHealth. Methods 63:276–281

14. Balsam J, Ossandon M, Bruck HA, Lubensky I, Rasooly A (2013) Low-cost technologies for medical diagnostics in low-resource settings. Expert Opin Med Diagn 7:243–255

15. Sapsford KE, Taitt CR, Loo N, Ligler FS (2005) Biosensor detection of botulinum toxoid A and staphylococcal enterotoxin B in food. Appl Environ Microbiol 71:5590–5592

16. Golden JP, Floyd-Smith TM, Mott DR, Ligler FS (2007) Target delivery in a microfluidic immunosensor. Biosens Bioelectron 22:2763–2767

17. Kostov Y, Sergeev N, Wilson S, Herold KE, Rasooly A (2009) A simple portable electroluminescence illumination-based CCD detector. Methods Mol Biol 503:259–272

18. Sapsford KE, Sun S, Francis J, Sharma S, Kostov Y, Rasooly A (2008) A fluorescence detection platform using spatial electroluminescent excitation for measuring botulinum neurotoxin A activity. Biosens Bioelectron 24:618–625

19. Higgins JA, Nasarabadi S, Karns JS et al (2003) A handheld real time thermal cycler for bacterial pathogen detection. Biosens Bioelectron 18:1115–1123

20. Sun S, Francis J, Sapsford KE, Kostov Y, Rasooly A (2010) Multi-wavelength Spatial LED illumination based detector for in vitro detection of botulinum neurotoxin A activity. Sens Actuators B 146:297–306

21. Sun S, Ossandon M, Kostov Y, Rasooly A (2009) Lab-on-a-chip for botulinum neurotoxin a (BoNT-A) activity analysis. Lab Chip 9:3275–3281

22. Rasooly R, Balsam J, Hernlem BJ, Rasooly A (2015) Sensitive detection of active Shiga toxin using low cost CCD based optical detector. Biosens Bioelectron 68:705–711

23. Rasooly R, Do PM (2010) Shiga toxin Stx2 is heat-stable and not inactivated by pasteurization. Int J Food Microbiol 136:290–294

24. Balsam J, Bruck HA, Kostov Y, Rasooly A (2012) Image stacking approach to increase sensitivity of fluorescence detection using a low cost complementary metal-oxide-semiconductor (CMOS) webcam. Sens Actuators B 171–172:141–147

25. Balsam J, Ossandon M, Bruck HA, Rasooly A (2012) Modeling and design of micromachined optical Soller collimators for lensless CCD-based fluorometry. Analyst 137:5011–5017

26. Sapsford KE, Francis J, Sun S, Kostov Y, Rasooly A (2009) Miniaturized 96-well ELISA chips for staphylococcal enterotoxin B detection using portable colorimetric detector. Anal Bioanal Chem 394:499–505

27. Sun S, Yang M, Kostov Y, Rasooly A (2010) ELISA-LOC: lab-on-a-chip for enzyme-linked immunodetection. Lab Chip 10:2093–2100

28. Yang M, Kostov Y, Bruck HA, Rasooly A (2009) Gold nanoparticle-based enhanced chemiluminescence immunosensor for detection of Staphylococcal enterotoxin B (SEB) in food. Int J Food Microbiol 133:265–271

29. Yang M, Kostov Y, Bruck HA, Rasooly A (2008) Carbon nanotubes with enhanced chemiluminescence immunoassay for CCD-based detection of Staphylococcal enterotoxin B in food. Anal Chem 80:8532–8537

30. Yang M, Kostov Y, Rasooly A (2008) Carbon nanotubes based optical immunodetection of Staphylococcal enterotoxin B (SEB) in food. Int J Food Microbiol 127:78–83

31. Irawan R, Tjin SC, Yager P, Zhang D (2005) Cross-talk problem on a fluorescence multi-channel microfluidic chip system. Biomed Microdevices 7:205–211

32. Hawkins KR, Yager P (2003) Nonlinear decrease of background fluorescence in polymer thin-films – a survey of materials and how they can complicate fluorescence detection in microTAS. Lab Chip 3:248–252

Chapter 16

Smartphone-Enabled Detection Strategies for Portable PCR–Based Diagnostics

Aashish Priye and Victor M. Ugaz

Abstract

Incredible progress continues to be made toward development of low-cost nucleic acid-based diagnostic solutions suitable for deployment in resource-limited settings. Detection components play a vitally important role in these systems, but have proven challenging to adapt for operation in a portable format. Here we describe efforts aimed at leveraging the capabilities of consumer-class smartphones as a convenient platform to enable detection of nucleic acid products associated with DNA amplification via the polymerase chain reaction (PCR). First, we show how fluorescence-based detection can be incorporated into a portable convective thermocycling system controlled by a smartphone app. Raw images captured by the phone's camera are processed to yield real-time amplification data comparable to benchtop instruments. Next, we leverage smartphone imaging to achieve label-free detection of PCR products by monitoring changes in electrochemical reactivity of embedded metal electrodes as the target DNA concentration increases during replication. These advancements make it possible to construct rugged inexpensive nucleic acid detection components that can be readily embedded in a variety of portable bioanalysis instruments.

Key words PCR, Smartphone, Point of care, Label free detection, Mobile health care, Image analysis

1 Introduction

The emerging demand for sophisticated medical diagnostic tools that can be readily deployed in resource limited settings has focused renewed attention on the need for advanced bio-analytical detection techniques. Next-generation instruments based on these technologies promise to catalyze a revolutionary shift away from dedicated laboratory-based approaches toward field-deployable systems that are compact, portable, inexpensive, and easy to use. But existing methods employed to perform post-analytical evaluation predominantly involve optical analysis (e.g., microscopy, fluorometry, colorimetry, cytometry) whose cost and complexity severely constrain the development of portable systems. Smartphones are poised to provide a versatile platform that can enable many of these limitations to be overcome. In addition to their

Avraham Rasooly and Ben Prickril (eds.), *Biosensors and Biodetection: Methods and Protocols Volume 1: Optical-Based Detectors*, Methods in Molecular Biology, vol. 1571, DOI 10.1007/978-1-4939-6848-0_16, © Springer Science+Business Media LLC 2017

ubiquity for voice communication, today's mobile devices embed a robust array of sensors (microphones, accelerometers), wireless data transmission protocols (Wi-Fi, Bluetooth), navigation and positioning tools (global positioning system (GPS)), and perhaps most significantly high-resolution CMOS (complementary metal--oxide semiconductor) optical sensors associated with their built-in cameras. The high sensitivity of these CMOS imaging systems, coupled with advanced processing capabilities, makes smartphones an attractive standalone platform to support development of portable analytical systems.

Incredible progress has already been made toward leveraging these advantages to position smartphones as an alternative to benchtop microscopes in biosensing applications [1–5]. Many of these efforts have focused on harnessing the small footprint and versatility of smartphones to produce custom devices tailored toward microscopy [6, 7], spectroscopy [8], single molecule analysis [9], colorimetry [10–13], paper-based microfluidics [14, 15], and label-free detection [16, 17]. In this chapter we describe recent efforts by our research group to harness the CMOS imaging capabilities embedded in ordinary consumer-class smartphones for development of real-time fluorescence- and electrochemical-based detectors for bioanalytical applications involving DNA replication via the polymerase chain reaction.

The polymerase chain reaction (PCR) is considered a gold-standard nucleic acid-based detection assay owing to its favorable combination of specificity and sensitivity. The assay protocol involves cyclically heating and cooling reagents through distinct temperature zones (denaturing (95 °C), annealing (55–65 °C), and extension (72 °C)) to produce exponential amplification of a specific target sequence within a template nucleic acid strand (Fig. 1a). Such thermal cycling protocols are conventionally carried out in "thermocycler" instruments that are quite slow (timescales of ~1 h or more) and power intensive. Alternatively, thermal convection-driven techniques can overcome these limitations by applying a temperature gradient to drive microliter-scale convective flow that transports reagents through the denaturing, annealing and extension temperature zones passively [18–24] (Fig. 1a). Thus, convective PCR offers pseudo isothermal operation by housing the reagents in a confined cylindrical enclosure subjected to a temperature gradient (Fig. 1b).

Despite the inherent simplicity of convective PCR format, the resulting micro-scale flows are quite intricate [19, 20, 25]. A spectrum of flow states (static fluid, periodic, quasi-periodic, chaotic, and transient flow) can be accessed by tuning the cylinder geometry (aspect ratio; h/d; d is the diameter of the cylinder) and the thermal driving force, i.e., Rayleigh number ($Ra = [g\beta(T_2 - T_1)h^3]/\nu\alpha$, where β is the fluid's thermal expansion coefficient, g is the gravitational acceleration, T_1 and T_2 are the temperatures of the top (cold)

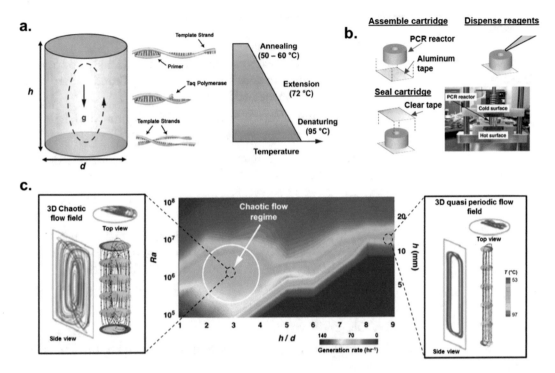

Fig. 1 Microscale convective flow states enable convective PCR. (**a**) Thermocycling is performed in a cylindrical reactor whose *top* and *bottom surfaces* are maintained at different fixed temperatures, establishing a flow field that circulates PCR reagents sequentially through denaturing, annealing, and extension conditions. (**b**) Reagents are loaded in cylindrical wells machined into polycarbonate rods and sealed, after which the reactor is subjected to a vertical temperature gradient. (**c**) A 3D computational fluid dynamics model coupled with PCR kinetics yields predictions of DNA replication time scales expressed in terms of a characteristic generation rate (i.e., DNA doubling events per hour) in a *Ra-h/d* parametric space. The parametric map reveals a zone in the chaotic flow regime where accelerated DNA replication is stably achievable over a span of two orders of magnitude in *Ra*. Insets originating from the map depict that the convective flow field has a strong chaotic component in lower aspect ratio geometries as compared to higher aspect ratio geometries (quasi-periodic flow trajectories). Trajectories were obtained from 3D CFD simulations performed over 5 min of convective flow with a vertically imposed temperature gradient of 40 °C

and bottom (hot) surfaces, respectively, h is the height of the cylindrical reactor, α is the thermal diffusivity, and ν is the kinematic viscosity). Furthermore, PCR kinetics are strongly coupled with the underlying convective flow state [19]. For example, PCR is executed most robustly in a broad chaotic flow regime (spanning two orders of Ra) in Ra–h/d parametric map where the fluid elements exhibit 3D chaotic trajectories as opposed to at higher aspect ratio geometries where the fluid elements exhibit more 2D periodic trajectories (Fig. 1c). These results provide crucial insight into design principles for convective PCR reactors as they can function universally in the chaotic flow regime encompassing all possible combinations of temperature and reactor volume associated with realistic PCR conditions.

2 Materials

2.1 Convective PCR Experiments

1. Convective PCR reactors/tubes: holes are machined by drilling into polycarbonate rods (Amazon Supply; Brand name: Small Parts; Part #: PlasticRod137).

2. Bovine serum albumin (cat. no. A2153; Sigma-Aldrich).

3. Rain-X Anti-Fog (SOPUS Products).

4. SYBR Green PCR Master Mix (2×; Cat. no. 4309155, Life Technologies).

5. AmpliTaq Gold DNA polymerase (included in the SYBR Green PCR Master Mix).

6. Forward and Reverse primers (10 μM; 5'-CTGAGGCCGGGTTATTCTTG-3' and 5'-CGACTGGC-CAAGATTAGAGA-3' Integrated DNA Technologies).

7. λ-phage template DNA (1 μg/mL; Cat. no. N3011S, New England Biolabs).

8. PCR grade water.

9. Aluminum tape (cat. no. PCR-AS-200; Axygen, Inc.) and clear tape.

10. A smartphone with PCR to go app.

11. Clip on mini-microscope attachment for the phone (60× mini pocket microscope for iPhone 4; supply).

12. Portable battery pack (Vinsic Tulip battery charger).

13. Optical filter (THORLABS; Ø25 mm OG515 Colored Glass Filter; Part # FGL515).

14. Microcontroller (Atmega328).

15. Ceramic resistor (Wire wound 10 Ω; Mouser electronics).

16. Temperature sensor (TMP35; Mouser electronics).

17. 5 mm round RGB LED, 5 mW.

18. RN41 Bluetooth module (Microchip).

19. Resistors and capacitors.

20. Polydimethylsiloxane (PDMS).

2.2 Electrode Dissolution Experiments

1. Glass microscope slides (cat. no. 12-550-A3, Fisher).

2. Dry transfer film (Press-n-Peel Blue, Techniks, Inc.).

3. Metal etchant (Gold Etchant TFA, Transene).

4. Acetone.

5. Cyanoacrylate adhesive.

6. Polycarbonate cylindrical cells (**item 1** in Subheading 2.1).

7. Conductive tape (xyz-axis Electrically Conductive Tape, 3 M).

8. DC power supply (E3612A, HP/Agilent).

9. Aluminum tape.

---

3 Methods

3.1 Smartphone-Enabled Analysissee Smartphones, PCR of Real-Time PCR Fluorescence: Design Considerations

PCR amplification is typically detected by a fluorescent signal produced by an intercalating dye or fluorophore-labeled probe whose signal can be used to monitor the target sequence concentration over time. Adaptation of PCR-based analysis to a portable format has proven challenging due to the electrical power requirements associated with temperature cycling, and a number of innovations have been made to overcome these issues [26–29]. But while these approaches simplify the thermal cycling apparatus, quantitative analysis of the amplified products still relies heavily on expensive optical detection components. A typical PCR detection unit incorporates an excitation source, optical focusing lenses, and a detector (e.g., photo multiplier tube (PMT), charged-coupled device (CCD) camera, photo-diode) to enable fluorescence quantification. Fluorescence data recorded from the PCR sample as a function of time are then analyzed and processed to yield a real-time PCR amplification curve.

To explore feasibility of performing portable real-time PCR fluorescence analysis, we developed a smartphone application (app) that integrates with a portable thermocycling instrument developed in our laboratory [30]. Instrument portability is enhanced by adopting a convective PCR format whereby a temperate gradient established between opposing surfaces establishes a microscale convective flow field that can be harnessed to drive DNA replication [18, 19, 31]. In this way, PCR can be actuated isothermally by maintaining a single heater at a constant temperature, drastically reducing electrical consumption to a level that can be supplied by standard 5 V USB sources that power consumer mobile devices. A smartphone is able to wirelessly integrate with the heating and detection module via Bluetooth, creating a seamless interface between the peripheral device components (heater, LED illumination source) and a smartphone app (Fig. 2a). Temperature settings can be controlled through the app, with thermal management achieved by using off the shelf ceramic resistors that convert electrical current to heat via joule heating (essentially a simple mini-hot plate). A temperature sensor regulates the applied current via a feedback loop programmed to maintain the temperature at 95 °C (the PCR denaturing setpoint) using an Arduino microcontroller (Fig. 2b). A polydimethylsiloxane (PDMS) casing around the resistor provides thermal insulation and directs heat transfer to the reactor.

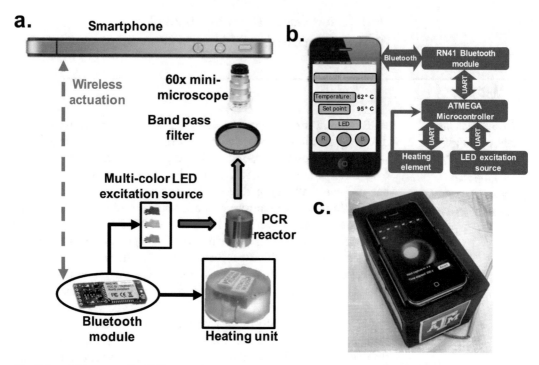

Fig. 2 A portable convective PCR setup. A smartphone enabled portable convective PCR device. (**a**) Convective PCR reactor is placed on an isothermal mini hotplate and is illuminated with an LED excitation source to emit fluorescence signal from the reactor which is then captured by the phone camera for further analysis. (**b**) The phone interfaces with the device's peripheral components (isothermal heater and LED source) wirelessly via Bluetooth module coupled with an Arduino microcontroller via the RX and TX lines (Universal asynchronous receiver/transmitter) (**c**) Device in operation

1. A simple convective thermocycling device consists of interchangeable polycarbonate reaction chambers (Fig. 1b) and a ceramic mini hot plate that functions as an isothermal heater (Fig. 2a).

2. Cylindrical reactor wells are embedded by machining holes in polycarbonate rods, with different combinations of hole diameter and plastic sheet thickness employed to achieve the desired aspect ratios (height/diameter = h/d).

3. The height of the reactor is chosen such that at steady state, the temperature of the top surface is maintained at the annealing temperature (55–65 °C) and the diameter is chosen such that geometry lies in the chaotic flow regime to enable robust convective PCR. Based on these design considerations the final reactor geometry in the results presented here are: $h/d = 5$, $h = 10$ mm and $d = 2$ mm.

4. The bottom surface of the reactor is heated using a mini hot plate constructed from two wire wound ceramic resistors (10 Ω) connected in series. The heaters were encapsulated using poly (dimethyl siloxane) (PDMS) to provide insulation.

A temperature probe (tmp35) senses the temperature of the heater, and the signal is sent to one of the analog pins in the microcontroller (Arduino UNO R3). This analog reading is converted into temperature which was continuously monitored to operate the heaters isothermally at 95 °C by actuating the pulse width modulation (PWM) pin of the ATMEGA microprocessor. The PWM pin outputs 100% of its signal until the temperature of the heater reaches the setpoint, and reduces to 85% thereafter. If the temperature exceeds 97 °C, the PWM signal reduces to zero. In our testing, this simple scheme was sufficient to maintain the heater at 95 ± 2 °C.

5. The reactor is placed on the isothermal mini hot plate surface and a 5 mW blue LED excitation source connected to low dropout (LDO) regulator illuminates the reactor from the side (Fig. 2a).

6. The isothermal heater and LED excitation source can be operated wirelessly through the smartphone which can communicate to a Bluetooth module linked to the Arduino microcontroller. The incorporated Bluetooth module reports the isothermal heater temperature and LED state. The smartphone app can establish a RX/TX connection (universal asynchronous receiver/transmitter at a baud rate of 9600 bits/s with no parity setting) via the inbuilt iOS core Bluetooth framework and send commands to the services advertised by the peripheral device to control the heater and the LED source.

7. The final circuit and all the components are soldered on a printed circuit board which is housed in a plastic case designed using FreeCAD and printed using a MakerGear M2 3D printer.

8. Reactors are first rinsed with a 10 mg/mL aqueous solution of bovine serum albumin followed by Rain-X Anti-Fog, and dried (*see* **Note 1**).

9. Convective PCR experiments are performed to replicate a 237 base pair target from a λ-phage DNA template using the following protocol.

Reagent	Volume (μL)
SYBR Green PCR Master Mix (2×)	50
Forward primer (10 μM)	10
Reverse prime (10 μM)	10
λ-phage template DNA (1 μg/mL)	1
Water, molecular biology grade	29

10. The lower surface of the polycarbonate reactor is sealed using aluminum tape, after which PCR reagent mixture (from **step 9**) is pipetted inside and the top surface sealed with another layer of clear tape (*see* **Note 2**).

11. After loading the reagents, the convective PCR reactor is placed on the preheated ceramic heater based mini hot plate maintained at 95 °C for 20 min.

12. The smartphone app is programmed to actuate an LED excitation source that illuminates the sample, and the top of the PCR reactor is imaged using the smartphone camera to quantify fluorescence at regular time intervals.

3.2 PCR to Go: A Fluorescence Analysis Smartphone App

The *PCR to Go* application has been developed for both iOS (*Xcode* 5.0) and Android (Android studio) platforms. It provides a user friendly GUI that sends commands to the detector/heater system and seamlessly switches between various in-app functions.

1. Once the app is launched it automatically pairs with the peripheral Bluetooth module of the detector system. The user can then set the heater and LED operation parameters.

2. The analysis can either be carried out in end point or real time detection modes. In each case, the camera exposure time and focal plane settings are manually fixed with the sample placed ~5 cm away from the camera lens (*see* **Note 3**).

3. A clip on mini microscope attachment (TOMTOP Mini Pocket Microscope Magnifier Jewelers Loupe for iPhone 4 4S with LED Light (Amazon.com)), supplemented with the ability to digitally zoom images via the app's affine transformation framework, enables a total of 60× magnification to resolve an area of ~1 mm^2 without pixel degradation (Fig. 3).

4. In real-time detection mode, the user inputs additional parameters such as the frequency of image acquisition and the number of images to be captured. The excitation LED is illuminated for 5 s during which the app synchronizes image capture.

5. Raw image data from the CMOS sensor are stored as bitmap data files containing 4 bits per pixel (one each for red, blue, green, and alpha values) corresponding to standard RGB color space (sRGB).

6. The user can then define a region of interest within the acquired image using touch gestures.

7. When the analyze button is tapped, the app extracts the corresponding sRGB byte matrix from the region of interest. The alpha channel, which measures transparency, is discarded as it is constant in all images.

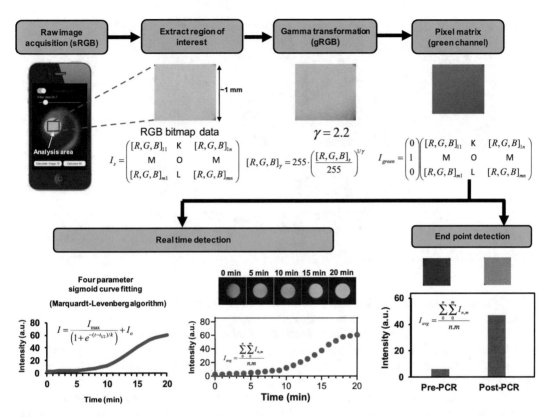

Fig. 3 Smartphone-based fluorescence analysis workflow. RGB bitmap data from the cropped raw images first undergoes a gamma transformation. The *green channel* intensity of the pixel matrix is then extracted to analyze the average fluorescence intensities. In end point detection mode pre- and post-PCR images are compared, while in real time detection mode averaged intensities from a series of acquired images are fit using a nonlinear sigmoid function to obtain a qPCR curve

8. A gamma transformation is applied to the RGB matrix with a gamma value of 2.2 to account for the nonlinear intensity scale of the CMOS sensor [32].

9. The red and blue channels' 8 bit intensity values are discarded, yielding a green channel pixel matrix that best represents the fluorescence signal.

10. The built-in nonlinear solver is invoked to fit the averaged green channel intensity over time using a four parameter sigmoid function via the Marquardt Levenberg algorithm [33, 34].

$$ I = \frac{I_{\max}}{\left(1 + e^{-(t - t_{1/2})/k}\right)} + I_o $$

where I_{max} and I_o represent the maximum and background fluorescence intensities, respectively. The quantities $t_{1/2}$ and k

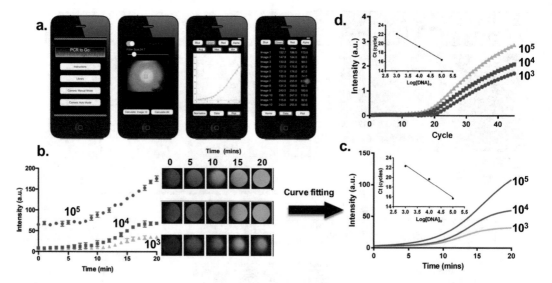

Fig. 4 Smartphone-based quantitative fluorescence detection (**a**) The *PCR to Go* app presents a simple GUI to control image acquisition, processing, and data analysis (https:/itunes.apple.com/us/app/pcr-to-go/id909227041?mt=8). (**b**) Smartphone-based quantification of a 237 bp target sequence from a λ-phage DNA template is demonstrated for three initial copy numbers ($[DNA]_0 = 10^5$, 10^4, and 10^3 copies/μL). (**c**) Sigmoidal fits are applied to the smartphone acquired real time data, and a standard curve is constructed using reaction times when fluorescence exceeds a threshold value of 20 units (inset, C_T = 9.4, 11.8, and 13.3 min for starting DNA concentrations of $[DNA]_0 = 10^5$, 10^4, and 10^3 copies/μL, respectively). (**d**) A benchtop real time PCR instrument (LightCycler 96, Roche) generates comparable results with a nearly identical standard curve

express the time required for intensity to reach half of the maximum value and slope of the curve, respectively, and are used as fitting parameters.

11. The fitting algorithm also allows background fluorescence subtraction yielding a constant baseline. The initial guess for the curve fitting iterations is determined from the parameters of the first image. Critical threshold reaction times are determined from the point where the sigmoidal fit exceeds a user-defined threshold value.

12. Finally, an in-app plotting feature enables results to be viewed graphically or exported as a .csv file for further analysis, along with a time stamp and GPS location of the run for archival.

To demonstrate the utility of our smartphone app, we applied it to analyze fluorescence during a PCR reaction run using a conventional SYBR Green reagent chemistry (Fig. 4a). Results were obtained by replicating a 237 bp target sequence from a λ-phage DNA template over a series of initial copy numbers ranging from 10^3 to 10^5 (Fig. 4b). An in-app algorithm fits the recorded fluorescence data into a four parameter sigmoidal curve and extracts a critical reaction time to exceed a threshold value of 20 fluorescence units—a time scale analogous to the critical cycle number obtained when real-time PCR is performed in a conventional thermocycler (Fig. 4c). These data can be used to construct a standard curve whose slope yields an effective doubling time of 35.8 s. For comparison, we evaluated the same series of reactions in a benchtop real-time PCR instrument (LightCycler 96, Roche), (Fig. 4d) and converted the convective data in Fig. 4c (expressed in units of reaction time) to equivalent cycle numbers by assuming that one conventional replication cycle occurs during the 35.8 s doubling time.

Both approaches agree remarkably well, yielding virtually identical standard curves (insets Fig. 4c, d). Our adaptation of smartphone-based fluorescence detection therefore enables quantitative analysis at template concentrations down to 1000 copies/µL, a level sufficient for many diagnostic scenarios (i.e., depending on the severity of an infection), and particularly impressive considering the simplified instrument format. It is also envisioned that, in practice, each PCR reactor will include a calibration target on the surface and multiple reactor cells that enable parallel control reactions to be simultaneously performed.

3.3 Detection Via Electrochemical Dissolution

An alternative approach to leverage smartphone imaging, potentially useful for detection of unlabeled biomolecules, involves monitoring electrochemical reactivity of metal electrodes and its dependence on the concentration of DNA in solution. We previously demonstrated this approach in a device that exploited the electrochemical reactivity of chromium to enable visual detection of PCR products [17]. This design consisted of a microfabricated electrode array patterned on a silicon substrate and bonded to an etched glass microchannel, yielding a cross section with electrodes spanning the floor of an enclosed chamber (Fig. 5a). Electrode dissolution is monitored by recording the change in reflectivity with time under oblique illumination with ordinary white light, from which the average intensity within a region of interest centered on the electrode can be evaluated. In this way, changes in dissolution rate in response to the bulk solution's chemical composition can be readily distinguished (Fig. 5b). We demonstrated feasibility of performing this detection using a smartphone camera outfitted with an inexpensive mini-microscope attachment (Fig. 5c). Fabrication of these devices is briefly summarized below:

1. Silicon wafers (P(100), 500 µm thick, 15 cm diameter, 5000 Å oxide layer; University Wafer) are cleaned in a reactive ion etcher.

2. Wafers are spin coated with hexamethyldisilazane (J.T. Baker) followed by a positive photoresist (Shipley 1827; Rohm & Haas).

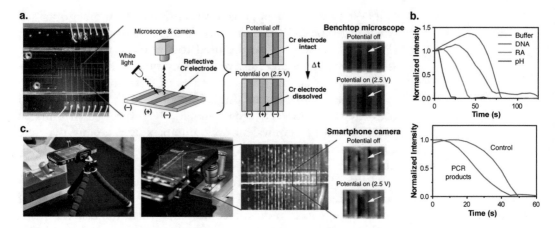

Fig. 5 Label-free detection via electrode dissolution. (**a**) Microdevice design incorporating a glass microchannel (275 × 45 μm cross section) bonded to a Si substrate patterned with a Cr microelectrode array. When a 2.5 V potential is applied, the Cr electrodes electrochemically dissolve at a rate dependent on the chemical composition of the bulk solution (*arrows*). (**b**) The evolution of reflected light intensity from the active electrode displays composition-dependent behavior. The upper panel compares dissolution kinetics of histidine buffer (50 mM, pH 7.6), a 100 bp dsDNA ladder (20 μg/mL in 50 mM histidine), a reducing agent (0.06% w/v $Na_2S_2O_4$ in 50 mM histidine), and at low pH (50 mM histidine, pH 4.4). Pre- and post-reaction PCR reagent mixtures are compared in the lower panel. (**c**) Electrodes are large enough to enable dissolution to be directly imaged using an ordinary smartphone (Apple iPhone 4S). Scale: all electrodes are 50 μm wide horizontally

3. Wafers are patterned, and developed (MF-319 developer; Rohm & Haas).

4. Electrodes are fabricated by depositing a 600 Å chromium layer by thermal evaporation.

5. Glass microchannels (275 × 45 μm cross section) are etched on glass wafers (borofloat, 500 μm thick, 15 cm diameter; Precision Glass and Optics) using hydrofluoric acid.

6. Assembled devices are wire bonded to a printed circuit board so that electrodes can be individually addressed.

Building on this proof of concept, we developed a greatly simplified design based on electrochemical dissolution of copper in a format that can be directly integrated with convective PCR reactors. Here, copper electrodes experience electrochemical corrosion at non-neutral pH (Fig. 6a), with degradation rates strongly dependent on bulk solution composition. To demonstrate this approach in the context of the above mentioned convective PCR platform, we patterned 100 nm-thick addressable copper electrodes on the top surface of the PCR reactor. This arrangement enables degradation kinetics to be straightforwardly monitored because the electrode surface (500 μm-wide, 1 mm inter electrode spacing) is large enough to easily view using the smartphone camera (Fig. 6a). This can be seen by comparing degradation kinetics before and after replication of a 237 bp target sequence from a λ-phage DNA

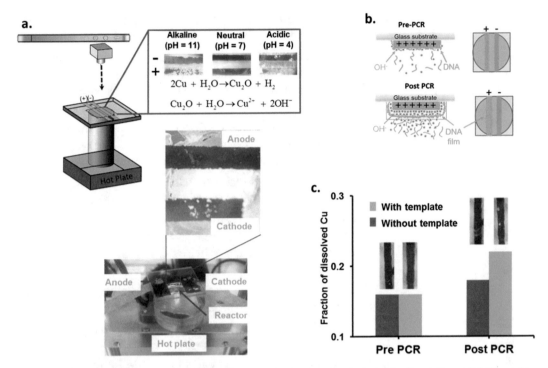

Fig. 6 Integrated label-free convection PCR detection by imaging local electrode dissolution. (**a**) Addressable Cu electrodes are patterned on a glass substrate affixed to the *top surface* of a cylindrical PCR reactor, enabling electrochemical dissolution to be viewed from *above* (drawing not to scale). Anodic dissolution occurs slowly at neutral pH, but is accelerated under alkaline conditions (images of the anode taken after a 3 V potential was applied for 1 min). (**b**) In the pre-PCR solution (*above*), the anode surface is not visibly changed when a 3 V potential is applied. Dissolution becomes favored (*below*) as more negatively charged DNA molecules are generated during PCR, forming an electrophoretically compacted film at the anode. (**c**) Images of the electrodes can be analyzed to quantify anodic dissolution before and after PCR, and compared against appropriate controls. Scale: all electrodes are 500 μm wide*See* Polymerase chain reaction (PCR)

template. Electrode dimensions were selected such that small potentials (3 V, much lower values can be applied by reducing the inter-electrode spacing) generate electric fields high enough to electrophoretically trap negatively charged DNA at the anode [35, 36]. The resulting DNA film imposes a membrane-like barrier against transport of electrochemically generated OH^- ions into the surroundings, leading to a local increase in pH that acts to accelerate the rate of copper dissolution (Fig. 6b). DNA replication can then be monitored in situ by analyzing video recordings of the electrodes to quantify the dissolved mass of Cu before and after PCR, revealing a faster dissolution rate owing to increase in target DNA concentration (Fig. 6c).

1. Copper film coatings (100 nm thick) were sputter deposited on glass microscope slides.

2. Electrodes (500 µm wide, 1 mm inter-electrode spacing) were patterned on the copper coated slides using dry transfer film.

3. The patterned electrode on the glass slide was immersed in metal etchant (Gold Etchant TFA, Transene) for ~1 min after which the remaining dry transfer film was stripped using acetone.

4. Patterned glass slides were affixed to the upper surface of the cylindrical acrylic cells using cyanoacrylate adhesive, and electrical connections were made using conductive tape.

5. Wires are soldered directly to pads on the patterned copper, enabling the individual electrode pair to be addressed.

6. Potentials were applied using a DC power supply. The bottom surfaces of the cells were sealed with aluminum tape and the assembly was heated from below using the same apparatus described above.

7. A smartphone was used to continuous monitor the electrode dissolution process using video recordings.

8. The acquired video was converted to image sequences, and analyzed using *ImageJ* software.

9. Dissolution was quantified within a region of interest overlaying the anode surface by converting the images to 8-bit grayscale and applying a black/white threshold of 160 out of 255.

10. The total white pixel count was then calculated to determine the mass of copper dissolved before and after convective PCR (Fig. 5c).

4 Notes

1. For the convective PCR runs we found that additional rinsing steps (**step 8**, Subheading 3.1) helped minimize adsorption at the sidewalls that may otherwise inhibit the reaction while also enhancing surface wettability so that reagents could be loaded without trapping air pockets inside the reactor. It is also essential to ensure that air bubbles are not introduced during sample loading as they can not only hinder the optical detection from the top but larger bubbles can also disrupt the convective flow resulting in failed PCR. Most of the time they are introduced at the time of sealing the reactor so proper care should be taken at this step. Sealing the reactors under pressure may also help to suppress any initial bubble formation. Additionally, injection molding can be applied to construct reactors, as opposed to the current machining process. Molding processes are envisioned to be used for mass production, and are likely to significantly reduce sidewall roughness that generates bubble nucleation

sites. These modifications suggest strong potential to significantly reduce bubble formation while convective PCR and thus improve operational reliability.

2. Prior to beginning **step 10** in Subheading 3.1, it is important that all the other pre-PCR steps be performed. For example, the enzyme we used in the SYBR master mix requires a hot start step. We ensured that enzyme activation was performed by heating the sample for 10 min at 95 °C in a conventional thermocycler (T-Gradient, Biometra).

3. A key challenge in smartphone-based optical detection is accounting for variability of ambient lighting. Inherent to all smartphones is a dynamic color balancing functionality optimized for photography that introduces variability in sensor parameters applied to adjust to the dynamic ambient conditions. This mode of operation is not ideal for fluorescence detection, and previous studies have explored approaches including the use of color calibration targets and reference samples to reduce the resulting image-to-image variability [37, 38]. Our system employs a 3D printed "dark box" to block ambient light and provides consistent illumination form the built-in LEDs. Additionally, our analysis app overrides the preset camera parameters so that fixed focal length and exposure times can be employed. Together, these approaches eliminate the most significant sources of fluorescence intensity variations.

References

1. Ozcan A (2014) Mobile phones democratize and cultivate next-generation imaging, diagnostics and measurement tools. Lab Chip 14:3187–3194

2. Laksanasopin T et al (2015) A smartphone dongle for diagnosis of infectious diseases at the point of care. Sci Transl Med 7:273re1

3. Zhu H, Isikman SO, Mudanyali O, Greenbaum A, Ozcan A (2013) Optical imaging techniques for point-of-care diagnostics. Lab Chip 13:51–67

4. Erickson D et al (2014) Smartphone technology can be transformative to the deployment of lab-on-chip diagnostics. Lab Chip 14:3159–3164

5. Xu X et al (2015) Advances in smartphone-based point-of-care diagnostics. Proc IEEE 103:236–247

6. Breslauer DN, Maamari RN, Switz NA, Lam WA, Fletcher DA (2009) Mobile phone based clinical microscopy for global health applications. PLoS One 4:e6320

7. Tseng D et al (2010) Lensfree microscopy on a cellphone. Lab Chip 10:1787–1792

8. Yu H, Tan Y, Cunningham BT (2014) Smartphone fluorescence spectroscopy. Anal Chem 86:8805–8813

9. Wei Q et al (2014) Imaging and sizing of single DNA molecules on a mobile phone. ACS Nano 8:12725–12733. doi:10.1021/nn505821y

10. Lee S, Oncescu V, Mancuso M, Mehta S, Erickson D (2014) A smartphone platform for the quantification of vitamin D levels. Lab Chip 14:1437–1442

11. Oncescu V, Mancuso M, Erickson D (2014) Cholesterol testing on a smartphone. Lab Chip 14:759–763

12. Oncescu V, O'Dell D, Erickson D (2013) Smartphone based health accessory for colorimetric detection of biomarkers in sweat and saliva. Lab Chip 13:3232–3238

13. Shen L, Hagen JA, Papautsky I (2012) Point-of-care colorimetric detection with a smartphone. Lab Chip 12:4240–4243

14. San Park T, Li W, McCracken KE, Yoon J-Y (2013) Smartphone quantifies Salmonella from paper microfluidics. Lab Chip 13:4832–4840

15. Fronczek CF, San Park T, Harshman DK, Nicolini AM, Yoon J-Y (2014) Paper microfluidic extraction and direct smartphone-based identification of pathogenic nucleic acids from field and clinical samples. RSC Adv 4:11103–11110

16. Gallegos D et al (2013) Label-free biodetection using a smartphone. Lab Chip 13:2124–2132

17. Huang Y-W, Ugaz VM (2013) Smartphone-based detection of unlabeled DNA via electrochemical dissolution. Analyst 138:2522–2526

18. Krishnan M, Ugaz VM, Burns MA (2002) PCR in a Rayleigh-Benard convection cell. Science 298:793–793

19. Priye A, Hassan YA, Ugaz VM (2013) Microscale chaotic advection enables robust convective DNA replication. Anal Chem 85:10536–10541

20. Muddu R, Hassan YA, Ugaz VM (2011) Chaotically accelerated polymerase chain reaction by Microscale Rayleigh–Bénard convection. Angew Chem Int Ed 50:3048–3052

21. Yao D-J, Chen J-R, Ju W-T (2007) Micro-–Rayleigh-Bénard convection polymerase chain reaction system. JM3 6:043007–043009

22. Braun D, Goddard NL, Libchaber A (2003) Exponential DNA replication by laminar convection. Phys Rev Lett 91:158103

23. Hennig M, Braun D (2005) Convective polymerase chain reaction around micro immersion heater. Appl Phys Lett 87:183901

24. Braun D (2004) PCR by thermal convection. Mod Phys Lett B 18:775–784

25. Priye A, Muddu R, Hassan YA, Ugaz VM (2011) Royal Society of Chemistry, Cambridge

26. Hühmer A, Landers J (2000) Noncontact infrared-mediated thermocycling for effective polymerase chain reaction amplification of DNA in nanoliter volumes. Anal Chem 72:5507–5512

27. Pal R et al (2005) An integrated microfluidic device for influenza and other genetic analyses. Lab Chip 5:1024–1032

28. Kopp MU, De Mello AJ, Manz A (1998) Chemical amplification: continuous-flow PCR on a chip. Science 280:1046–1048

29. West J et al (2002) Application of magnetohydrodynamic actuation to continuous flow chemistry. Lab Chip 2:224–230

30. Priye A et al (2016) Lab-on-a-drone: toward pinpoint deployment of smartphone-enabled nucleic acid-based diagnostics for mobile health care. Anal Chem 88:4651–4560

31. Ugaz VM, Krishnan M (2004) Novel convective flow based approaches for high-throughput PCR thermocycling. JALA 9:318–323

32. Skandarajah A, Reber CD, Switz NA, Fletcher DA (2014) Quantitative imaging with a mobile phone microscope. PLoS One 9:e96906

33. Liu W, Saint DA (2002) Validation of a quantitative method for real time PCR kinetics. Biochem Biophys Res Commun 294:347–353

34. Rutledge R (2004) Sigmoidal curve-fitting redefines quantitative real-time PCR with the prospective of developing automated high-throughput applications. Nucleic Acids Res 32:e178

35. Huang Y-W, Shaikh F, Ugaz VM (2011) Tunable synthesis of encapsulated microbubbles by coupled electrophoretic stabilization and electrochemical inflation. Angew Chem Int Ed 50:3739–3743

36. Shaikh F, Ugaz VM (2006) Collection, focusing, and metering of DNA in microchannels using addressable electrode arrays for portable low-power bioanalysis. Proc Natl Acad Sci U S A 103:4825–4830

37. Sari YA, Ginardi RH, Sarno R (2013) Assessment of color levels in leaf color chart using smartphone camera with relative calibration. Inform Syst 2:4

38. Yetisen AK, Martinez-Hurtado J, Garcia-Melendrez A, da Cruz Vasconcellos F, Lowe CR (2014) A smartphone algorithm with interphone repeatability for the analysis of colorimetric tests. Sens Actuators B 196:156–160

Chapter 17

Streak Imaging Flow Cytometer for Rare Cell Analysis

Joshua Balsam, Hugh Alan Bruck, Miguel Ossandon, Ben Prickril, and Avraham Rasooly

Abstract

There is a need for simple and affordable techniques for cytology for clinical applications, especially for point-of-care (POC) medical diagnostics in resource-poor settings. However, this often requires adapting expensive and complex laboratory-based techniques that often require significant power and are too massive to transport easily. One such technique is flow cytometry, which has great potential for modification due to the simplicity of the principle of optical tracking of cells. However, it is limited in that regard due to the flow focusing technique used to isolate cells for optical detection. This technique inherently reduces the flow rate and is therefore unsuitable for rapid detection of rare cells which require large volume for analysis.

To address these limitations, we developed a low-cost, mobile flow cytometer based on streak imaging. In our new configuration we utilize a simple webcam for optical detection over a large area associated with a wide-field flow cell. The new flow cell is capable of larger volume and higher throughput fluorescence detection of rare cells than the flow cells with hydrodynamic focusing used in conventional flow cytometry. The webcam is an inexpensive, commercially available system, and for fluorescence analysis we use a 1 W 450 nm blue laser to excite Syto-9 stained cells with emission at 535 nm. We were able to detect low concentrations of stained cells at high flow rates of 10 mL/min, which is suitable for rapidly analyzing larger specimen volumes to detect rare cells at appropriate concentration levels. The new rapid detection capabilities, combined with the simplicity and low cost of this device, suggest a potential for clinical POC flow cytometry in resource-poor settings associated with global health.

Key words POC, Webcam, Flow cytometry, Rare cells, Fluidics, Fluorescence detection

1 Introduction

In many clinical applications, environmental analysis, and basic research, flow cytometers have become important diagnostic tools. One current limitation for flow cytometry is that the associated low flow rates makes it practical for analyzing small specimen volumes. This is not appropriate for detecting rare cells, including circulating tumor cells (CTCs), and therefore prevents flow cytometers from being used for rare cells analysis. Thus there is a need for new flow cytometry techniques for rare cell detection that are

Avraham Rasooly and Ben Prickril (eds.), *Biosensors and Biodetection: Methods and Protocols Volume 1: Optical-Based Detectors*, Methods in Molecular Biology, vol. 1571, DOI 10.1007/978-1-4939-6848-0_17, © Springer Science+Business Media LLC 2017

capable of handling larger specimen volumes. If these techniques can be made more affordable and portable it would also be possible to successfully introduce them in resource-poor settings. To this end, a new imaging point-of-care (POC)*see* Point-of-care (POC) flow cytometer for detection of rare cells [1–3] based on streak imaging is described in this protocol.

1.1 Point-of-Care Cytometry

The development of POC analytical tools has largely focused on lab-on-a-chip (LOC) microfluidic technologies due to their small size, simplicity, and low manufacturing cost [4, 5]. In addition, smartphone cameras are being used effectively as optical detection systems for mobile POC devices based on LOC platforms [6–14]. For example, an optofluidic fluorescent imaging cytometer using a smartphone camera with a spatial resolution of 2 μm has been recently described [9, 13]. However, the flow rate for this system is 1 μL/min, limiting analyses to small volumes. In addition to smartphones, standalone inexpensive webcam imaging systems may also overcome limitations of smartphone cameras. For example, smartphone cameras often have limited video capabilities (e.g., many phones are limited to 30 fps) and have less versatile optical systems (e.g., inability to change lenses or directly set imaging parameters). To overcome these limitations we developed a small, mobile, and low-cost flow cytometer based on webcam imaging capable of large volume analysis and rare cell detection for a variety of POC applications.

The main components of current flow cytometers consist of: [1] a fluidic system for carrying fluorescently labeled biological material (e.g., cells), and [2] an optical system for fluorescence detection. Most current flow cytometers utilize microfluidic sheathing to focus the cells or particles to a channel whose width permits only a single cell to pass, enabling narrow-field detection using photomultipliers or other narrow-field photodetectors. This technique is known as "hydrodynamic focusing." In one recent device a passive hydrodynamic flow focusing was achieved using chevron grooves imbedded on the walls of the microchannel [15, 16]. This enabled the sheath fluid to completely surround the sample stream so that cells could be individually interrogated by the narrow-field fluorescent detector. However, one limitation of the focused stream is a low flow rate due to the high hydrodynamic resistance and pressure constraints of the cell, ultimately limiting the device to small volumes or long analysis times. In addition, the commonly used photomultipliers used for detection are delicate, expensive, and difficult to transport.

In general, current flow cytometers using hydrodynamic focusing flow cells are not suitable for POC applications or for rare cell detection due to their size, expense, lack of durability, small sample capacity, and high power requirement. An alternative approach to the focused microfluidic sheathing used in conventional flow cytometers has been recently developed using a wide-field flow cell [1–3] for large volume analysis. By expanding the cross section of the flow field by orders of magnitude over conventional hydrodynamic focusing flow cells, low flow rates can be used to interrogate larger specimen volumes over the same period of time while enabling easier detection by reducing the linear velocity of target cells and increasing the detection area.

With lower flow velocities, wide-field flow cell cytometry can use less-sensitive alternatives to photomultipliers for optical detection such as imaging devices based on complementary metal–oxide–semiconductor (CMOS) or charge-coupled device (CCD) detectors. Because of the larger flow field it is necessary to use these imaging devices to interrogate large numbers of cells in parallel. However, due to their inherent portability CMOS or CCD cameras have already been employed as relatively simple, low cost devices for optical detection over large areas in several array assays [17–20]. Their main advantage is that they have sufficient sensitivity to analyze light from an area large enough to cover the entire surface of a LOC or array [21–23]. This has made CMOS or CCD-based detectors an ideal choice for POC detectors.

1.2 Streak Imaging Flow Cytometry

Using CCD or CMOS imaging sensors for flow cytometry opens up new possibilities for image analysis. One approach is to image moving cells at long exposure times such that at a given flow rate a cell moves a distance X in time t to produce a line of pixels described as a "streak image." The dimensions of the streak are directly related to the size of the cell and its velocity. Unlike conventional imaging aimed of capturing precise sharp image of the cell using low flow rate and short exposure, streak imaging enables the use of higher flow rates and long exposure to screen larger volumes of specimens.

Figure 1a illustrates a cell image captured as a line generated using a flow rate of 20 mL/min with a frame rate of 20 FPS. Unlike a spot that would be generated during regular imaging, the line pattern associated with exposing many pixels during streak imaging enables easy identification, especially for low quality detectors with high background noise. The minimum streak length required is such that at least one pixel measures the maximum possible signal in order to maximize signal. This streak length is equal to twice the length of the cell image in pixels before streaking (L) plus one pixel (i.e., for a cell image of 3×3 pixels, a streak length of 7 is required). This effect is shown graphically in Fig. 1b, where a cell of length 3 pixels is seen to move a distance of 4 pixels, thereby producing an image streak with length 7 pixels with a single bright pixel in its center. An actual cell streak image is shown in Fig. 1c for a similar cell image size and similar flow conditions, along with a plot of actual pixel brightness along the center of the cell streak in Fig. 1d. Cell streaking such as this results in at least one pixel measuring the maximum possible signal from the cell.

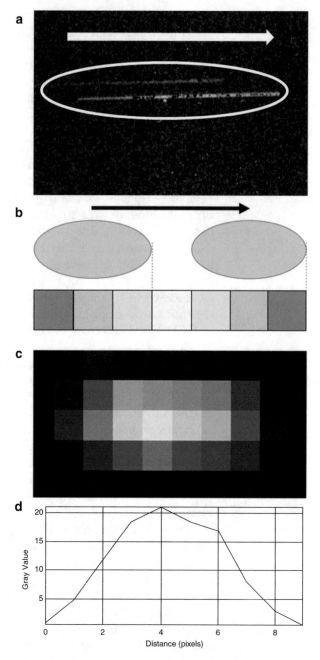

Fig. 1 Streak imaging of Cells—(**a**) An image of a cell captured at a flow rate of 20 mL/min with a frame rate of 20 FPS (the *arrow* shows the flow direction). (**b**) Schematic of a cell traversing a number of pixels showing a maximum brightness achieved in the pixel at the image center. (**c**) An actual cell image is shown under these approximate flow conditions (125 µL/min, 20 FPS), along with (**d**) a plot of its brightness along the center line of pixels

Recently, a new computational enumeration approach for accurate counting of cells in streak mode using low resolution cameras has been developed (Miguel Ossandon unpublished results). The image analysis algorithm identifies potential cells based on the streak intensity, length and relative location of the streaks in consecutive frames. The algorithm is based on: (1) finding streaks, (2) identifying candidate cells, and (3) filtering out spurious cells to identify true cells. This approach may enable automation of streak imaging cytometry.

2 Materials

2.1 Cell Culture and Fluorescence Staining

1. THP-1 human monocytes cells, ATCC #TIB-202 (ATCC Manassas, VA).

2. For cell culturing, RPMI-1640 medium supplemented with 10% (v/v) heat-inactivated FBS, 1% antibiotic–antimycotic solution, and 2 mM glutamine is used.

3. SYTO-9 dye (Molecular Probes Eugene, OR).

4. The cells are cultured in $CO2$ incubator.

2.2 Optical Components for Imaging Flow Cytometer

1. As a detector, Sony PlayStation® Eye webcam is used as an inexpensive color CMOS photodetector (Amazon, Seattle WA).

2. An alternative camera with improved sensitivity is a C mount 1.3 Mp gray scale CCD camera CMLN 13S2M-CS (Point Grey Research, Richmond BC Canada).

3. A C-mount CCTV lens (Pentax 12 mm f/1.2) is used to replace the webcam lens to improve imaging (Spytown, Utopia NY).

4. Green emission filter HQ535/50M (Chroma Technology Corp Rockingham VT).

5. 1 W 450 nm blue laser is used for fluorescent excitation illumination (Hangzhou BrandNew Technology Co., Zhejiang, China).

2.3 Flow Cell Fluid Delivery, and Imaging

1. 3 mm clear acrylic sheet (Piedmont Plastics, Beltsville MD).

2. Quartz glass microscope slide.

3. 3 M 9770 Adhesive transfer Tape (Piedmont Plastics, Beltsville MD).

4. Nylon rod (McMaster-Carr, Robbinsville, NJ).

5. Fusion100 syringe pump is used for fluid delivery (Chemyx, Stafford TX).

6. Epilog Legend CO_2 65 W laser cutter (Epilog, Golden CO) used for flow cell fabrication.

2.4 Computer Control and Data Analysis

1. The webcam sensor is connected to a 32-bit Windows-based laptop computer via a USB2 port.

2. Drivers and software allowing the webcam to be controlled on a personal computer are developed and freely distributed by Code Laboratories, Inc. (Henderson NV).

3. The camera control software (CL-Eye Test) is used to set camera parameters (exposure time, frame rate, and gain) and to capture and save video in uncompressed AVI format.

4. Video files are analyzed using ImageJ software (freely distributed by NIH, http://rsb.info.nih.gov/ij/).

5. Data analysis and plotting is carried out in Microsoft Excel (Redmond, WA).

3 Methods

3.1 Cell Culture and Labeling

Fluorescently stained THP-1 human monocytes are used as a model to simulate rare CTCs. Though monocytes themselves are not rare, they are diluted to levels similar to those associated with rare cells.

1. Cells are cultured with RPMI-1640 medium supplemented with 10% (v/v) heat-inactivated FBS, 1% antibiotic–antimycotic solution, and 2 mM glutamine (*see* **Note 1**) in a 5% CO_2 environment at 37 °C.

2. Cells are removed from an active culture, pelleted by centrifugation and resuspended in deionized water at an approximate concentration of 10^6 cells/mL (staining protocol for Syto-9 dye advises against the use of phosphate buffer solutions).

3. To stain cells, 10 μL of Syto-9 dye (3.34 mM stock concentration) is added to 1 mL of suspended cells and allowed to rest at room temperature in the dark for 20 min. Washing of cells to remove excess dye is not necessary due to the low intrinsic quantum yield when not bound to nucleic acids (<0.01), and because of the later dilution of this stock solution by factors of 10^4–10^6. One known factor which contributed substantially to the decreasing counting efficiency at lower concentrations is the cell staining protocol. The protocol recommended by the dye manufacturer is followed. However, dye which has entered the cell but not yet bound to DNA is noted to diffuse out of cells over time, leading to diminishing cell brightness. Because all dilutions are prepared simultaneously and then analyzed sequentially from highest to lowest concentration, lower dilutions exhibited cells with lower fluorescent intensity. Allowing the cells to equilibrate at a low concentration for several hours would allow this diffusion process time to complete.

4. After staining, cells are diluted to approximately 1–10 cell/μL (measured by microscopy) to allow for accurate manual counting. Cell concentration is measured by placing 3 μL sample droplets on a microscope slide and counting cells in the droplet in real time under laser excitation using the same imaging platform employed to image the flow cells in these experiments. This is repeated many (N > 20) times. An average cell concentration of 3.73 cells/μL with a standard error of 0.3 is a typical population estimate resulting from these measurements. From this relatively high concentration, lower concentration samples of 100, 10, and 1 cell/mL are generated by single-step dilution. For each dilution, 268 μL of stock solution (with measured concentration of 3.73 cells/μL) is diluted into a volume of deionized water to yield the final target concentration (10, 100, and 1000 mL of deionized water for concentrations of 100, 10, and 1 cell/mL, respectively). Based on a normal sampling distribution with standard deviation of 0.3 cells/μL, the 95% confidence range for each concentration is 84–116, 8.4–11.6, and 0.84–1.16 cells/mL for concentrations of 100, 10, and 1 cell/mL, respectively. Pipetting volume error is measured to be less than 1%.

3.2 Wide-Field Flow Cytometer for POC Applications

A wide-field flow cytometer is developed that is simple, lightweight, and inexpensive. The components are available commercially, and the key to the design is the integration of the elements in a compact configuration that is isolated from ambient light to enhance imaging of the fluorescence signal, and which allows for easy placement and removal of the wide-field flow cell.

1. The flow cytometer for fluorescence detection (Fig. 2) consists of four main elements: (1) a webcam used as an optical detector, (2) a laser excitation source for illumination, (3) a wide-field flow cell, and (4) an optical stage to hold each module in alignment for stable and clear imaging.

2. For fluorescence detection, a green emission filter with a center wavelength of 535 nm and bandwidth of 50 nm is used with a 1 W 450 nm blue laser (*see* **Note 1**). The optical elements must be vertically centered (*see* **Note 2**).

3. The optical detector consists of the electronics of a webcam which is disassembled in order to remove the original lens, and a 12 mm f/1.2 CCTV lens for improved light collection and image magnification (*see* **Note 3**).

4. The fluidics module consists of a flow cell and a programmable syringe pump. The flow cytometer platform (Fig. 2) consists of a stationary platform and a moveable stage for focusing using a screw mechanism.

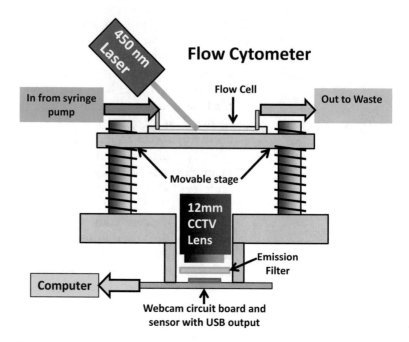

Fig. 2 Schematic of mobile wide-field flow cytometer for POC—The flow cytometer consists of four modules: (1) a webcam as an optical detector, (2) a blue 450 nm excitation source for illumination, (3) a wide-field flow cell, and (4) a stage to focus the image and to hold each module in alignment. The sensing element consists of the electronic elements of a webcam, a 12 mm f/1.2 CCTV lens, a green emission filter. It is connected to a computer to collect and analyze data. The excitation source is a 450 nm 1 W blue laser. The sample handling module consists of the wide-field flow cell, a programmable syringe pump, and a waste container

5. The flow cytometer platform is constructed using 0.5 in. thick clear acrylic sheet and nylon rod which is used as a rail for focusing. The webcam electronic circuitry and optics are attached to the stationary platform while the flow cell is attached to the translating stage. It is important to enclose the device to reduce background light (*see* **Note 4**).

3.3 Wide-Field Flow Cell Fabrication

To maximize residence time of cells in the interrogation window and maximize the number of fluorescent photons captured (i.e., to increase sensitivity), a new high throughput flow cell is designed using a wide flow channel that increases volumetric flow rates at lower flow velocities through an orders of magnitude higher cross-sectional area than used for hydrodynamic focusing.

1. The flow cell (Fig. 3) consists of a 4–20 mm wide flow channel that maximizes the volume of fluid in the interrogation window. Fluorescent labeled samples are injected via an inlet through the channel, excited by a laser in the channel, and

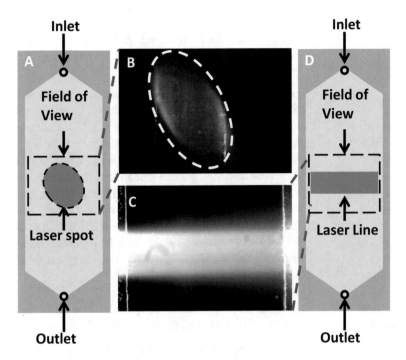

Fig. 3 Wide-field flow cell—(**a**) A schematic of the wide-field flow cell with key elements illuminated with spot laser. (**b**) An image from the flow cell illuminated with spot laser. (**c**) A schematic of the wide-field flow cell with key elements illuminated with line laser illumination. (**d**) An image from the flow cell illuminated with line laser illumination

then imaged by the webcam before being collected at the outlet.

2. The wide-field flow cell is fabricated using an Epilog Legend CO_2 65 W laser cutter (*see* **Note 5**) using techniques similar to those described in our previous work [5, 24–28] (*see* **Note 6**).

3. The flow cell consisted of three functional layers: (1) a plain glass or quartz microscope slide lower layer, (2) a middle layer laser machined from 3 M 9770 double-sided adhesive transfer tape to define the geometry of the fluid channel, and (3) a glass or quartz microscope slide on the top layer that has two holes drilled for the inlet and outlet ports aligned with the ends of the fluid channel layer (*see* **Note 6**).

4. Fluid interfacing is accomplished by bonding 1 cm^2 acrylic plates over each port hole. The acrylic plates have through holes laser machined, into which 18 G needles are bonded. Tygon 1.3 mm ID tubing is used to connect directly from the needles to either a waste container or an input syringe. The experiments consist of injecting fluorescently labeled samples into the wide channel via an inlet, excite them with the laser, and then image them with the webcam before they are collected at the outlet (*see* **Note 6**).

5. Channel depth is set to keep the flow field within the depth of field of the lens being used. A Pentax CCTV 12 mm f/1.2 is operated at approximately f/2.4 to reduce field curvature and improve depth of field.

6. A Fusion100 syringe pump connected to the flow cell is used for flow rate control. The maximum flow rate achievable through this flow cell is 10 mL/min. The adhesive may fail at higher flow rates due to the dynamic pressures required. The flow rate in experiments presented in this chapter is limited to 500 μL/min due to the maximum frame rate of the webcam employed for sensing. For very faint cells the flow rate should be decreased in order to improve sensitivity. Sensitivity also increases if exposure time is increased to allow cell images to form short streaks, with length greater than or equal to the length of the cell image plus one pixel (detailed in Subheading 3.7).

3.4 Optical System for Webcam-Based Cytometer Platform

The optical system consists of: (1) the blue laser for fluorophore excitation, (2) a webcam detector, and (3) an excitation filter.

1. **Laser Excitation:** For fluorescent imaging of a wide-field, a 1 W laser is used for illumination in two modes of illumination, spot illumination and line illumination:

   (a) In spot illumination the laser illuminates the flow cell at an angle of incidence of approximately 45° (*see* **Note 2**) at this angle the laser illuminate a bright elliptical spot which covered the width of the flow cell (Fig. 3b). The main advantage of spot illumination is that it is high intensity, which is suitable for higher flow rates.

   (b) In line illumination, the laser illuminate the edge of the slide an angle of incidence of approximately 45° and the light illuminate the flow cell through the slide (Fig. 3d)

   (c) The 1 W consumer grade laser is fairly expensive (~$300) for a device designed for use in a low-resource setting. To reduce the cost of the laser, a lower power laser with line generator optics could be used to further focus the laser spot. This could allow the critical parameter of photon flux to remain unchanged while reducing overall power; however, this approach may reduce the area analyzed. Alternatively, high power LEDs could be utilized. This would require the addition of an excitation filter as well as collimating optics, leading to a more complex and potentially more expensive configuration.

2. **Webcam detector and emission filter:** A Sony PlayStation® Eye webcam is used as the photodetector in this platform. The Sony PlayStation Eye device was disassembled and the main circuit board (with attached image sensor and USB cable) is

removed. A C-mount CCTV lens (Pentax 12 mm f/1.2) is used to replace the stock camera lens (*see* **Note 2**). For fluorescence detection, a green emission filter with center wavelength 535 and 50 nm bandwidth is used for detecting fluorescent emission. To improve sensitivity, other video devices which employ CCDs with electronics for high frame rate video imaging can be used, such as the Point Grey Research CMLN 13S2M-CS.

3.5 Configuration of Imaging Flow Cytometer

The new imaging flow cytometer, the webcam integrated the sensor into the bottom of the platform below the flow cell, as shown in Fig. 2. It is connected to a 32-bit Windows-based laptop computer via a USB2 port. The drivers and software controlling the webcam are developed and freely distributed by Code Laboratories. The camera control software (CL-Eye Test) enables setting of several camera parameters (exposure time, frame rate, and gain), and the capture of video in uncompressed AVI format. The imaging parameters for the camera (exposure time and frame rate) depend on the brightness of the fluorescently labeled cells and the flow rate (*see* **Note 7**). An example of fluorescence labeled cells is the single video frame of THP-1 stained with SYTO-9 dye in a wide-field flow cell at a flow rate of 500 μL/min, as shown in Fig. 4.

3.6 Background Signal and Noise Reduction

Frames may contain a signal of interest (e.g., image of a cell), interfering background signal (e.g., autofluorescence), and random noise. To enhance detection, the constant component of the interfering signal can be subtracted. For background subtraction, the median value and the maximum value of each pixel in a video file of the relevant color channel can be calculated. Figure 5 illustrates the green channel video frames of a video clip containing 2000 frames (10.7 s of video) showing a single cell passing through the flow cell shown in different positions along its path. The median image of all the frames, shown in Fig. 5a, represents the average background signal with no information from the cells. The maximum image, shown in Fig. 5b, indicates the highest signal recorded at each pixel during the video with the cell position in each frame marked. In order to improve cell detection, especially for lower SNR images than shown in this example, the median image is subtracted from the maximum image (Fig. 5c), which allowed for improved visualization of cell movement by removing background autofluorescence from the flow cell.

3.7 Noise Reduction in Streak Mode Images

Figure 6a shows an actual cell streak image (circled) with flow direction indicated. Figure 6b shows a close-up of the cell streak image containing background noise and showing individual pixels and. In order to reduce noise each column of pixels is averaged along the streak length *n* to produce a single averaged row of pixels, labeled avg(n) in Fig. 6. Figures 6d–i and ii shows a plot of pixel

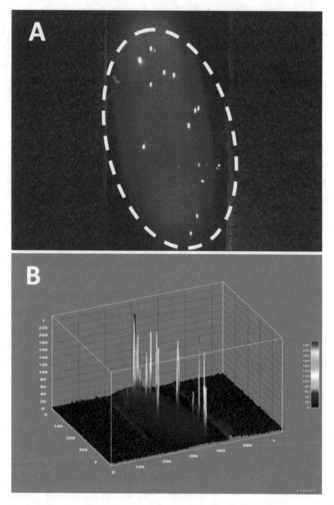

Fig. 4 Cell imaged using wide-field flow cytometer. (**a**) Single video frame of fluorescent labeled THP-1 human monocytes in wide-field flow cell at 500 µL/min, and (**b**) 3D visualization of A

values before and after averaging, respectively, indicating a 3× improvement in SNR. SNR is calculated as follows:

$$SNR = \frac{\mu_{\text{signal}}}{\sigma_{\text{noise}}}$$

Where μ_{signal} is the peak value in the plot, and σ_{noise} is the standard deviation of the pixel values on either side of the peak. When calculating σ_{noise}, a five pixel exclusion zone on either side of a peak is used to prevent the inclusion of any signal components in the noise measurement. Twenty-five pixel windows on either side of this exclusion zone are used to measure local noise levels (i.e., a total of fifty pixels around each peak).

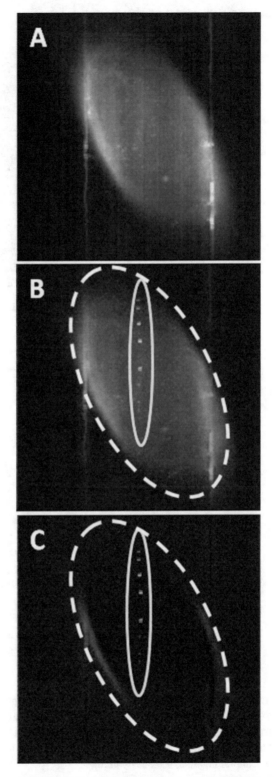

Fig. 5 Background subtraction to improve imaging. (**a**) Median pixel value from 2000 frames showing average background autofluorescence from the flow cell. (**b**) Maximum pixel value from 2000 video frames showing a single cell moving through the laser spot. (**c**) Result of subtracting image B from image C, allowing for improved visualization of cell movement and faint cell images

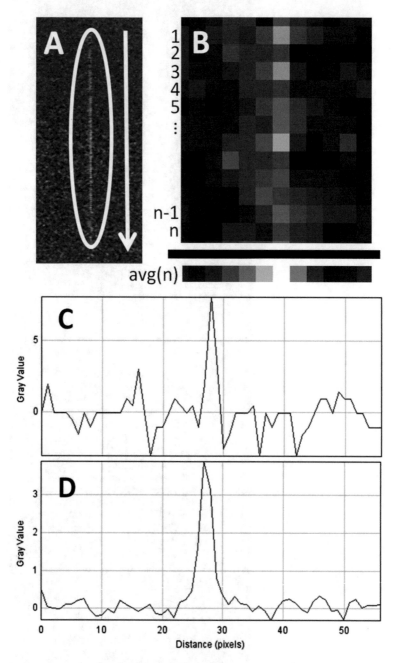

Fig. 6 Noise reduction in streak mode images. (**a**) A cell streak image (*circled*) with flow direction indicated. (**b**) Close-up of cell streak image showing individual pixels and background noise. In order to reduce noise, each column of pixels is averaged over the streak length n to produce a single averaged row of pixels, labeled avg (n). (**c**) A plot of pixel values before (i) and after (ii) averaging, showing a 3× improvement in SNR. The plot in (i) is for the row with the brightest pixel value in (**d**).

3.8 Color Channel Extraction

When using a color camera, spectral filtering can be employed to improve SNR by reducing non-specific background light (e.g., excitation light). The webcam CMOS sensor employs a typical Bayer color filter array in which one half of all pixels have a green filter, 25% have a red filter, and 25% have a blue filter. The emission profile of Syto-9 dye used in this work has a large overlap in the green spectrum resulting in this color channel having the highest SNR. The blue and red channels typically showed SNRs of one half and one tenth those of the green channel, respectively. For this reason the red and blue channels are discarded and only the green channel is analyzed. The "Split Channels" function in ImageJ is used to divide a color image into its constituent channels.

3.9 Rare Cell Counting Using Streaking Mode Imaging

Counting of rare cell events is simulated using dilutions of fluorescently labeled THP-1 monocytes. Dilutions are prepared as described in Subheading 1, and then measured using the wide-field flow cytometer in Fig. 2. Figure 7a shows results using the flow cell depicted in Fig. 3a. Dilutions of 100, 10, and 1 cell per mL respectively yielded average concentrations of 84, 7.9, and 0.56 cells/mL, with 95% confidence bounds of 61–107, 6.9–9, and 0.16–0.96 cells/mL. Error bars in Fig. 7 represent a 95% confidence interval.

Following this set of experiments, flow cell geometry and system operating parameters (flow rate, frame rate) are optimized

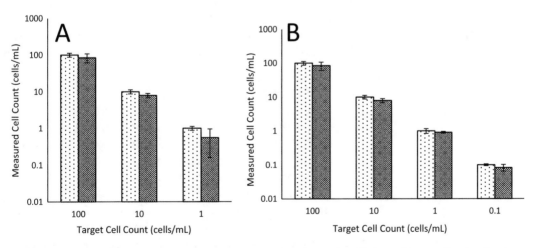

Fig. 7 Rare cell counting using streaking mode—(**a**) Results of counting dilutions of rare cells using line laser illumination. The large error bars evident at the level of 1 cell/mL prompted further optimization of the flow cell geometry and operating parameters. (**b**) Results of counting dilutions of rare cells using an updated flow cell design to improve image SNR by reducing cell velocity. The field-of-view (FOV) was also improved. Following this optimization, performance of the wide-field flow cytometer improved substantially at the level of 1 cell/mL, allowing measurement expansion to as low as 0.1 cell/mL. Measurements made via microscopy (sparsely filled columns) vs. wide-field flow cytometry (*dense fill*) were not significantly different at the 95% confidence level

based on the methods described in Subheading 3.6. This resulted in the flow cell pictured in Fig. 7b with increased channel width and an increased field-of-view. The excitation laser spot is modified to form a line covering the width of the channel. The modifications to the flow cell resulted in cells moving with lower velocity, which translates to higher image SNR. This allowed for substantially improved results at low cell dilutions as seen in Fig. 7c. Dilutions of 1 and 0.1 cell/mL are prepared and measured via microscopy (95% confidence bounds: 0.84–1.16 and 0.094–0.106, respectively). These dilutions are then measured in cell streaking mode with the optimized wide-field cytometer, yielding average counts of 0.91 and 0.083 cells/mL, each with a 95% confidence bound of 0.85–0.97 and 0.065–0.102 cells/mL, respectively. In both cases the two means are not significantly different at the 95% confidence level.

Our technology has many promising clinical applications, including detection of CTCs for POC. This may enable better disease prognosis, earlier detection of metastasis-capable malignancy, guidance in selection of treatments, and evaluation of treatment efficacy to help prevent patients from being exposed to ineffective treatments. The simplicity and low cost of the webcam-based wide-field flow cytometer suggests that this configuration may have potential for developing clinically relevant POC flow cytometry for rare cell detection in resource-poor settings.

3.10 Factors for Optimization of Streak Imaging Flow Cytometer

In order to optimize the performance of a wide-field cytometer, the following method for setting various device parameters can be used. This will result in maximized cell image SNR, maximized sample throughput, and accurate flow sampling.

1. The distance between the imaging lens and the flow cell should be set such that images of cells are projected onto at most a 3×3 array of pixels. Distance should not be so great that a cell image is less than one pixel in size. At those distances, photon flux per pixel begins to diminish and SNR drops quickly. When cells are imaged on more than one pixel, photon flux per pixel is a constant and independent of the distance between flow cell and lens. The largest possible distance should be chosen in order to maximize the field-of-view (FOV). Set the flow cell width so that it covers the FOV of the image sensor at this lens-to-flow cell distance. This allows the largest possible flow cell width to be used, thereby minimizing flow velocity and maximizing SNR at a given flow rate.

2. The relationship between average cell velocity and flow rate should be determined empirically. The maximum velocity achievable will be determined later and is a function of cell image signal-to-noise ratio (SNR) (e.g., higher SNR allows higher velocity). This in turn depends on the cell staining

characteristics (e.g., quantum yield of the dye, number of fluorophores bound, and excitation field intensity), the camera used (e.g., camera sensitivity, characteristic noise levels, and maximum frame rate), and the lens used (e.g., focal ratio).

3. Exposure time should be set such that cell image brightness is maximized. For a given average flow velocity and average cell brightness (i.e., photon emission rate) there will be a maximum cell image brightness that can be produced. This is based on the number of photons than can strike a pixel as the cell image passes over it.

4. Volumetric flow rate should be set such that the entire distribution of cells is significantly above the noise background of the system (this typically corresponds to an SNR > 5). At higher flow rates image SNR will decrease. Above a certain flow rate, cells at the faint end of the distribution will be indistinguishable from background noise and the average cell count will decrease. In order to set flow rate, a stock solution of fluorescent cells should be prepared and counted beginning at a high flow rate and decreasing until the average cell count per sample reaches a constant value. As previously mentioned, this maximum flow rate will depend on several factors including the inherent brightness of cell staining and the excitation photon flux. The parameters of exposure time and frame rate depend directly on flow rate, so these parameters will need to be updated as flow rate is varied in order to maintain similar levels of device sensitivity across various tests.

4 Notes

1. Make sure the lens, filters, and flow cell are vertically centered and aligned and that the laser illuminates the image area of the flow cell.

   Cell cultures with very high density may result in cell anucleation (ghost cells), cells with low fluorescence signal when labeled with nucleic acid stain.

2. Make sure the arrows on the coated filters are facing the camera, and that if the filters being used have angular dependence (e.g., interference filters) the excitation source beam is confined to the correct range of angles.

3. A fast focal ratio (e.g., f/1.2) will enable shorter exposure time, but focusing will be more difficult as the depth of field will be reduced. Extension tubes are added between the lens and sensor to allow for close focusing and enlargement of cell images. This can result in noticeable image field curvature (i.e., image corners out of focus while image center is in focus), so spacing and imaging distance must be optimized. Focal ratio can be reduced to reduce field curvature.

4. The imager enclosure must be sealed to block external light sources. A long exposure can be used to detect light leaks.

5. For flow cell fabrication, the laser power and speed for cutting polymers has to be determined empirically. The minimum laser power is recommended in order to reduce overheating or excessive burning of the material.

6. For strong bonding, all air bubbles between the surfaces and the adhesive tape should be pressed out manually. The adhesive layer should be attached to the acrylic side first to ensure that proper alignment exists between the inlet and outlet ports. Bubbles between these layers can be pressed out before the protective paper layer on the back of the adhesive tape is removed. After this, with the protective paper removed, the glass layer can be bonded. Bubbles between the glass and adhesive layer can be removed by carefully applied pressure, or by use of a heated press. Care must be taken to ensure even pressure because of the risk of glass breakage. To avoid this, the glass layer can be replaced by another layer of acrylic. However, the high autofluorescent emission rate of acrylic will result in reduced SNR and diminished dynamic range of the sensor. For improved sensitivity, two quartz microscope slides can be used in place of the acrylic layer and the glass slide. Inlet and outlet holes must be cut using a glass cutting drill bit or other appropriate method. High pressure ports can be constructed by pressing an 18G needle into a laser cut hole in a small square of acrylic sheet, sealing with cyanoacrylate glue, and bonding over the inlet and outlet holes in the quartz slide using double sided adhesive transfer tape with a center hole cut out.

7. For image quantification, the webcam performance will vary depending both on the device and the application used to collect the images from the camera. Prior to preparing a sample for measurement it is important to understand the performance of the camera and application being used. First, block any light from reaching the camera sensor by taping a layer of aluminum foil over the lens aperture. Take two images and open them in ImageJ or a comparable software package. If a color sensor is used, use only the green channel for analysis. Convert the images to 32-bit float format, and subtract one image from the other. Finally, use the software package to produce a histogram of all pixel values. This histogram should have an approximately Gaussian profile. If the original images have pixel values that are predominantly zero valued, you must change the settings of the camera capture software so that the black end of the sensor histogram is not being artificially clipped. This clipping behavior is typically set by an adjustment labeled brightness. If the images captured by the sensor with

no light arriving cannot be made to show values above zero with an approximately Gaussian profile, it is likely that the device is not suitable for performing sensitive fluorescence measurements of faint objects.

References

1. Balsam J, Bruck HA, Rasooly A (2015) Cell streak imaging cytometry for rare cell detection. Biosens Bioelectron 64:154–160

2. Balsam J, Bruck HA, Rasooly A (2015) Mobile flow cytometer for mHealth. Methods Mol Biol 1256:139–153

3. Balsam J, Bruck HA, Rasooly A (2014) Webcam-based flow cytometer using wide-field imaging for low cell number detection at high throughput. Analyst 139(17):4322–4329

4. Sun S, Ossandon M, Kostov Y, Rasooly A (2009) Lab-on-a-chip for botulinum neurotoxin a (BoNT-A) activity analysis. Lab Chip 9 (22):3275–3281

5. Sun S, Yang M, Kostov Y, Rasooly A (2010) ELISA-LOC: lab-on-a-chip for enzyme-linked immunodetection. Lab Chip 10(16):2093–2100

6. Wei Q, Qi H, Luo W, Tseng D, Ki SJ, Wan Z et al (2013) Fluorescent imaging of single nanoparticles and viruses on a smart phone. ACS Nano 7(10):9147–9155

7. Coskun AF, Nagi R, Sadeghi K, Phillips S, Ozcan A (2013) Albumin testing in urine using a smart-phone. Lab Chip 13 (21):4231–4238

8. Navruz I, Coskun AF, Wong J, Mohammad S, Tseng D, Nagi R et al (2013) Smart-phone based computational microscopy using multi-frame contact imaging on a fiber-optic array. Lab Chip 13(20):4015–4023

9. Zhu H, Ozcan A (2013) Wide-field fluorescent microscopy and fluorescent imaging flow cytometry on a cell-phone. JoVE 74:e50451.

10. Zhu H, Sencan I, Wong J, Dimitrov S, Tseng D, Nagashima K et al (2013) Cost-effective and rapid blood analysis on a cell-phone. Lab Chip 13(7):1282–1288

11. Zhu H, Sikora U, Ozcan A (2012) Quantum dot enabled detection of Escherichia coli using a cell-phone. Analyst 137(11):2541–2544

12. Zhu H, Yaglidere O, Su TW, Tseng D, Ozcan A (2011) Wide-field fluorescent microscopy on a cell-phone. Conf Proc IEEE Eng Med Biol Soc 2011:6801–6804

13. Zhu H, Mavandadi S, Coskun AF, Yaglidere O, Ozcan A (2011) Optofluidic fluorescent imaging cytometry on a cell phone. Anal Chem 83 (17):6641–6647

14. Zhu H, Yaglidere O, Su TW, Tseng D, Ozcan A (2011) Cost-effective and compact wide-field fluorescent imaging on a cell-phone. Lab Chip 11(2):315–322

15. Golden JP, Kim JS, Erickson JS, Hilliard LR, Howell PB, Anderson GP et al (2009) Multi-wavelength microflow cytometer using groove-generated sheath flow. Lab Chip 9 (13):1942–1950

16. Howell PB Jr, Golden JP, Hilliard LR, Erickson JS, Mott DR, Ligler FS (2008) Two simple and rugged designs for creating microfluidic sheath flow. Lab Chip 8(7):1097–1103

17. Taitt CR, Anderson GP, Ligler FS (2005) Evanescent wave fluorescence biosensors. Biosens Bioelectron 20(12):2470–2487

18. Ngundi MM, Qadri SA, Wallace EV, Moore MH, Lassman ME, Shriver-Lake LC et al (2006) Detection of deoxynivalenol in foods and indoor air using an array biosensor. Environ Sci Technol 40(7):2352–2356

19. Moreno-Bondi MC, Taitt CR, Shriver-Lake LC, Ligler FS (2006) Multiplexed measurement of serum antibodies using an array biosensor. Biosens Bioelectron 21(10):1880–1886

20. Ligler FS, Sapsford KE, Golden JP, Shriver-Lake LC, Taitt CR, Dyer MA et al (2007) The array biosensor: portable, automated systems. Anal Sci 23(1):5–10

21. Kostov Y, Sergeev N, Wilson S, Herold KE, Rasooly A (2009) A simple portable electroluminescence illumination-based CCD detector. Methods Mol Biol 503:259–272

22. Sapsford KE, Sun S, Francis J, Sharma S, Kostov Y, Rasooly A (2008) A fluorescence detection platform using spatial electroluminescent excitation for measuring botulinum neurotoxin A activity. Biosens Bioelectron 24 (4):618–625

23. Sun S, Francis J, Sapsford KE, Kostov Y, Rasooly A (2010) Multi-wavelength Spatial LED illumination based detector for in vitro detection of Botulinum Neurotoxin A Activity. Sens Actuators B 146(1-8):297–306

24. Rasooly A, Bruck HA, Kostov Y (2013) An ELISA Lab-on-a-Chip (ELISA-LOC). Methods Mol Biol 949:451–471

25. Rasooly A, Kostov Y, Bruck HA (2013) Charged-coupled device (CCD) detectors for Lab-on-a Chip (LOC) optical analysis. Methods Mol Biol 949:365–385

26. Balsam J, Bruck HA, Rasooly A (2013) Capillary array waveguide amplified fluorescence detector for mHealth. Sens Actuators B 186:711–717

27. Balsam J, Rasooly R, Bruck HA, Rasooly A (2014) Thousand-fold fluorescent signal amplification for mHealth diagnostics. Biosens Bioelectron 51:1–7

28. Balsam J, Ossandon M, Bruck HA, Lubensky I, Rasooly A (2013) Low-cost technologies for medical diagnostics in low-resource settings. Expert Opin Med Diagn 7(3):243–255

Chapter 18

Rapid Detection of Microbial Contamination Using a Microfluidic Device

Mustafa Al-Adhami, Dagmawi Tilahun, Govind Rao, Chandrasekhar Gurramkonda, and Yordan Kostov

Abstract

A portable kinetics fluorometer is developed to detect viable cells which may be contaminating various samples. The portable device acts as a single-excitation, single-emission photometer that continuously measures fluorescence intensity of an indicator dye and plots it. The slope of the plot depends on the number of colony forming units per milliliter. The device uses resazurin as the indicator dye. Viable cells reduce resazurin to resorufin, which is more fluorescent. Photodiode is used to detect fluorescence change. The photodiode generated current proportional to the intensity of the light that reached it, and an op-amp is used in a transimpedance differential configuration to ensure amplification of the photodiode's signal. A microfluidic chip is designed specifically for the device. It acts as a fully enclosed cuvette, which enhances the resazurin reduction rate. In tests, the *E. coli*-containing media are injected into the microfluidic chip and the device is able to detect the presence of *E. coli* in LB media based on the fluorescence change that occurred in the indicator dye. The device provides fast, accurate, and inexpensive means to optical detection of the presence of viable cells and could be used in the field in place of more complex methods, i.e., loop-mediated isothermal amplification of DNA (LAMP) to detect bacteria in pharmaceutical samples (Jimenez et al., J Microbiol Methods 41(3):259–265, 2000) or measuring the intrinsic fluorescence of the bacterial or yeast chromophores (Estes et al., Biosens Bioelectron 18(5):511–519, 2003).

Key words Contamination detection device, Resazurin, Resorufin, *E. coli* detection, Microfluidic device, Thermal bonding of PMMA

1 Introduction

Each year thousands of people die and millions are infected due to food, water, or medicine contamination [1]. In order to prevent most of these poisonings a rapid, precise, and sensitive microbial detection method is highly desirable [2, 3]. There are many common methods to detect pathogens and other biologics [4]. Most of these methods require well-equipped and environmentally stable laboratories as well as highly trained staff to handle devices and reagents or antibodies [2]. These methods are hard to apply in the

Avraham Rasooly and Ben Prickril (eds.), *Biosensors and Biodetection: Methods and Protocols Volume 1: Optical-Based Detectors*, Methods in Molecular Biology, vol. 1571, DOI 10.1007/978-1-4939-6848-0_18,
© Springer Science+Business Media LLC 2017

field and especially in the biotechnology industry where the final product has to always be sterile to prevent infection of the patients [1, 5]. The recent complex biologics are hard and sometimes impossible to sterilize with the traditional methods of sterilization [6]. Therefore, it has been of increasing importance to develop a mechanically robust and easily handled device that can detect contamination rapidly. Catching the contamination at early stage can save both time and labor for the manufacturer and, more importantly, increases the safety of the final product. For example, having near-real-time feedback in bioreactors can be very critical since each batch takes many days to grow before harvest; it would be of great interest to abort the process early in case of contamination [7].

Another important application of rapid microbial detection is in the quality control processes for biopharmaceuticals. Since biologics are highly susceptible for contamination by adventitious agents such as mycoplasma, there is a need for risk mitigation procedure like testing to confirm the absence of any unwanted contaminants [8]. In this way, contaminated products can be caught early, which minimizes the risk of having them produced and then sold in pharmacies. Furthermore, if the device is low-cost and low-effort, it could be used during the biological process to test manufactured drugs for contamination [9].

Methods of contamination detection considered to be quick usually refer to a time frame of one day. For example [9], Jimenez et al. discuss the use of PCR analysis for detecting low levels of bacteria and mold contamination in pharmaceutical samples; in this case, the protocol states that the method can detect presence of viable cells only in their exponential phase of growth, which might take up to 24 h. The method mentioned using PCR analysis for contamination detection above is very labor-intensive and takes 27 h to achieve results. It can detect contamination of colony forming units as low as 10 CFU. It is not clear though whether the detected cells are dead or alive at the time of detection. There are other reported methods that are faster, for example measuring the intrinsic fluorescence of the bacterial or yeast chromophores [12]. This approach is fast and sensitive but it is limited to a slightly higher number of colony forming units and it is very specific which requires a high level of knowledge about the environment in which the contamination is detected as to compensate for background signal [13].

Cell viability assays are widely used in drug discovery applications to determine the ability of organs, cells, or tissues to live on their own and develop. These assays are very useful for the rapid detection and quantification of microorganisms [3, 10–13]. For example, there are many assays like tetrazolium reduction, and protease activity assays that measure different aspects of the viable cell activity. They require an incubated population of viable cells to convert a substrate to a colored product. On the other hand, ATP

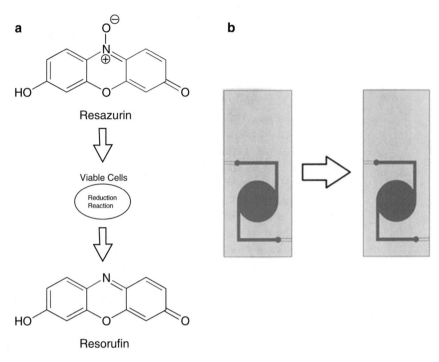

Fig. 1 (**a**) The reduction of resazurin to resorufin chemical formula. (**b**) The color transformation when resazurin is converted to resorufin

assays use a different approach in which adding a reagent ruptures the cells and therefore it does not need any incubation time. More recently, resazurin has been found to be an inexpensive alternative that can provide a fluorescent readout at good sensitivity (*see* **Note 1**) [14, 15].

Resazurin is a blue dye which itself is weakly fluorescent [14]. However, viable cells retain the ability to reduce resazurin into resorufin, which is highly fluorescent [15]. Nonviable cells rapidly lose metabolic capacity which is why they do not reduce this indicator dye. Resazurin-based assays are used for viability detection of cells other than bacterial cells, e.g., human cells for clinical transplantation [16], stem cells [17], CD4 T cells [18], and malarial gametocytocidal assay [19]. Also, resazurin-based assay can be used for the screening of bacteria for the radiation sensitivity (*see* **Note 2**) [20].

As shown in Fig. 1, the conversion of resazurin to resorufin changes the color of the dye from blue to red, which is accompanied with significant increase of the absorption green–yellow–green region of visible light ($\lambda_{max} = 570$ nm). The fluorescence emission in the orange–red region ($\lambda_{max} = 590$ nm) is enhanced as well. These wavelengths are quite far from the absorption and emission of the fluorophores in most media, which allows the assay to be used directly in the cell culture media under test. The conversion process gives a linear curve over a wide range of cell concentration [21]. The resazurin assay is as sensitive as thymidine assay [22] for detecting cell proliferation [23].

In this study, we present a highly sensitive, accurate, fast, and a USB-powered portable device that can detect the presence of viable cells in a given sample. We demonstrate that the device can detect colony forming units as low as 10 CFU/mL in a given *E. coli* sample in 30 min or less (*see* **Note 3**). The device results are then validated using a standard plating method. This low-cost, tablet powered device can be used for the field measurements of viable-cell-induced fluorescent assay.

The main purpose of the device is to detect viable cells in the tested samples. The device does not identify the species or the growth phase of the cells that are detected. The main application of the device is to do initial screening on sample which would be followed by more traditional plate-based tests where the culture can be grown and identified. However, having the ability to rapidly screen a variety or samples irrespectively of the contaminating species is of great value in the pharmaceutical and food industries. The device provides means for reliable detection of contamination in less than 30 min. Each reading is compared with a standard plate count method. The device has proven to be accurate and fast, which makes it suitable for rapid detection of contamination applications (*see* **Note 4**). When used quantitatively, it is necessary to develop of custom calibration curve such as that in Fig. 9. It will depend on the type of species to be detected.

2 Materials

1. Green Light Emitting Diode (NSPG-500, Nichia).

2. Resistors R_A—1 kΩ, RA—1 kΩ, R_C—10 Ω, capacitor C—330 pF.

3. Operational amplifier OPA 354 (Texas Instruments) for the LED driver.

4. Excitation filter 532.0-35-75 (Intor, Soccorro, NM).

5. Emission filter 590.0-40-75 (Intor, Soccorro, NM).

6. Photodiode S12232-01 (Hamamatsu).

7. Two operational amplifiers OPA2301 (Texas Instruments) for the photodetector amplifier.

8. Resistors R_M—100 kΩ, R_T—3 MΩ, C_L—220 nF, R_K—619 Ω, R_P—12 kΩ, C_H-100 pF.

9. Analog-to-digital converter (16 bit ADC, ADS8318, Texas Instruments).

10. Digital-to-analog converter (12 bit DAC, DAC6571, Texas Instruments).

11. Microcontroller (MSP430F2272 Texas Instruments).

12. *E. coli* NM303.

13. Luria–Bertani (LB)-agar plate to verify the viability of the Escherichia coli NM303. Luria–Bertani broth is purchased in the form of powder from MP Biomedicals (Santa Ana, CA). The agar is purchased from Fisher Scientific.

14. Resazurin sodium salt powder is from Acros Organics (Fair Lawn, NJ).

15. Plates are purchased from BD Biosciences (San Jose, CA).

16. Other reagents like NaCl, Na_2HPO_4, and NaH_2PO_4.

17. PBSr from Gibco.

18. Syringe Filters.

19. PMMA sheets purchased from Astra products (Thickness = 0.2 and 1.5 mm).

20. Ethyl alcohol (90% concentration).

3 Methods

3.1 Portable Kinetics Fluorometer

1. The kinetics fluorometer is a single-excitation, single emission photometer that can detect fluorescence and then plots it. The block schematics of the device is presented in Fig. 2. The green denotes the excitation light path (light source, filter, aperture, direction of the excitation light), while the red denotes the emission light path). Note that excitation and emission light paths are oriented at 90° at each other to each other [26].

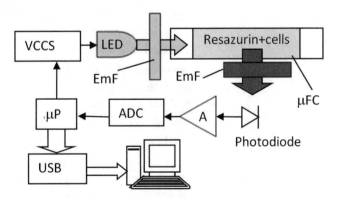

Fig. 2 Block schematics of the fluorometer. *VCCS* voltage controlled current source, *LED* light emitting diode, *uFc* microfluidic cassette, *ExF, EmF* Excitation and emission filters, *PD* photodiode, *A* amplifier, *ADC* analog to digital converter, *uP* microprocessor, *USB* universal serial bus

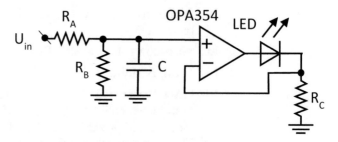

Fig. 3 Schematics of the voltage-controlled current source for LEDs. The extraction module consists of an LED with emission maximum 525 nm and voltage controlled current source which sets a constants current through the LED

2. The excitation module consists of an LED with emission maximum 525 nm and voltage controlled current source which sets a constant current through the LED. The schematics of the module is shown in Fig. 3.

3. The current through the LED is controlled via the voltage applied at U_{in}. With the suggested values, an input voltage in the range between 0 and 1 V will result in current through the LED between 0 and 50 mA. The capacitor decreases the spurious transients on the LED which result in unwanted electromagnetic high-frequency noise. The driver is good for modulating the current through the LED for frequencies up to 3 MHz.

4. An excitation filter must be mounted in front of the LED in order to filter all the light that enters the cassette. Ideally, the filter and the LED should be mounted in a holder that keeps them mechanically aligned. The holder must be made of non-fluorescent materials

5. Photodiode is employed to detect the fluorescence intensity by generating current proportional to the intensity of the light reaching them. The photodiode should be mounted at an angle of 90 degrees to the excitation beam, with the emission filter mounted in front of it.

6. The overall positioning of the LED, photodiode, the filters and the cassette is presented in Fig. 4.

7. The first stage of the amplifier converts the current in voltage with 10^7 transimpedance gain and second stage provides additional amplification of 20 (Fig. 5).

8. The second stage is separated from the first using a blocking capacitor, which prevents the dc offsets from the output of the first stage being amplified (Fig. 5). The high-pass filter formed between C_L and R_K also significantly attenuates the ever-present 60 Hz. The maximum amplification is in the band between 5 and 20 kHz.

Fig. 4 Spatial positioning of the optics and optoelectronics. *ExF* Excitation Filter, *EmF* Emission Filter, *PD* Photo diode, *LED* Light emitting diode

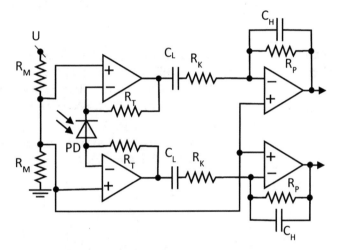

Fig. 5 Schematics of the differential photoreceiver Photodiodes do not exhibit internal amplification as do photomultipliers or avalanche photodiodes, therefore the photocurrent is converted to voltage using an op amp in transimpedence differential configuration and additional amplification is required

9. The output of the second stage is digitized using a fast analog-to-digital converter(ADS8318, TI) with a maximum through-put of 500 kilosamples per second. This is a 16-bit ADC with SPI output. It is connected to the SPI port of microcontroller (MSP430F2272, TI).

10. The ADC conversion is initiated by the microcontroller by toggling the CONVST (conversion start) pin of the microcontroller. The microcontroller enters conversion phase, which lasts for 1.4 μs. After the conversion end, ADC enters acquisition phase, during which the result is read by the microcontroller via the SPI protocol. It also performs the LED control and times the acquisition of the intensities [22]. 997 readings are initiated and collected and averaged by the microcontroller

as a single data point. The averaged value is transferred to the computer via serial-to-USB bridge (FT232R, FTDI chip).

11. The computer control program is written in Labview. It allows to set the number of the averaged readings (997 in this example) and controls the brightness of light source (LED) via DAC (DAC7571, TI). The computer initiates the measurements at a prescribed interval (i.e., once a second) and stores the received data points as an array. A second array with the time of measurements is also stored. Every time a new data point is obtained, it is added to the data array and the time at which it has been acquired is added to the time array. These values are approximated with a straight line. It is fitted to the data using the least squares method. The slope of this line is the resazurin reduction rate.

3.2 Microfluidic Cassettes

1. A PMMA sheet of 1.5 mm in thickness is first cut using any available equipment (CNC mill, laser cutter). In order to remove the internal stresses from the plastic sheet, it is annealed using a PID controlled oven [25].

2. After the sheet is annealed, its surface is flattened with sand paper [26], treated with 90% ethanol and immediately covered with a 0.2 mm thick PMMA sheet acting as the cover sheet [27].

3. The ethanol treated sheets are then clamped with a flat surface clamp and placed in the oven at 55 °C for 5 min (Fig. 6). After the oven, the assembly is taken out to cool down and the clamps are removed [28].

4. After the microfluidic chip is fabricated, holes are drilled on the side of the device. These holes are then filled with silicone and left to cure to act like a septum for the microfluidic chip. Figure 1b shows one of the chip designs (*see* **Note 4**) [11].

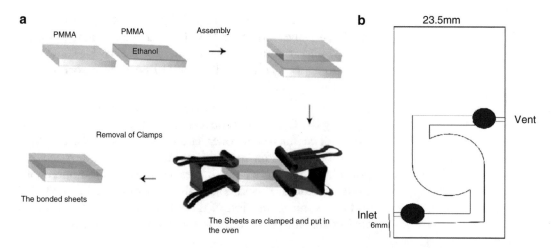

Fig. 6 (**a**) Step sequence to bond two PMMA sheets. (**b**) Schematics of the microfluidics cassette

3.3 Preparation and Plating of E. coli Cells

1. Initially, 100 mL primary culture is prepared using 10 μL of *E. coli* NM303 cells, which is grown at 37 °C in a shaker at 150 rpm (Labline Instruments, Melrose park, IL) for 8 h until the optical density of primary culture is measured to be 2.5 at 600 nm (model 8453, Agilent Technologies, Santa Clara, CA).

2. The primary seed culture (1%) is used to inoculate into 200 mL secondary culture are grown at 37 °C in a shaker at 150 rpm to reach an optical density of 0.4 at 600 nm [3].

3. This sample is used for making serial dilutions (in 50 mL tubes) from 1:10 to $1:10^5$. The $1:10^5$ diluted sample is used to make further dilutions to make a final concentration of viable cells which is calibrated against a standard plate count. An appropriate volume is calculated (from the $1:10^5$ aliquot) for the enumeration of 1000 CFU/mL to 10 CFU/mL and plated on LB-agar plate.

4. It is tested for viable cell number using both standard plate count method and the resazurin reduction test. The block-schematics of the protocol are presented in Fig. 7.

5. One milliliter aliquots (in triplicates) of sample containing cells from the serially diluted tubes are plated on LB-agar plate. The plates are incubated at 37 °C for 24 h. CFU are counted from the each plate.

3.4 Preparation of Resazurin Dye

1. 2.3 mg of resazurin is mixed in a 1 mL of filtered water to create a master mix. 100 μL of master mix is added to a 1.5 mL centrifuge tube and diluted with 900 μL of filtered PBSr.

2. 980 μL PBS is then mixed with 20 μL of resazurin stock solution to create reaction mix. The master mix is then returned to the freezer to be used later.

3. 1 mL of the cell culture with the respective number of CFU is deposited in an Eppendorf tube. Then, 3.3 μL of the freshly made reaction mix is added to it and. The final mix is vortexed and 300 μL are injected into the microfluidic cassette (*see* **Note 6**). Two needles are used on the both sides of the cassette. The first needle is used to inject the mix, while the second needle serves as a vent for the air. After withdrawing both needles, the cassette is inserted into the device cassette holder. The leftover reaction mix is stored in a refrigerator to be used within 12 h, otherwise it is discarded.

3.5 Measurement of the Resazurin Reduction

1. Once the cassette is in the device, the control program is started. It continuously measures and displays the fluorescence intensity. The control program also calculates the running value of the slope of the fluorescence intensity change. If the slope is close to zero (Fig. 8, squares), there is no

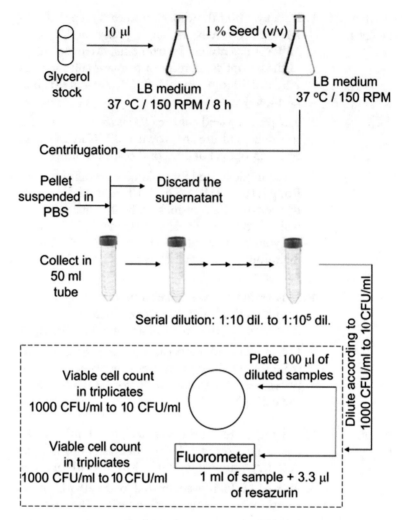

Fig. 7 Experimental protocol for low CFU detection. The *dashed box* represents the actual measurement (reprinted from the PDA Journal of Pharmaceutical Science and technology)

contamination. If the slope is significant (Fig. 8, triangles), the media is contaminated.

2. To build a calibration curve, several dilutions of bacteria are plated on LB-agar plates and grown for 24 h at 37 °C. The plates are analyzed for colony-forming units (CFU) with respect to each dilution.

3. In parallel the bacterial dilutions are analyzed for resazurin reduction over a time period of 3 min.

4 Notes

1. An even lower limit of detection can be achieved if the sample is pre-concentrated. This can be achieved by aspirating the

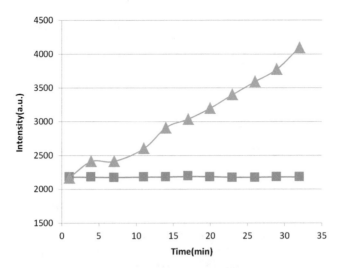

Fig. 8 Fluorescene intensity changes in contaminated vs. non-contaminated media. *Triangles*: Fluorescene intensity increase in media contaminated with 10,000 CFU. *Squares*: fluroscene intensity stays practically constant in sterile LB media. Both measurements performed at room temperature (25 °C). (a.u. refers to arbitrary unit)

sample into the syringe through a 0.2 μm filter. As a result, the cells will be trapped on the filter. For example, a 100 mL sample can be filtered fairly quickly. Then, a small percentage (i.e., 10% or even 1%) of the filtered media is used to wash the cells from the filter and into the Eppendorf tube. If now the same protocol is followed, ideally 10 or 100 times lesser CFU could be detected. However, the approach requires additional validation due to the possibility of for some of the cells to get stuck on the filter or to be lysed during filtration.

2. The indicator (resazurin) is a blue dye which itself is weakly fluorescent. Viable cells reduce resazurin into resorufin, which is highly fluorescent. Nonviable cells do not have metabolic capacity and do not reduce indicator dye. By examining the increase of fluorescence intensity with time, we are able to correspond the slope with the number of viable cells. It is worth mentioning that the magnitude of fluorescence change is small. Therefore, the device is working at relatively high amplifications of the fluorescence signal (*see* **Note 6**). Strong ambient light may interfere with device operation, even though it is shielded (the cassette holder is made of black plastic). Care should be taken to avoid placing it close to room lights.

3. To confirm the normal operation of the device, different CFU/mL values are monitored and the slopes are determined as in Fig. 8. The reduction rate slope of resazurin is correlated to the number of bacterial cells in a given sample. The bacterial cells (*E. coli*) concentrations ranged from 10 CFU/mL to 10,000 CFU/mL. All sample readings are analyzed in

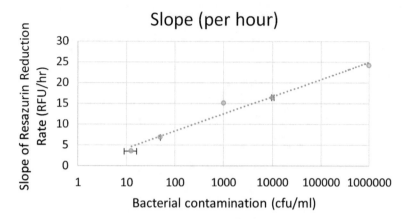

Fig. 9 Correlation between the CFU number and the slope of fluorescence increase. Samples below 10,000 CFU/mL were verified using agar plates while the other numbers were estimated

triplicates (some of the standard deviations as shown in Fig. 9 are too small to be visible). As seen in the figure, there is a good correlation between the resazurin reduction over the specified range and the log number of viable cells.

4. The system can be used in the field for contamination detection like the contamination of media. The portable fluorometer is powered from a USB port of a computer and has been shown to work with laptop or tablet featuring full USB port. It is designed to house a specifically designed microfluidic cassette (Fig. 1).

5. The cassettes are fully enclosed and have long input channels to provide mixing. The design helps to decrease possible ingress of air in the solution, which may interfere with the measurement. The actual input and output shape of the chamber are designed to feature gradual increase and decrease of the channel width. This prevented the formation of bubbles. The light enters the cassette sideways from the edge, and the wide-area photodiode picks up the fluorescence from the sample.

6. A control and visualization program written in Labview provides the timing for the fluorescence acquisition. It also can calculate the slope value with every incoming reading and saves the data for further analysis on the computer's internal drive.

7. The protocol for field use of the device is developed with a goal of simplicity. Only a syringe and an Eppendorf tube in addition to the cassette and the portable fluorometer are used. The use of a vortex mixer is optional, as the sample and the indicator dye could be mixed by hand.

References

1. Pharmtech.com (2015) An overview of rapid microbial-detection methods|Pharmaceutical technology. N.p. 2015. Web. 3 June 2015

2. Hoehl MM et al (2012) Rapid and robust detection methods for poison and microbial contamination. J Agric Food Chem 60 (25):6349–6358

3. Hobson NS, Tothill I, Turner AP (1996) Microbial detection. Biosens Bioelectron 11 (5):455–477

4. Fda.gov (2015) Archived BAM method: rapid methods for detecting foodborne pathogens. N.p. Web. 23 July 2015

5. Vogel SJ, Tank M, Goodyear N (2013) Variation in detection limits between bacterial growth phases and precision of an ATP bioluminescence system. Lett Appl Microbiol 58 (4):370–375 Web

6. Folsome CE (1964) Functional transformation in mammalian cell culture systems. Nature 202 (4936):1023–1024 Web

7. Celsis.com (2010) Quality control—microbial testing: rapid microbiological methods in lean manufacturing. N.p. Web. 23 July 2015

8. Pettit AC, Kropski JA, Castilho JL, Schmitz JE, Rauch CA, Mobley BC, Wang XJ, Spires SS, Pugh ME (2012) The index case for the fungal meningitis outbreak in the United States. N Engl J Med 367(22):2119–2125

9. Gurramkonda C et al (2014) Fluorescence-based method and a device for rapid detection of microbial contamination. PDA J Pharm Sci Technol 68(2):164–171 Web

10. Jimenez L, Smalls S, Ignar R (2000) Use of PCR analysis for detecting low levels of bacteria and mold contamination in pharmaceutical samples. J Microbiol Methods 41(3):259–265

11. Ncbi.nlm.nih.gov 2015 Polymerase chain reaction (PCR). N.p. Web. 23 July 2015

12. Estes C, Duncan A, Wade B, Lloyd C, Ellis W Jr, Powers L (2003) Reagentless detection of microorganisms by intrinsic fluorescence. Biosens Bioelectron 18(5):511–519

13. Brown NA (1990) Cell based assays. Developmental toxicity assays in vitro. Anal Proc 27 (9):246 Web.

14. Bionity.com (2015) Alamarblue® assay for assessment of cell proliferation using the Fluostar OPTIMA. N.p. Web. 23 July 2015

15. Boyce ST, Anderson BA, Rodriguez-Rilo HL (2006) Quantitative assay for quality assurance of human cells for clinical transplantation. Cell Transplant 15(2):169–174

16. Boyce ST, Anderson BA, Rodriguez-Rilo HL (2006) Quantitative assay for quality assurance of human cells for clinical transplantation. Cell Transplant 15(2):169–174

17. Nagaoka M, Hagiwara Y, Takemura K, Murakami Y, Li J, Duncan SA, Akaike T (2008) Design of the artificial acellular feeder layer for the efficient propagation of mouse embryonic stem cells. J Biol Chem 283 (39):26468–26476

18. Longhi MP, Wright K, Lauder SN, Nowell MA, Jones GW, Godkin AJ, Jones SA, Gallimore AM (2008) Interleukin-6 is crucial for recall of influenza-specific memory CD4 T cells. PLoS Pathog 4(2):e1000006

19. Tanaka TQ, Williamson KC (2011) A malaria gametocytocidal assay using oxidoreduction indicator, alamarBlue. Mol Biochem Parasitol 177(2):160–163

20. Hudman DA, Sargentini NJ (2013) Resazurin-based assay for screening bacteria for radiation sensitivity. Springerplus 2(1):55

21. Fields RD, Lancaster MV (1993) Dual-attribute continuous monitoring of cell proliferation/cytotoxicity. Am Biotechnol Lab 11 (4):48–50

22. Ahmed SA, Gogal RM Jr, Walsh JE (1994) A new rapid and simple non-radioactive assay to monitor and determine the proliferation of lymphocytes: an alternative to [3H]thymidine incorporation assay. J Immunol Methods 170 (2):211–224

23. Al-Nasiry S, Geusens N, Hanssens M, Luyten C, Pijnenborg R (2007) The use of Alamar Blue assay for quantitative analysis of viability, migration and invasion of choriocarcinoma cells. Hum Reprod 22(5):1304–1309

24. Kostov Y et al (2014) Portable system for the detection of micromolar concentrations of glucose. Measurement Science and Technology 25 (2):025701 Web

25. Henderson RM, Selock N, Rao G 2012 Robust and easy microfluidic connections in acrylic. Chips and Tips. http://blogs.rsc.org/chipsandtips/2012/04/23/robust-and-easy-macrofluidic-connections-in-acrylic. Accessed 16 Sept 2015

26. Zhu X, Liu G, Guo Y, Tian Y (2007) Study of PMMA thermal bonding. Microsystem Technology 13:403–407

27. Tran H, Wu W, Lee N (2013) Ethanol and UV-assisted instantaneous bonding of PMMA assemblies and tuning in bonding reversibility. Sens Actuators B 181:955–962

28. Ng S, Tjeung R, Wang Z, Lu A, Rodriguez I, Rooij N (2007) Thermally activated solvent bonding of polymers. Microsystem Technology 14:753–759

Chapter 19

Resonance Energy Transfer-Based Nucleic Acid Hybridization Assays on Paper-Based Platforms Using Emissive Nanoparticles as Donors

Samer Doughan*, M. Omair Noor*, Yi Han, and Ulrich J. Krull

Abstract

Quantum dots (QDs) and upconverting nanoparticles (UCNPs) are luminescent nanoparticles (NPs) commonly used in bioassays and biosensors as resonance energy transfer (RET) donors. The narrow and tunable emissions of both QDs and UCNPs make them versatile RET donors that can be paired with a wide range of acceptors. Ratiometric signal processing that compares donor and acceptor emission in RET-based transduction offers improved precision, as it accounts for fluctuations in the absolute photoluminescence (PL) intensities of the donor and acceptor that can result from experimental and instrumental variations. Immobilizing NPs on a solid support avoids problems such as those that can arise with their aggregation in solution, and allows for facile layer-by-layer assembly of the interfacial chemistry. Paper is an attractive solid support for the development of point-of-care diagnostic assays given its ubiquity, low-cost, and intrinsic fluid transport by capillary action. Integration of nanomaterials with paper-based analytical devices (PADs) provides avenues to augment the analytical performance of PADs, given the unique optoelectronic properties of nanomaterials. Herein, we describe methodology for the development of PADs using QDs and UCNPs as RET donors for optical transduction of nucleic acid hybridization. Immobilization of green-emitting QDs (gQDs) on imidazole functionalized cellulose paper is described for use as RET donors with Cy3 molecular dye as acceptors for the detection of SMN1 gene fragment. We also describe the covalent immobilization of blue-emitting UCNPs on aldehyde modified cellulose paper for use as RET donors with orange-emitting QDs (oQDs) as acceptors for the detection of HPRT1 gene fragment. The data described herein is acquired using an epifluorescence microscope, and can also be collected using technology such as a typical electronic camera.

Key words Quantum dots, Upconverting nanoparticles, Resonance energy transfer, Paper-based bioassays, Nucleic acid hybridization

*These authors contributed equally to this work.

Avraham Rasooly and Ben Prickril (eds.), *Biosensors and Biodetection: Methods and Protocols Volume 1: Optical-Based Detectors*, Methods in Molecular Biology, vol. 1571, DOI 10.1007/978-1-4939-6848-0_19,
© Springer Science+Business Media LLC 2017

1   Introduction

1.1   Quantum Dots in Resonance Energy Transfer

Quantum dots (QDs) are colloidal semiconductor nanocrystals with diameter in the range of 2–10 nm. [1] The two commonly reported structural types of QDs are core QDs (e.g., CdSe, CdTe, and CdS) and core/shell QDs (e.g., CdSe/ZnS, CdSe/CdS, and $CdSe_xS_{1-x}/ZnS$) [1]. The core/shell structural type is prevalent in bioassay development as the shell passivates the optical properties of QDs that originate from its core. QDs exhibit robust and unique optical properties that originate from quantum confinement effects. These properties include narrow (full-width-at-half-maximum in the range of 25–40 nm), symmetric and size tunable emission spectra, greater photostability and brightness than organic fluorophores, high quantum yields and a large extinction coefficient over a broad range of wavelengths that stretches from the UV region to their first exciton peak in the visible region [1]. These properties are well suited for the use of QDs in optical multiplexing for the development of bioassays. QDs are commercially available with numerous surface chemistries allowing for easy conjugation of biorecognition elements for use in bioassays [1].

QDs are popular RET donors due to their high quantum yield and tunable photoluminescence (PL) spectra. The surface area of QDs allows for immobilization of multiple recognition elements, where a single QD can participate in multiple RET events [1]. Our group has paired green-emitting QDs (gQDs) and red-emitting QDs (rQDs) with various molecular dyes as RET acceptors in nucleic acid hybridization assays [2, 3]. While QDs are popular RET donors, their use as RET acceptors has been limited to time gated measurements [4], chemiluminescence resonance energy transfer (CRET) [5] and bioluminescence resonance energy transfer (BRET) [6]. The broad absorbance band of QDs that stretches into the UV region of the spectrum makes them susceptible to direct excitation in the process of exciting the RET donor. We have recently paired QDs as RET acceptors with UCNP donors [7, 8]. UCNPs are excited in the near-IR and IR region and emit in the UV to near-IR region of the spectrum. This permits the excitation of the RET donor without the direct excitation of the QDs. We have previously shown the utility of UCNP/QD RET pair for bioassay development using analytes such as proteins [7] and nucleic acids [8].

1.2   Upconverting Nanoparticles as Donors in Resonance Energy Transfer

UCNPs are luminescent nanocrystals that are typically tens of nanometers in size. They exhibit anti-Stokes emission based on the process of upconversion. Low-energy pump photons are accumulated in multiple long-lived excited-states of lanthanide ions that are supported in an inert crystal lattice. The subsequent relaxation to the ground electronic state results in the emission of radiation of higher energy than the excitation light [9]. A common class of UCNPs is lanthanide doped inorganic crystals, such as $NaYF_4$ and Y_2O_3. The unique $4f^n\,5d^{0-1}$ electronic orbitals of Ln^{3+} ions, which are shielded by the $5s^2$ and $5p^6$ sub-shell electrons, present energy states that can be long lived (up to 0.1 s) and are thus ideal for UC processes. There are five major upconversion processes observed in lanthanide doped UCNPs: excited state absorption (ESA), energy transfer upconversion (ETU), cooperative upconversion (CUC), photon avalanche (PA), and energy migration upconversion (EMU) [9]. Of these processes, ETU is the most efficient and is based on co-doping the inorganic lattice with two lanthanide species, a sensitizer (donor) and an activator (acceptor). For selection criteria of dopants, refer to Ref. 9. The sensitizer ions are capable of absorbing photons in the NIR or IR region of the spectrum and effectively transfer the energy non-radiatively to an activator ion for UC luminescence. The sequential excitation of one activator and its subsequent relaxation to the ground electronic state results in an anti-Stokes emission [9]. Herein, we use Yb^{3+} and Tm^{3+} as sensitizer and activator, respectively, to obtain emission peaks in the UV and blue regions of the spectrum. By changing the identity and concentration of the dopant ions, different emission profiles can be obtained. Lanthanide doped UCNPs are protected with an inert shell of the same material as the host lattice to prevent non-radiative relaxation pathways resulting from interactions with the high-energy vibrational modes of surface ligands and collisions with solvent molecules [9].

Our group has used UCNPs on paper as direct labels [10] and as RET donors with molecular dyes [11] and QDs [8] as acceptors in nucleic acid hybridization assays. Blue emitting $NaYF_4$: 0.5% Tm^{3+}, 30% $Yb^{3+}/NaYF_4$ core/shell UCNPs have been paired with orange emitting QD (oQD) acceptors [8]; green and red emitting $NaYF_4$: 2%Er^{3+}, 18% $Yb^{3+}/NaYF_4$ core/shell UCNPs have been paired with Cy3 and Cy5.5 dyes, respectively [11]. The use of a near-IR excitation source minimizes background signals due to the suppression of light scatter and autofluorescence [9].

1.3 Solid-Phase Assays

Solid-phase resonance energy transfer (RET)-based bioassays are characterized by immobilization of a donor nanoparticle (e.g., QDs or UCNPs) on the surface of a solid support that is modified with a suitable surface chemistry to facilitate its immobilization [1]. The immobilized donor nanoparticle surface is then conjugated to a biorecognition element to allow selective interaction with a solution-phase analyte. The selective binding event modulates the luminescence of donor nanoparticle, which is achieved by pairing an appropriate acceptor (a fluorescent dye or another nanoparticle) with the donor nanoparticle. The selective binding interaction modulates the efficiency of energy transfer between the donor and the acceptor, where the resulting emission from an acceptor serves as an analytical signal [1].

Solid-phase RET-based bioassays that make use of nanoparticles as an active component of a transduction interface offer various advantages in terms of assay development and analytical performance when compared with the corresponding solution-phase assays. Immobilization of nanoparticles on a solid support eliminates the need to maintain their colloidal suspension, which allows implementation of solution conditions (e.g., solvent composition and reaction conditions) that may be beneficial to the analytical performance of the assay [1, 12]. On the other hand, exposure of such solution-phase conditions to a colloidal suspension of nanoparticles can potentially compromise their colloidal stability, the latter being an integral requirement for the performance of solution-phase nanoparticle assays. Immobilization of a biorecognition element is greatly simplified in the case of solid-phase assays, where the immobilized nanoparticles can be exposed to a high concentration of reagents to facilitate bioconjugation. Excess reagents can be rinsed from the surface without the need for time-consuming purification steps that are typically encountered in a multistep synthetic protocol. As a result, bioconjugation can be readily achieved [1, 12]. A solid support that is modified with a selective chemistry can be used to capture targets of interest from a complex matrix. The selectivity of biomolecular interaction can be improved by exposing the substrate to washes in order to suppress the contribution of interferents prior to measurement [1, 12]. Solid-phase assays are also useful for the development of reusable assays. The biomolecule–target complex can be dissociated by changing the environment (e.g., by means of manipulation of temperature, ionic strength or introduction of denaturing agents), while retaining the immobilized selective chemistry on the surface of a solid support for subsequent interrogation of another sample solution for the analyte of interest [1, 12]. It should be noted that reusability is an important feature of biosensors. Solid-phase nanoparticle assays are also amenable to integration with a number of near-field optical transduction techniques, which include evanescent wave excitation [13, 14] and plasmonics [15]. These techniques not only improve the versatility of nanoparticle-based solid-phase assays, but they can also have a pronounced effect on the analytical performance of the assays.

Another significant advantage of solid-phase assays employing nanoparticle mediated RET for signal transduction is the improvement in assay sensitivity that arises from enhancement of energy transfer efficiency at an interface. When donors and acceptors are immobilized at sufficiently high density at an interface there are additional energy transfer pathways between donors and acceptors, where multiple donors can interact with a single acceptor in addition to the interaction of one donor with multiple acceptors [1]. At an interface, there are no discrete donor–acceptor pairs. Instead, there is a two-dimensional plane of donors and acceptors. As a result, the probability of energy transfer to a given acceptor increases for solid-phase RET-based assays. This enhancement is sufficient to compensate for the loss of available surface area of a nanoparticle for selective interaction, which is experienced when one face of a nanoparticle is blocked by immobilization to an interface [1].

1.4 Paper as a Support for Development of Diagnostic Assays

There is a growing interest in the development of decentralized diagnostic assays that can be applied at the point-of-care and point-of-need settings in order to improve patient care and to make health care more accessible. In this regard, paper-based assays have attracted considerable attention in recent years given the advantageous attributes of paper substrates. These attributes include: (1) low-cost and widespread commercial availability of paper substrates with different pore sizes and flow rates [16]; (2) cost-effective, high-throughput and ease of patterning of paper-based analytical devices (PADs) using methods such as wax printing, stamping, and drawing [17, 18]; (3) PADs can be operated independent of any supporting equipment given the autonomous fluid flow offered by the hydrophilic cellulosic fibers of paper substrates; (4) well-established methods for surface modification of cellulose for biomolecule immobilization; (5) an elegant approach to eradicate biohazard by means of incineration of paper; (6) requirement of a small reagent/sample volume; and (7) compatibility with biological samples [19]. A number of these attributes are in congruence with the ASSURED (affordable, sensitive, specific, user-friendly, rapid and robust, equipment-free, and deliverable to end-users) guidelines set by the World Health Organization in order to develop diagnostics for the developing world [19]. Examples of PADs include dipsticks assays, lateral flow devices, paper-based 96-zone and 384-zone plates, microfluidic paper-based analytical devices (µPAD), three-dimensional PADs and origami PADs [20].

1.5 Paper-Based Solid-Phase Nucleic Acid Hybridization Assays Using the gQD/Cy3 and UCNP/QD RET Pairs

In this chapter, we describe methods for solid-phase nucleic acid hybridization assays on a paper-based platform using QDs and UCNPs as donors in RET-based transduction scheme. The paper substrates are patterned using a wax printing method and subsequently chemically derivatized with functional groups that are suitable for the immobilization of the donor nanoparticles. In case of the gQD/Cy3 (donor/acceptor) RET pair (Fig. 1a), the paper substrates are chemically modified with imidazole groups to allow immobilization of gQDs that are pre-modified with oligonucleotide probes. Subsequent hybridization of the target strand and the Cy3-labeled reporter strand brought the Cy3 acceptor dye in close proximity to the surface of immobilized gQDs to allow RET-

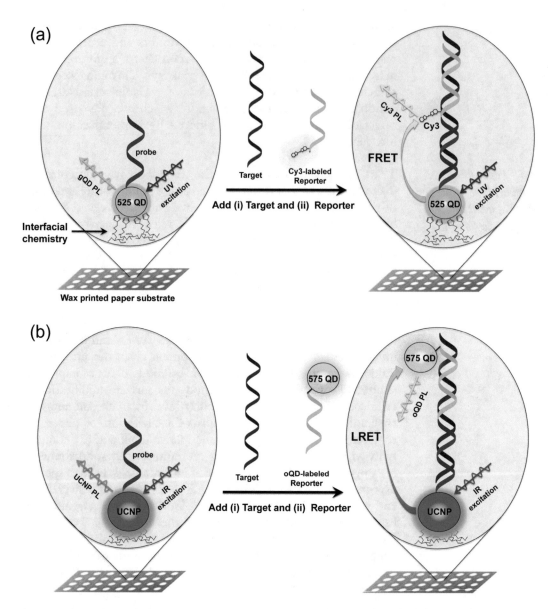

Fig. 1 Illustration of interfacial chemistry for paper-based solid-phase transduction of nucleic acid hybridization using the (**a**) gQD/Cy3 RET pair and the (**b**) UCNP/QD RET pair. The paper substrates were patterned using wax printing and the circular hydrophilic paper zones were derivatized with (**a**) imidazole groups for the immobilization of QD–probe oligonucleotide conjugates and (**b**) aldehyde moieties for immobilization of UCNPs. Sequential addition and hybridization of target and the (**a**) Cy3-labeled or the (**b**) QD-labeled reporter oligonucleotides provided the necessary proximity for RET sensitized emission from the acceptor (**a**) Cy3 dye or (**b**) oQD upon excitation of donor (**a**) gQDs (emission maximum at 525 nm) or (**b**) UCNP with (**a**) a UV or (**b**) IR excitation source. Upon a selective hybridization event, the emission PL shifts from the donor to the acceptor and the modulation of the donor and the acceptor PL serve as an analytical signal

sensitized emission from the Cy3 acceptor dye, which served as an analytical signal upon excitation of gQDs. In case of the UCNP/oQD RET pair (Fig. 1b), the paper substrates are modified with aldehyde groups to allow covalent immobilization of blue-emitting UCNPs. The immobilized UCNPs are subsequently bioconjugated to an oligonucleotide probe. The hybridization of the target strand and the reporter strand that is bioconjugated to an oQD resulted in a close proximity of the UCNP and the oQD, where the resulting RET-sensitized emission from the oQD served as an analytical signal upon excitation of UCNPs. This chapter also entails experimental and data analysis methods for a ratiometric transduction of nucleic acid hybridization using the aforementioned RET pairs. Readout of the donor and acceptor emissions is described using both an epifluorescence microscope and a low-cost electronic camera employing colored digital imaging. Given the advantageous attributes of paper substrates for the development of low-cost diagnostic assays, the nucleic acid hybridization assays described herein can potentially find applications in field-portable and remote diagnostic applications.

2 Materials

2.1 Reagents

1. (4-(2-hydroxyethyl)-1-piperazineethanesulfonic acid) (HEPES) buffer: 100 mM HEPES (pH 7.2).

2. 1-(3-aminopropyl)imidazole (API, ≥97%).

3. 1.7 mL microcentrifuge tubes.

4. 2.5 M NaCl solution prepared using cartridge-purified water (the Milli-Q water).

5. Anhydrous ethanol.

6. Avidin from egg white.

7. Borate buffer (BB1): 50 mM borate (pH 9.2).

8. Borate buffered saline (BB2): 50 mM borate (pH 9.2, 100 mM NaCl).

9. Chloroform, reagent grade.

10. Dithiothreitol (>99%).

11. Ethyl acetate, reagent grade.

12. Hydrophobic green-emitting QDs (gQDs) and orange-emitting QDs (oQDs). We use commercially available alkyl-ligand coated ternary alloyed $CdSe_xS_{1-x}/ZnS$ QDs (gQDs with peak PL at 525 nm and oQDs with peak PL at 575 nm) from Cytodiagnostics Inc. (Burlington, ON, Canada).

13. Hexanes, reagent grade.

14. L-glutathione, reduced (GSH).

15. Lithium chloride (LiCl, anhydrous, ACS reagent, ≥99%).

16. Maleimidohexanoic acid—G(Aib)GHHHHHH from Can-Peptide (Pointe-Claire, QC).

17. Methanol, reagent grade.

18. NHS-PEG$_4$-biotin from Thermo Scientific (Rockford, IL, USA).

19. Octadecene, reagent grade.

20. Oleic acid, reagent grade, 90%.

21. Phosphate-buffered saline (PBS).

22. Phosphorylethanolamine (PEA).

23. Sodium (meta)periodate (NaIO$_4$, ≥99%).

24. Sodium cyanoborohydride (NaCNBH$_3$, reagent grade, 95%).

25. Sodium hydroxide.

26. Sterile ultrapure Milli-Q water (specific resistance ≥18 MΩ cm).

27. Tetramethylammonium hydroxide (TMAH): 25% w/w solution in methanol.

28. Thulium (III) acetate hydrate (99.9%).

29. Tris(2-carboxyethyl)phosphine hydrochloride (TCEP) powder (≥98%).

30. Whatman® cellulose chromatography papers (Grade 1, 20 cm × 20 cm).

31. Ytterbium(III) acetate hydrate (99.9%).

32. Yttrium(III) acetate hydrate (99.9%).

33. Oligonucleotide sequences listed in Table 1 (*see* **Note 1**). The sequences are from Integrated DNA Technologies (Coralville, IA, USA).

2.2 Oligonucleotide Sequences for the Hybridization Assays

See Table 1.

2.3 Instrumentation and Equipment

1. Nikon Eclipse L150 epifluorescence microscope (Nikon, Mississauga, ON, Canada) custom fitted with the following components: 4× Nikon WD Plan Fluor objective lens (NA = 0.13); a filter cube comprising ZET405/20× as an excitation filter, Z405rdc as a dichroic mirror and a HQ430lp as an emission filter (Chroma Technologies Corp., Bellows Falls, VT, USA); a 25 mW diode laser excitation source with an output of 402 nm (Radius 402, Coherent Inc. Santa Clara, CA, USA) and a diode array spectrometer (QE65000, Ocean Optics Inc., Dunedin, FL, USA) as a detector.

Table 1
Probe, target and reporter oligonucleotide sequences used in the hybridization assays with the gQD/Cy3 and UCNP/QD RET pairs

Name	Sequence (5′ to 3′ direction)
gQD/Cy3 RET	
SMN1 probe	DTPA-5′-ATT TTG TCT GAA ACC CTG T-3′
SMN1 FC target	5′-TCC TTT ATT TTC CTT ACA GGG TTT CAG ACA AAA T-3′
SMN1 Cy3 reporter	Cy3-5′-AAGGAAAATAAAGGA-3′
UCNP/QD RET pair	
HPRT1 probe	5′-Biotin-CAA AAT AAA TCA AGG TCA-3'
HPRT1 FC target	5′-GAT GAT GAA CCA GGT TAT GAC CTT GAT TTA TTT TG-3'
HPRT1 reporter	5′-TAACCTGGTTCATCATC-Thiol-3'

Abbreviations: *FC* fully complementary, *DTPA* dithiol phosphoramidite, *Cy3* cyanine 3

2. Nikon Eclipse L150 epifluorescence microscope (Nikon) custom fitted with the following components: 40× Nikon WD Plan Fluor objective lens (NA = 0.60); a filter cube comprising a ZET964/86 bp excitation filter (Chroma), a ZT1064rdc-sp dichroic mirror (Chroma) and a 455/35 (Nikon) or 570/20 (Chroma) emission filters; a 2.5 W tunable 980 nm collimated diode laser (Laserglow Technologies, Toronto, ON, CAN) and a H5784-20 photomultiplier tube (Hamamatsu, Bridgewater, NJ, USA).

3. Handheld ultraviolet (UV) lamp (UVGL-58, LW/SW, 6W, The Science Company®, Denver, CO, USA).

4. HP8452A Diode-Array Spectrophotometer (HP Corporation, Palo Alto, CA, USA).

5. iPad mini (Apple, Cupertino, CA, USA).

6. Low-binding polypropylene microcentrifuge tubes, 1.7 mL capacity.

7. Micropipettes and tips.

8. Neutral density (ND) filters (ND 4, ND 8 and ND 16).

9. Orbital shaker.

10. 0.2 μm Polyethersulfone (PES) syringe filters.

11. Xerox ColorQube 8570DN solid ink wax printer (Xerox Canada, Toronto, ON, Canada).

12. Amicon Ultra-0.5 mL 100 kDa centrifugal filters (Millipore Corporation, Billerica, MA, USA).

3   Methods

3.1   UCNP Synthesis

3.1.1   Synthesis of Core NaYF$_4$: 0.5% Tm^{3+}, 30% Yb^{3+} UCNPs

1. Add 0.4562, 0.2534 and 0.0042 g of Y(CH$_3$CO$_2$)$_3$·xH$_2$O, Yb(CH$_3$CO$_2$)$_3$·4H$_2$O, Tm(CH$_3$CO$_2$)$_3$·xH$_2$O and a stir bar into a 100 mL three-neck round bottom flask.

2. Add 30 mL of octadecene and 12 mL of oleic acid into the round bottom flask.

3. Place the round bottom flask on a heating mantle controlled by a temperature precision controller.

4. Insert the precision controller thermometer in the central neck of the three-neck round bottom flask through a septum. Attach a two-neck collection round bottom flask via a bent adaptor to one of the side necks of the reaction flask. Seal the remaining necks of the reaction and the collection flasks with septa (*see* **Note 2**).

5. Stir the mixture gently under vacuum at 115 °C for 30 min. The solution should turn clear and colourless at this point.

6. Cool the mixture to 50 °C under a gentle stream of argon. Insert the in-line via a needle in the unused neck of the three-neck round bottom flask. Insert the out-line via a needle in the unused neck of the collection round bottom flask. The argon flow should be maintained for the remainder of the synthesis.

7. Prepare a 20 mL methanol solution containing 0.20 g NaOH and 0.30 g NH$_4$F via sonication. Shake the solution occasionally to help dissolve the solids (*see* **Note 3**). Add the methanol solution to the reaction mixture. The solution will turn cloudy.

8. Allow the reaction to stir for 30 min at 50 °C before heating it to 75 °C to evaporate the methanol. The solution should become clear at this point (*see* **Note 4**).

9. Increase the temperature to 300 °C rapidly and maintain it for 1 h. Wrap the reaction flask with glass fibre to help keep the solution warm (*see* **Note 5**).

10. Allow the reaction mixture to cool to room temperature under argon.

11. Add an equal volume of absolute ethanol to the reaction mixture and centrifuge the solution at 4500 rpm to collect the UCNPs.

12. Resuspend the UCNPs in hexanes and repeat **step 11**. Repeat twice.

13. Store the washed core UCNPs in hexanes in a glass vial at 4 °C for subsequent growth of shell.

3.1.2 Synthesis of NaYF₄: 0.5% Tm³⁺, 30% Yb³⁺/ NaYF₄ Core/Shell UCNPs

1. Add 0.5738 g of $Y(CH_3CO_2)_3 \cdot xH_2O$ into a 100 mL three-neck round bottom flask.

2. Follow **steps 2–5** in Subheading 3.1.1.

3. Cool the mixture to 80 °C under argon and add the core UCNPs from **step 13**.

4. Cool the reaction temperature to 50 °C after all the hexane has evaporated.

5. Prepare a 20 mL methanol solution containing 0.14 g NaOH and 0.26 g NH_4F via sonication. Add the mixture to the reaction vessel. The reaction mixture will turn cloudy (*see* **Note 3**).

6. Allow the reaction to stir for 30 min before heating the mixture to 75 °C to evaporate the methanol. The solution should be clear at this point (*see* **Note 4**).

7. Increase the temperature to 300 °C rapidly and maintain it for 1 h (*see* **Note 5**).

8. Allow the reaction mixture to cool to room temperature under argon.

9. Add an equal volume of absolute ethanol and centrifuge at $2173 \times g$ to collect the UCNPs.

10. Resuspend the UCNPs in hexanes and repeat **step 9**. Repeat twice.

11. Store the oleic acid capped core/shell UCNPs in hexanes or toluene in a glass vial at 4 °C for subsequent modification.

3.2 Preparation of Water Soluble QDs and UCNPs

3.2.1 Preparation of Water Soluble UCNPs

1. Mix 2 mL of hexanes containing 100 mg of oleic acid capped core/shell UCNPs from **step 11** in Subheading 3.1.2, 400 mg of phosphorylethanolamine (PEA) and 1 mL of tetramethylammonium hydroxide (TMAH) in 10 mL of absolute ethanol in a capped glass vial.

2. Stir the reaction vigorously overnight at 70 °C in an oil bath.

3. Allow the reaction to cool to room temperature and recover PEA capped UCNPs by centrifugation at $1315 \times g$.

4. Resuspend the UCNPs in ethanol via sonication and mix the solution with an equal volume of hexanes before centrifugation at 4500 rpm.

5. Repeat **step 4** two times.

6. Resuspend the washed PEA capped UCNPs in 10 mL of water and pass the solution through a 0.2 μm polyether sulfone (PES) syringe filter to remove aggregates. If desired, evaporate water using a rotary evaporator to concentrate UCNPs.

7. Store the PEA capped UCNPs in excess PEA at 4 °C.

1. Add 75 µL of 10 µM alkyl QDs into 2 mL of chloroform in a glass vial.

2. Dissolve 0.2 g of reduced L-Glutathione (GSH) in 600 µL of TMAH.

3. Add the solution of QDs drop-wise to the GSH solution while swirling.

4. Allow the cloudy mixture to sit overnight in the dark.

5. Extract water soluble GSH coated QDs using BB2 in 100 µL fractions. Place the collected fractions in a 1.7 mL microcentrifuge tube.

6. Add an equal volume of ethanol and collect the QDs by centrifugation at $6869 \times g$ for 5 min.

7. Resuspend the QDs in 200 µL of BB2.

8. Repeat **steps 6** and **7** two more times.

9. Resuspend the GSH-QDs in BB1.

10. Measure the absorbance spectrum of the QDs and use their first exciton peak and known molar extinction coefficient to determine the NP concentration.

11. Store the GSH-QD at 4 °C for subsequent use.

3.3 Bioconjugation of Oligonucleotides to QDs

3.3.1 Bioconjugation of Hexahistidine-Terminated Oligonucleotides to GSH-QDs

1. Incubate thiol-terminated oligonucleotides (27 nmol in 100 µL) with 500 equivalents of dithiothreitol (DTT) (27 µL of 0.5 M DDT solution in 1× PBS) and 100 µL of 1× PBS buffer for 1 h at room temperature to reduce disulfides moieties into sulfhydryl groups.

2. Extract excess DTT with 600 µL of anhydrous ethyl acetate four times.

3. Add 6-maleimidohexanoic acid—G(Aib)GHHHHHH (0.7 mg in 30 µL of DMSO) to the oligonucleotide solution in 20 times molar excess and allow the reaction mixture to shake for 12 h at room temperature (*see* **Note 6**).

4. Purify the modified oligonucleotides using a NAP-5 desalting column as per the manufacturer's instructions.

5. Quantify the purified hexahistidine-modified oligonucleotides by UV-vis spectroscopy ($\lambda_{\max} = 260$ nm).

6. Store the hexahistidine-modified oligonucleotides at −20 °C.

7. Incubate hexahistidine-modified oligonucleotides (HPRT1 reporter) with GSH-QDs at the desired QD:DNA ratio for 1 h in BB2 in a 1.7 mL microcentrifuge tube on an orbital shaker (*see* **Note 7**).

8. Purify the hexahistidine-modified oligonucleotides by centrifugation three times using an Amicon Ultra-0.5 mL 100 kDa

centrifugal filter as per the manufacturer's instructions to remove excess nucleic acids.

9. Store the washed nucleic acid conjugated QDs in BB2 at 4 °C.

3.3.2 Bio-Conjugation of DTPA-Terminated Oligonucleotides to GSH-QDs

1. Prepare 50 mM solution of TCEP by dissolving 7 mg of TCEP in 500 μL of borate buffer (BB1).

2. To a 1.7 mL microcentrifuge tube, pipette 169 μL of borate buffered saline (BBS).

3. To the same tube, pipette 82 μL of 50 mM TCEP solution that has been prepared in **step 1**.

4. Add 8.2 nmol of DTPA terminated oligonucleotide probe (SMN1 probe) to a solution prepared in **step 3** and place the tube on an orbital shaker for 15 min (*see* **Note 8**).

5. Pipette 200 pmol of GSH-gQDs to the solution in **step 4** (*see* **Note 9**).

6. Pipette additional volume of borate buffered saline (BB2) to the solution in **step 5** such that the final solution volume is 493 μL. For the example of QD–probe conjugates preparation provided in this section, an additional 169 μL of borate buffered saline (BB2) will need to be added.

7. Agitate the contents of the tube overnight on an orbital shaker.

8. Prepare a fresh 50 mM TCEP solution using **step 1**.

9. Pipette a 40 μL aliquot of 50 mM TCEP solution from **step 8** into the contents of the tube in **step 7**.

10. Incrementally pipette 100 μL of 2.5 M NaCl solution to the tube in **step 9**. We use an interval time of 10 min for each sequential addition of 10 μL of 2.5 M NaCl (*see* **Note 10**).

11. Place the tube in **step 10** on an orbital shaker for an overnight incubation and subsequently store the contents of the tube (QD–probe conjugates) inside a fridge at 4 °C until further use. The concentration of QD–probe conjugates in this solution is ca. 312 nM (total solution volume is 633 μL).

3.4 Wax Printing and Chemical Modification of Paper

3.4.1 Wax Printing of Paper Substrates

1. Using an appropriate software, draw a pattern of the paper device that is to be printed on the Whatman® cellulose chromatography paper substrates. We use AutoCAD 2012 software (*see* **Note 11**). For the purpose of the work described in this chapter, the dimensions of each paper device is 25 mm × 60 mm and a single sheet of 20 cm × 20 cm Whatman® chromatography paper substrate comprised 18 paper devices (arranged in a 6 × 3 array format). We use paper devices that contain 32 circular zones with a diameter of 5 mm (*see* **Note 12**), arranged in a 4 × 8 array format. We print on bare paper, and the zones are surrounded by a filling with black ink.

2. Print the paper substrates using a Xerox ColorQube 8570DN solid ink wax printer (*see* **Note 13**).

3. Preheat an oven or a hot plate to 120 °C.

4. Place the paper sheet (printed side up) inside the oven or on top of the hotplate for 2.5 min to confine the hydrophilic paper zones across the thickness of the paper substrate (*see* **Note 14**).

5. Allow the paper sheet to cool to room temperature.

6. Using scissors, trim the patterned paper sheet to individual paper devices.

3.4.2 Chemical Derivatization of Paper Zones with Aldehyde Functionality

1. Attach a binder clip to the end of a wax-printed paper substrate in order to suspend the paper device in air. We do so by inserting a micropipette tip in the arms of the binder clip such that the bottom of the micropipette tip rests inside one of the holders of the microcentrifuge tray rack (*see* **Note 15**).

2. Prepare 1.4 M solution of LiCl by dissolving 30 mg of LiCl in 500 μL of Milli-Q water.

3. Prepare 94 mM solution of $NaIO_4$ by dissolving 10 mg of $NaIO_4$ in 500 μL of Milli-Q water.

4. Mix the two solutions prepared in **steps 2** and **3** in equal volumes.

5. Spot a 5 μL aliquot of the solution prepared in **step 4** onto each paper zone and incubate the spotted paper devices inside an oven set at 50 °C for 30 min.

6. Repeat **step 5**.

7. Wash the paper devices three times with Milli-Q water (*see* **Note 16**).

8. Place the washed paper devices on any form of absorbent blotting paper in order to wick off the excess water.

9. Dry the paper devices inside a vacuum desiccator.

3.4.3 Chemical Derivatization of Aldehyde Modified Paper Zones with Imidazole Functionality

1. Suspend aldehyde modified paper substrates in air using a binder clip as mentioned previously.

2. Prepare 320 mM solution of $NaCNBH_3$ by dissolving 10 mg of $NaCNBH_3$ in 500 μL of 100 mM HEPES buffer (pH 8.0).

3. To 400 μL of 320 mM solution of $NaCNBH_3$ prepared in **step 2**, pipette 12 μL of API. The concentration of API in the resulting solution is 0.24 M.

4. Pipette a 5 μL aliquot of the solution prepared in **step 3** onto each paper zone that has been modified with the aldehyde functionality.

5. Incubate the spotted paper substrates under ambient conditions for 1 h.

6. Submerge each paper device inside a 50 mL conical centrifuge tube filled with BB1.

7. Place the conical centrifuge tubes on an orbital shaker for 15 min in order to wash the paper devices.

8. Remove the imidazole modified paper substrates from the conical tubes and place them on any form of an absorbent blotting paper in order to wick off the excess solution.

9. Dry the paper substrates inside a vacuum desiccator.

3.5 Immobilization of QDs and UCNPs on Paper

3.5.1 Immobilization of QD–Probe Oligonucleotide Conjugates on Imidazole Modified Paper Substrates

1. Suspend imidazole modified paper substrates in air using a binder clip as mentioned previously.

2. Spot a 3 μL aliquot of 312 nM solution of QD–probe conjugates solution onto each paper zone that has been modified with imidazole functionality and incubate for 1 h at room temperature in dark (*see* **Note 17**).

3. Wash the paper devices for 15 min with BB1 using an orbital shaker by submerging each paper device inside a 50 mL conical centrifuge tube containing BB1.

4. Wick off the excess solution from the paper substrates using any form of absorbent blotting paper and dry the QD–probe conjugates modified paper substrates inside a vacuum desiccator.

3.5.2 Immobilization of UCNPs on Aldehyde Modified Paper Substrates

1. Mix 200 μL of PEA-UCNPs (3 mg/mL) with 200 μL of a 0.1 mM sodium cyanoborahydride solution in HEPES buffer (100 mM, pH 7.2).

2. Suspend aldehyde modified paper substrates in air using a binder clip as mentioned previously.

3. Pipet 5 μL of the solution in **step 1** onto each aldehyde modified paper zone.

4. After incubating for 10 min., wash the paper substrate in 0.1% v/v aqueous Tween® 20 solution for 5 min in a conical centrifuge tube.

5. Wash the paper substrate with purified water for 2 min.

6. Place the washed paper substrate on any form of absorbent blotting paper to wick off the excess water.

7. Dry the paper substrate inside a vacuum desiccator.

3.5.3 Layer-By-Layer Assembly of the Assay Using Immobilized UCNPs on Paper Substrates

1. Suspend paper substrates with immobilized UCNPs in air using a binder clip as mentioned previously.

2. Pipet 3 μL of a freshly prepared 2 mM aqueous NHS-PEG-biotin solution onto each spot of the paper substrate containing immobilized UCNPs and allow to air dry.

3. Wash the paper for 2 min with purified water in a conical centrifuge tube.

4. Place the washed paper substrate on an absorbent blotting paper to wick off the excess water before placing it inside a vacuum desiccator to dry.

5. Pipet 5 μL of a 20 μM avidin solution in HEPES buffer (100 mM, pH 7.2) onto each paper zone. Allow the zones to air dry.

6. Wash the paper for 2 min with BB1 in a conical centrifuge tube.

7. Place the washed paper substrate on an absorbent blotting paper to wick off the excess water before placing it inside a vacuum desiccator to dry.

8. Pipet 5 μL of 10 μM biotinylated oligonucleotide probe (HPRT1 probe) solution in BB1 onto each paper zone. Allow the paper to air dry.

9. Repeat **steps 6** and **7**.

3.6 Calibration Curve of Target DNA

1. Prepare dilutions of the SMN1 FC or HPRT1 FC target in the concentration range of 20 nM to 15 μM using 50 mM borate buffered saline (pH 9.2, 100 mM NaCl) (*see* **Note 18**).

2. Prepare a dilution of SMN1 Cy3 reporter at 10–15 μM concentration or HPRT1 oQD reporter at 2 μM using borate buffered saline (BB2).

3. Spot 3 μL of borate buffered saline (BB2) onto the paper zones that belong to the top row of the paper device (*see* **Note 19**).

4. Spot a 3 μL aliquot of SMN1 FC or HPRT1 FC target at various concentrations onto each of the different rows of the paper device (excluding the first row, *see* point 3). In order to collect replicate measurements, the same concentration of the target can be spotted onto the paper zones that belong to the same row of the paper device (*see* **Note 20**).

5. Allow the targets/sample solutions to incubate on the paper device for 1 h (*see* **Note 21**).

6. Spot a 3 μL aliquot of 10–15 μM solution of SMN1 Cy3 reporter or 2 μM HPRT1 oQD reporter onto each of the paper zones that have been subjected to the target/sample hybridization and allow the reporter solution to incubate on the paper device for 30 min (*see* **Note 22**).

7. Wash the paper substrates for 5 min with BB2 using an orbital shaker by submerging each paper device inside a 50 mL conical centrifuge tube containing BB2. Add 1% Tween® 20 by volume to the wash solution when QDs are used as acceptors.

8. Wick off the excess solution from the paper substrates using an absorbent blotting paper and dry the paper substrates inside a vacuum desiccator prior to the data collection (*see* **Note 23**).

3.7 Data Acquisition

1. gQD/Cy3 RET Pair.

    (a) Using the epifluorescence microscope platform (402 nm laser excitation source and diode array spectrometer as a detector), measure the PL spectrum from each paper zone that has been modified with the selective chemistry (immobilized QD–probe conjugates) (*see* **Note 24**). Also, collect a background spectrum from imidazole modified paper zone (without immobilized QD–probe conjugates). Proceed to subheading 3.8, **step 1** for data analysis.

    (b) Representative data acquired using the epifluorescence microscope platform is shown in Fig. 2b and c.

    (c) For the hybridization assays using the gQD/Cy3 RET pair, the red-green-blue (RGB) color palette of a digital camera can also be used to acquire quantitative information. Using the long wavelength (365 nm) setting of a handheld UV lamp (UVGL-58, LW/SW, 6W, The Science Company®), illuminate the paper devices for the excitation of immobilized QDs on paper substrates and collect a colored digital image of the paper device under the default settings of the built-in Camera application of an iPad mini in a dark environment (*see* **Note 25**). Proceed to Subheading 3.8, **step 1** for data analysis.

    (d) Representative data from the iPad detection platform is shown in Fig. 2a and c.

2. UCNP/QD RET Pair.

    (a) Using the epifluorescence microscope platform with a 980 nm laser excitation source and the PMT detector, acquire the PL image from each paper zone. Appropriate optical emission filters are used to collect the UCNP and QD signals independently. Herein, we use 455/35 nm and 570/20 nm band-pass filters to collect UCNP and QD luminescence, respectively. The paper is scanned on the microscope stage and an image is obtained using LabView (*see* **Note 26**). Proceed to subheading 3.8, step 1 for data analysis.

    (b) Representative data from the epifluorescence detection platform is shown in Fig. 3.

3.8 Data Analysis

1. Analysis of PL spectra for gQD/Cy3 RET Pair.

    (a) Subtract background spectrum from each PL spectrum that has been collected from the paper zones with immobilized QD–probe conjugates.

    (b) Normalize each background corrected PL spectrum to the maximum PL intensity of gQDs, i.e., divide all the PL intensities by the maximum PL intensity of gQDs (the

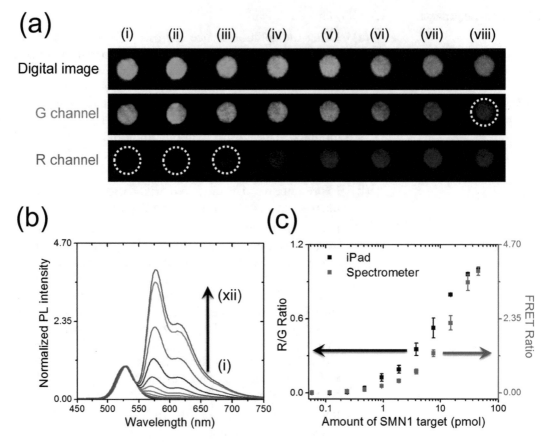

Fig. 2 Solid-phase QD-RET transduction of nucleic acid hybridization on paper substrates using the gQD/Cy3 RET pair. (**a**) Colored digital image and the corresponding pseudo-colored PL images of gQDs (G channel) and Cy3 (R channel) after R-G-B splitting of the colored digital image with increasing amount of SMN1 FC TGT. The amounts of SMN1 FC TGT in (*i*) to (*viii*) were 0, 0.94, 1.9, 3.8, 7.5, 15, 30, and 45 pmol, respectively. (**b**) Normalized PL spectra acquired using the epifluorescence microscope platform showing the assay response with increasing amount of SMN1 FC TGT. The amounts of SMN1 FC TGT in (*i*) to (*xii*) were 0, 0.057, 0.12, 0.23, 0.47, 0.94, 1.9, 3.8, 7.5, 15, 30, and 45 pmol, respectively. (**c**) Calibration curves showing the RET ratio response (*red*) and the R/G ratio response (*black*) with increasing amount of SMN1 FC TGT. Figure adapted with permission from Ref. 21. Copyright 2014 American Chemical Society

λmax for gQDs is in the range of 520–530 nm). The resulting spectra will depict profiles as shown in Fig. 2b.

(c)   Calculate RET Ratio for each background subtracted and normalized PL spectrum using Eq. 1. In Eq. 1, the wavelength range of 560–590 nm in the numerator of each term is used to integrate the Cy3 (acceptor) PL intensity, while the wavelength range of 510–540 nm in the denominator of each term is used to integrate the gQD (donor) PL intensity. The subscripts DA and D denote

(a)

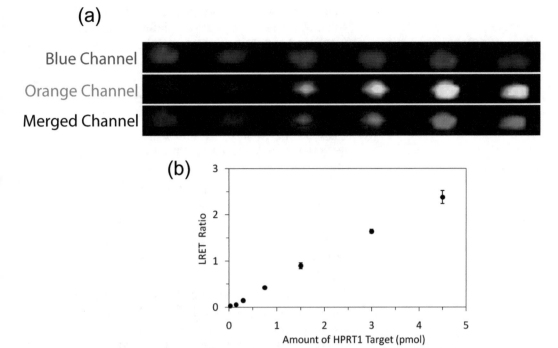

Blue Channel

Orange Channel

Merged Channel

(b)

Fig. 3 Solid-phase UCNP/QD (donor/acceptor) RET transduction of nucleic acid hybridization on paper substrates. (**a**) Pseudo-colored microscope images of UCNPs (*blue channel*), oQDs (*orange channel*) and an overlay of both images with increasing amount of HPRT1 FC target. The amounts of HPRT1 FC target ranged from 15 fmol to 6 pmol. (**b**) Calibration curve showing the RET ratio with increasing amount of HPRT1 FC target

measurements made in the presence and absence of the acceptor, respectively (*see* **Note 27**).

$$\text{FRET ratio} = \left(\frac{\sum\limits_{\lambda=560}^{\lambda=590} \text{PL}(\lambda)}{\sum\limits_{\lambda=510}^{\lambda=540} \text{PL}(\lambda)}\right)_{\text{DA}} - \left(\frac{\sum\limits_{\lambda=560}^{\lambda=590} \text{PL}(\lambda)}{\sum\limits_{\lambda=510}^{\lambda=540} \text{PL}(\lambda)}\right)_{\text{D}} \quad (1)$$

(d)   Plot the calculated RET ratios in **step 1c** as a function of the amount of target DNA that has been spotted onto each paper zone. An example of the resulting data set is shown in Fig. 2c.

2. Analysis of Digital Images.

(a)   Open the acquired colored digital images using the ImageJ software (National Institute of Health, Bethesda, MB, USA).

(b)   Click on the image tab and sequentially select the color and then split channels option from the dropdown menu in order to split each colored digital image into the corresponding red, green and blue color channels.

The channels of interest are the red and green color channels, which respectively interrogate the acceptor (Cy3) and the donor (gQDs) PL intensities as shown in Fig. 2a (*see* **Note 28**).

(c) Draw a circle around a paper zone in the split red and green color channels images by selecting the oval drawing tool (*see* **Note 29**).

(d) Click on the Analyze tab and select the measure option from the dropdown menu. This will open a new window displaying the area, mean, min and max values corresponding to the paper zone that has been selected. The value of interest is the mean value.

(e) Determine the mean value from each of the paper zones by dragging the same drawn circle from one paper zone to another and subsequently repeating **step 2d** for different paper zones.

(f) Calculate the R/G ratio corresponding to each of the paper zones using Eq. 2. In Eq. 2, I_R corresponds to the mean intensity of the paper zone in the red channel, while I_G corresponds to the mean intensity of the same paper zone in the green channel. The subscripts DA and D denote measurements made in the presence and absence of the acceptor, respectively (*see* **Note 30**).

$$\text{R/G ratio} = \left(\frac{I_R}{I_G}\right)_{DA} - \left(\frac{I_R}{I_G}\right)_{D} \tag{2}$$

(g) Plot R/G ratios as a function of the amount of target DNA that has been spotted onto each paper zone. An example of the resulting data set is shown in Fig. 2c.

3. Analysis of Microscope Images for the UCNP/QD RET Pair.

(a) Using the ImageJ software (National Institute of Health, Bethesda, MB, USA), open the images that have been acquired by scanning the paper using an epifluorescence microscope.

(b) Click on Plugins, Macros, Record (*see* **Note 31**).

(c) Draw circles around each paper zone for the resultant image acquired using 455/35 nm optical filter using the oval drawing tool. The circle drawn for one zone can be dragged over to other zones to ensure that the mean intensity determined from each zone represents the same area.

(d) After each circle is drawn, press the "M" key on the keyboard to measure an integrated intensity in the paper zone. This will display the area, and the mean, minimum, and maximum intensity values of the paper zone that has been selected. Use the mean value as the average signal from each zone.

(e) In the Macros window, save the resultant code.

(f) Open the image that has been scanned using the 560/20 nm filter.

(g) Go to Plugins, Macro, Run and select the code that has been saved. This will automatically give results from each zone on this image in the same order as the zones have been created in the previous image.

(h) Calculate a RET Ratio (Eq. 3) for each reaction zone by taking the ratio of the signal in the QD (acceptor) channel and UCNP (donor) channel.

$$\text{RET Ratio} = \left(\frac{I_o}{I_B}\right)_{DA} - \left(\frac{I_o}{I_B}\right)_{D} \tag{3}$$

(i) Plot the calculated RET ratios as a function of amount of DNA target that has been spotted onto each paper zone. Take an average of the three paper zones that have been subjected to the same target concentration. An example of the resulting data set is shown in Fig. 3.

4   Notes

1. SMN1 sequence is a genetic marker that is diagnostic of the neuromuscular disorder known as spinal muscular atrophy [22]. We have used the SMN1 sequences as model sequences to demonstrate the principles of solid-phase QD-RET nucleic acid hybridization assays. HPRT1 is a housekeeping gene found in mammalian cells. It is selected as a generic target to demonstrate the principles of solid-phase RET between UCNPs and QDs. Housekeeping genes are routinely used for control and calibration in biotechnological applications and genomic studies [8].

2. A small incision is made in the septum for the thermometer.

3. This solution can be prepared after **step 5** has been set up and will be ready to use in this step.

4. Mark the solvent level on the reaction flask before the addition of the methanol and wait for the solvent level to return to the same level. The methanol should evaporate and condense in the collection flask. Check the side flask to determine the end of condensation.

5. Be careful not to get any material trapped between the heating mantel and the reaction flask as it is a fire hazard.

6. Ensure that the 6-maleimidohexanoic acid solution is prepared before extracting the excess DTT. Add the 6-maleimidohexanoic acid solution to the reduced thiol DNA immediately after extraction to ensure that the sulfhydryl functionality is fully available.

7. A maximum of approximately 20 oligonucleotide strands can be loaded on one QD using this method.

8. As an example, if the concentration of SMN1 probe is 239.8 μM, then 34 μL of the probe solution will need to be pipetted. TCEP is used as a reducing agent to reduce the disulfide (DTPA) moiety associated with each oligonucleotide probe to a dithiol [23]. TCEP is added at 500 times molar excess of the DTPA probe concentration. Therefore, if a different amount of DTPA probe is used, then the volume of 50 mM TCEP solution should be adjusted accordingly.

9. As an example, if the concentration of GSH-gQDs is 5.1 μM, then ca. 39 μL of the QD solution will need to be pipetted. The amount of QDs added in this step is such that the stoichiometric ratio between the QDs and DTPA probe is 1 to 40, respectively. The bioconjugation of DTPA probe to CdSeS/ZnS (core/shell) QDs is driven by self-assembly via the coordination linkage of dithiol moiety to the Zn^{2+} atoms on ZnS shell of the QDs [24–26].

10. This step is referred to as the salt aging of QD–probe conjugates. Salt aging is used to increase the loading capacity of oligonucleotide probes to the surface of QDs by screening the electrostatic repulsion associated with the negative charge of the sugar-phosphate backbone of oligonucleotides [21]. We have previously reported that without the salt-aging step, an average of 17 oligonucleotide probes can be self-assembled to the surface of similar GSH-QDs when incubated with 40 times molar excess of the oligonucleotide probes [21]. However, with the salt-aging step, an average of 35–40 oligonucleotide probes can be self-assembled to the surface of the GSH-QDs for the aforementioned preparation [21].

11. Ensure that the pattern of the paper device fits within the 20 cm × 20 cm dimensionality of the Whatman® cellulose chromatography paper sheet.

12. After the wax melting step, the actual diameter of each paper zone will be smaller than what has been originally designed, i.e., less than 5 mm. An estimate of the size of the features that are wax printed on the paper substrates after the melting step can be made as described by Carrilho et al. [17].

13. A custom setting of the printer may be required in order to print the paper substrate.

14. This step melts and spreads the wax laterally and vertically, resulting in a formation of hydrophobic wax barriers across the thickness of the paper. Owing to the lateral spreading of the wax, the resulting diameter of each hydrophilic paper zone on the paper device will be approximately 3 mm [17].

15. This ensures that any solution that is spotted onto the paper zones does not come in contact with any other surface, which can otherwise result in reagent loss.

16. We wash paper devices by submerging each paper device inside a 50 mL conical centrifuge tube that is filled with Milli-Q water.

17. Dilutions of the stock QD–probe conjugates solution can also be prepared and spotted onto the imidazole modified paper zones. Quantitative information can also be acquired using various concentrations of the QD–probe conjugates solution (e.g., 8.7, 52, or 167 nM) provided that calibration data exist for these concentrations of immobilized QD–probe conjugates.

18. It is preferred that a serial dilution method is used in order to prepare various dilutions of SMN1 FC TGT. Also, ensure that the NaCl concentration in each of the target solutions is adjusted to 100 mM. We use 2.5 M NaCl solution in order to adjust NaCl concentration to 100 mM.

19. This allows for the collection of PL spectra and images from paper zones that have been modified with immobilized QD–probe conjugates but not exposed to the target sample.

20. There are four paper zones along each row of the paper device. This allows four replicate measurements to be collected by spotting the same target concentration along a particular row of the paper device. To evaluate nonspecific adsorption, a row of paper zones can also be spotted with SMN1 NC target.

21. During this incubation step, the target/sample solutions may dry on the paper device. However, this does not affect the experimental outcome.

22. We typically apply reporter that is in two- to threefold molar excess of the target concentration. For experiments involving single nucleotide polymorphism discrimination, the paper substrates are first washed with 15% (v/v) formamide solution in BB1 for 15 min and then with BB2 for 1 min prior to the reporter hybridization step.

23. We have previously reported that data collection from dry paper substrates offers at least an order of magnitude higher assay sensitivity and at least an order of magnitude lower limit

of detection as compared to the corresponding hydrated paper substrates [21].

24. We use a 4× objective lens (NA = 0.13) for the collection of PL spectra from paper substrates. Depending on the concentration of QD–probe conjugates that has been spotted onto the paper zones and the power of the laser excitation source, a neutral density (ND) filter may be required in order to attenuate the laser excitation intensity to prevent detector saturation. We use a ND 8 filter for experiments that rely on excitation with the 25 mW diode laser and QD–probe conjugates spotted onto the paper substrates at 312 nM (936 fmol) concentration. Due to ratiometric signal processing, fluctuations in the absolute PL intensities of donor and acceptor that are caused by factors such as variations in detector sensitivity, excitation source intensity and sample dilutions do not significantly affect the experimental outcome [27]. While QDs exhibit a strong and broad absorption band [1], which is a useful attribute to efficiently excite QDs with a broad range of excitation wavelengths, the selection of excitation wavelength close to 400 nm is done in order to minimize direct excitation of the Cy3 acceptor [1].

25. For the collection of PL images, place the paper substrates and the UV lamp on a horizontal surface and hold the iPad mini orthogonal to the horizontal surface. Depending on the concentration of QD–probe conjugates that has been spotted onto the paper zones, a placement of ND filter in front the iPad mini camera may be required to prevent pixel saturation. For QD–probe conjugates spotted onto the paper devices at 312 nM (936 fmol) concentration, we place paper devices 10 cm away from the UV lamp and use ND 16 filter. The selection of ND filter (ND 4, ND 8, or ND 16) will also be dependent on the separation distance between the lamp and the paper device.

26. We use a 40× objective lens (NA = 0.60) to collect the photoluminescence from the paper. The tunable power of the IR laser is set between 0.4 and 1 W depending on the concentration and brightness of immobilized UCNPs.

27. The second term in Eq. 1 is a correction factor and accounts for the crosstalk of the gQDs donor PL with the Cy3 acceptor PL.

28. After splitting the color channels, the images will be in the grey color scheme. Pseudo-coloring of the images can be done by clicking on the image tab and then selecting the Lookup Tables option from the dropdown menu, followed by selecting the desired color scheme. We use red color for displaying the red color channel and green color for displaying the green color channel.

29. Make sure the drawn circle fits within the dimensions of the paper zone. The same drawn circle can be dragged from one paper zone to another in order to analyze the mean PL intensity from each paper zone. This also ensures that the mean intensity determined from each of the paper zones represents the same unit area.

30. The second term in Eq. 2 is a correction factor and accounts for the crosstalk of the gQDs donor intensity in the red channel.

31. This allows the software to track the locations on the paper from where the signals are collected as an integrated area for the image collected with the 455/35 nm filter. This can then be applied to the image collected with the 560/20 nm filter.

Acknowledgments

The authors gratefully acknowledge the Natural Sciences and Engineering Research Council of Canada (NSERC) for financial support of their research. S.D. acknowledges NSERC for the provision of a graduate scholarship. M.O.N. is grateful to the Ontario Centres of Excellence (OCE) for provision of a Talent Edge postdoctoral fellowship. Y.H. acknowledges the Ministry of Training, Colleges and Universities for provision of an Ontario Graduate Scholarship (OGS).

References

1. Noor MO et al (2014) Building from the "Ground" up: developing interfacial chemistry for solid-phase nucleic acid hybridization assays based on quantum dots and fluorescence resonance energy transfer. Coord Chem Rev 263:25–52

2. Noor MO, Krull UJ (2013) Paper-based solid-phase multiplexed nucleic acid hybridization assay with tunable dynamic range using immobilized quantum dots as donors in fluorescence resonance energy transfer. Anal Chem 85 (15):7502–7511

3. Noor MO, Tavares AJ, Krull UJ (2013) On-chip multiplexed solid-phase nucleic acid hybridization assay using spatial profiles of immobilized quantum dots and fluorescence resonance energy transfer. Anal Chim Acta 788:148–157

4. Algar WR et al (2012) Quantum dots as simultaneous acceptors and donors in time-gated forster resonance energy transfer relays: characterization and biosensing. J Am Chem Soc 134 (3):1876–1891

5. Freeman R, Liu XQ, Willner I (2011) Chemiluminescent and Chemiluminescence Resonance Energy Transfer (CRET) detection of DNA, metal ions, and aptamer-substrate complexes using hemin/G-quadruplexes and CdSe/ZnS quantum dots. J Am Chem Soc 133(30):11597–11604

6. Kumar M et al (2011) A rapid, sensitive, and selective bioluminescence resonance energy transfer (BRET)-based nucleic acid sensing system. Biosens Bioelectron 30(1):133–139

7. Doughan S et al (2014) Solid-phase covalent immobilization of upconverting nanoparticles for biosensing by luminescence resonance energy transfer. ACS Appl Mater Interfaces 6 (16):14061–14068

8. Doughan S, Uddayasankar U, Krull UJ (2015) A paper-based resonance energy transfer nucleic acid hybridization assay using upconversion nanoparticles as donors and quantum dots as acceptors. Anal Chim Acta 878:1–8

9. DaCosta MV et al (2014) Lanthanide upconversion nanoparticles and applications in

bioassays and bioimaging: a review. Anal Chim Acta 832:1–33

10. Ju Q, Uddayasankar U, Krull U (2014) Paper-based DNA detection using lanthanide-doped LiYF4 upconversion nanocrystals as bioprobe. Small 10(19):3912–3917

11. Zhou F, Krull UJ (2014) Spectrally matched duplexed nucleic acid bioassay using two-colors from a single form of upconversion nanoparticle. Anal Chem 86(21):10932–10939

12. Algar WR, Tavares AJ, Krull UJ (2010) Beyond labels: a review of the application of quantum dots as integrated components of assays, bioprobes, and biosensors utilizing optical transduction. Anal Chim Acta 673(1):1–25

13. Algar WR, Krull UJ (2009) Interfacial transduction of nucleic acid hybridization using immobilized quantum dots as donors in fluorescence resonance energy transfer. Langmuir 25(1):633–638

14. Algar WR, Krull UJ (2009) Toward a multiplexed solid-phase nucleic acid hybridization assay using quantum dots as donors in fluorescence resonance energy transfer. Anal Chem 81(10):4113–4120

15. Robelek R et al (2004) Multiplexed hybridization detection of quantum dot-conjugated DNA sequences using surface plasmon enhanced fluorescence microscopy and spectrometry. Anal Chem 76(20):6160–6165

16. Martinez AW et al (2007) Patterned paper as a platform for inexpensive, low-volume, portable bioassays. Angew Chem Int Ed 46(8):1318–1320

17. Carrilho E, Martinez AW, Whitesides GM (2009) Understanding wax printing: a simple micropatterning process for paper-based microfluidics. Anal Chem 81(16):7091–7095

18. Martinez AW et al (2008) FLASH: a rapid method for prototyping paper-based microfluidic devices. Lab Chip 8(12):2146–2150

19. Martinez AW et al (2010) Diagnostics for the developing world: microfluidic paper-based analytical devices. Anal Chem 82(1):3–10

20. Yetisen AK, Akram MS, Lowe CR (2013) Paper-based microfluidic point-of-care diagnostic devices. Lab Chip 13(12):2210–2251

21. Noor MO, Krull UJ (2014) Camera-based ratiometric fluorescence transduction of nucleic acid hybridization with reagentless signal amplification on a paper-based platform using immobilized quantum dots as donors. Anal Chem 86(20):10331–10339

22. Watterson JH et al (2004) Rapid detection of single nucleotide polymorphisms associated with spinal muscular atrophy by use of a reusable fibre-optic biosensor. Nucleic Acids Res 32(2):1–9

23. Noor MO, Shahmuradyan A, Krull UJ (2013) Paper-based solid-phase nucleic acid hybridization assay using immobilized quantum dots as donors in fluorescence resonance energy transfer. Anal Chem 85(3):1860–1867

24. Pong BK, Trout BL, Lee JY (2008) Modified ligand-exchange for efficient solubilization of CdSe/ZnS quantum dots in water: a procedure guided by computational studies. Langmuir 24(10):5270–5276

25. Algar WR et al (2011) The controlled display of biomolecules on nanoparticles: a challenge suited to bioorthogonal chemistry. Bioconjug Chem 22(5):825–858

26. Sapsford KE et al (2013) Functionalizing nanoparticles with biological molecules: developing chemistries that facilitate nanotechnology. Chem Rev 113(3):1904–2074

27. Algar WR, Krull UJ (2008) Quantum dots as donors in fluorescence resonance energy transfer for the bioanalysis of nucleic acids, proteins, and other biological molecules. Anal Bioanal Chem 391(5):1609–1618

Chapter 20

Enhanced Performance of Colorimetric Biosensing on Paper Microfluidic Platforms Through Chemical Modification and Incorporation of Nanoparticles

Ellen Flávia Moreira Gabriel, Paulo T. Garcia, Elizabeth Evans, Thiago M.G. Cardoso, Carlos D. Garcia, and Wendell K.T. Coltro

Abstract

This chapter describes two different methodologies used to improve the analytical performance of colorimetric paper-based biosensors. Microfluidic paper-based analytical devices (μPADs) have been produced by a stamping process and CO_2 laser ablation and modified, respectively, through an oxidation step and incorporation of silica nanoparticles on the paper structure. Both methods are employed in order to overcome the largest problem associated with colorimetric detection, the heterogeneity of the color distribution in the detection zones. The modification steps are necessary to improve the interaction between the paper surface and the selected enzymes. The enhanced performance has ensured reliability for quantitative analysis of clinically relevant compounds.

Key words Oxidation process, Nanoparticle, Colorimetric detection, Clinical diagnostic, Urinalysis test

1 Introduction

According to the definition, a biosensor is an analytical device that combines two specific characteristics: an element able to interact specifically with a target and a transducer able to convert the recognition process into a measurable signal [1, 2]. The extensive use of biosensors can be understood since these devices are able to provide reliable, specific, and rapid responses. There are numerous examples in literature reporting the construction of biosensors dedicated to different applications such as clinical and biological monitoring [3–5], environmental pollutants [6], and food and water contamination [7, 8]. In a particular case of clinical and diagnostic monitoring, the development of biosensors has increased significantly, since it has been demonstrated that early diagnosis can dramatically change the clinical treatment type [1].

Avraham Rasooly and Ben Prickril (eds.), *Biosensors and Biodetection: Methods and Protocols Volume 1: Optical-Based Detectors*, Methods in Molecular Biology, vol. 1571, DOI 10.1007/978-1-4939-6848-0_20,
© Springer Science+Business Media LLC 2017

In order to understand the biological systems and mainly the specific biomolecular interaction involved in the process, different techniques have been proposed. The most commonly explored technologies are the surface plasmon resonance (SPR) [9] and electrochemical impedance spectroscopy (EIS) [10]. Both techniques offer high sensitivity and are commercially established. However, the instrumentation required in both techniques is expensive and it demands trained personal to operate the instrument accordingly. These disadvantages hinder promptly the implementation of both SPR and EIS in research groups with limited financial support [11]. Alternatively, different techniques or simpler tools are emerging as promising alternatives to monitor binding events. Some examples involve the use of chemosensors [12], supercapacitive admittance tomoscopy technique [13], and capacitively coupled contactless conductivity detection [11, 14].

In general, the fabrication of biosensors involves the thin-film deposition followed by surface functionalization. These procedures are laborious and raise the final cost of the biosensor. Therefore, as an alternative that could maintain the main characteristics of biosensors (specific interaction with target analyte and measurement of signal response), microfluidic paper-based analytical devices (μPADs) have emerged as a new generation of powerful platforms serving as a low-cost substrate for biosensing applications. Since the pioneering report published by Whitesides group in 2007 [15], many research groups have demonstrated that μPADs could offer important advantages in the design of novel biosensing systems, especially for clinical diagnostics. The main advantages of μPADs are the low-cost, low sample consumption, minimal waste disposal, ease-of-use, global affordability, disposability, biocompatibility, and capability to be used in remote areas [3, 15–17]. All of these features make μPADs attractive to be implemented in low-income communities, principal targets to improve public health care [1].

Different techniques have been employed to produce paper-based biosensors including photolithography [15, 18, 19], inkjet etching [20, 21], wax printing [22], laser cutting [23, 24], and stamping process [4, 25]. All fabrication methods allow the production of microfluidic channels on paper surface, which due to its porous structure facilitates solution transport through capillary action. In addition to the fluidic channels, specific regions known as detection zones are designed and integrated with channels to proceed the chemical reaction responsible for the specific recognition process. In order to provide the response, suitable transduction methods are coupled with paper-based biosensors including colorimetric [3–5, 24], electrochemical [26–28], fluorescence [29], chemiluminescence [30, 31], and mass spectrometry [32, 33] detectors. Among the different detection systems, colorimetric detection represents one of the most popular tools used as transduction mode on μPADs. Color is usually developed through enzymatic or complexometric reactions between analyte and chromogenic agent. The presence of target analyte in sample may further be qualitatively identified by the naked eye inside the detection zones. Moreover, the concentration level can be quantified based on the color intensity, which is measured by the pixel intensity recorded in a digital image captured with scanner [3–5], digital camera [34], or cell phone cameras [35–37].

Despite the wide applicability of colorimetric biosensors on µPADs, many reports have presented drawbacks that still need to be addressed to ensure reliability for quantitative measurements. The most remarkable is related the lack of color homogeneity or uniformity generated in the detection zones during the reaction process between the analyte and indicator, and typically catalyzed by a specific enzyme. The heterogeneous color is due to the poor interaction between the support material (paper) and the enzyme or chromogenic reagents.

To overcome this problem, this chapter describes two different methods to enhance the analytical performance of colorimetric biosensors on µPADs. The improvement is achieved through the (1) oxidation process and (2) incorporation of silica nanoparticles onto paper structure. Both modification processes are adopted to provide better support for enzyme adsorption and therefore improve the color homogeneity and uniformity inside the detection zones. For each procedure, the optimal conditions of the modification steps are investigated. The µPADs modified by the oxidation process are evaluated for the detection of glucose and uric acid. On the other hand, the µPADs incorporated with silica nanoparticles are explored on glucose, lactate, and glutamate assays. In both instances, a flatbed scanner is used to quantify each analyte using the histogram tools in the graphic software. The feasibility of both modification methodologies are evaluated by urinalysis tests using an artificial urine samples.

2 Materials

2.1 Paper-Based Analytical Device

1. Computer equipped with graphic software Corel Photo-Paint™ and Adobe Photoshop®.

2. Paraffin.

3. Metal Stamp designed in stainless steel by a local shop (MS Máquinas, Goiânia, GO, Brazil).

4. CO_2 Laser ablation system model Mini 24, 30 W obtained from Epilog Laser Systems (Golden, CO, USA).

2.2 Modification Process (Oxidation and Incorporation with SiO₂)

1. 0.5 M sodium periodate solution prepared in water.

2. Silicon dioxide nanopowder (SiO_2—15 nm) obtained from Sigma-Aldrich (St. Louis, MO, USA).

3. 5% (v/v) solution of 3-aminopropyltrielthoxysilane (APTES) (Sigma-Aldrich; St. Louis, MO, USA) in ethanol.

4. Nanoparticle suspension: 3.3 mg of SiO_2 nanoparticles suspended in 10 mL of 5% APTES solution.

2.3 Colorimetric Detection

1. Flatbed Scanner.

2. Histogram tools from graphic software.

2.4 Bioassay Solutions

1. Solution 1: potassium iodide (0.6 M) and trehalose (0.3 M) dissolved in 100 mM phosphate buffer at pH 6.0.

2. Solution 2: 4-aminoantipyrine (4-AAP) and sodium 3,5-dichloro-2-hydroxybenzenesulfonic acid (DHBS), both in 6.6 mM, dissolved in 100 mM phosphate buffer at pH 6.0.

3. Solution 3: 15 mM 3,3′,5,5′-tetramethylbenzidine (TMB) dissolved in ethanol and wrapped with aluminum foil to protect from light.

4. Solution 4: A 5:1 mixture of glucose oxidase (120 U/mL) and horseradish peroxidase (30 U/mL) is prepared in 100 mM phosphate buffer at pH 6.0.

5. Solution 5: A 1:1 mixture of lactate oxidase (100 U/mL) and horseradish peroxidase (339 U/mL) dissolved in 100 mM phosphate buffer at pH 6.0.

6. Solution 6: A 1:1 mixture of L-glutamate oxidase (4.16 U/mL) and horseradish peroxidase (339 U/mL) prepared in 100 mM phosphate buffer at pH 7.4.

7. Solution 7: A 1:1 mixture of uricase oxidase (80 U/mL) and horseradish peroxidase (339 U/mL) dissolved in phosphate buffer solution pH 6.0.

8. All reagents employed on colorimetric assays are acquired from Sigma-Aldrich (St. Louis, MO, USA).

2.5 Artificial Urine Sample

1. Artificial urine is prepared at pH 6.0 and consists of 2 mM citric acid (Sigma-Aldrich; St. Louis, MO, USA), 25 mM sodium bicarbonate (Sigma-Aldrich; St. Louis, MO, USA), 170 mM urea (Sigma-Aldrich; St. Louis, MO, USA), 2.5 mM calcium chloride (E.M. Science; Gibbstown, NJ, USA), 90 mM sodium chloride (Mallinckrodt Baker; Center Valley, PA, USA), 2 mM magnesium sulfate (Mallinckrodt Baker; Center Valley, PA, USA), 10 mM sodium sulfate (Sigma-Aldrich; St. Louis, MO, USA), 7 mM sodium phosphate monobasic anhydrous (Fisher Scientific; Waltham, MA, USA), 7 mM sodium phosphate dibasic anhydrous (Fisher Scientific; Waltham, MA, USA), and 25 mM ammonium chloride (Sigma-Aldrich; St. Louis, MO, USA). This solution is stored in the refrigerator (4 °C) until use.

3 Methods

3.1 Methodologies to Produce μPADs

3.1.1 Stamping Process

1. Stamped μPADs are fabricated according to a procedure previously developed by our group [38] and schematically described in Fig. 1.

2. Initially, a sheet of filter paper is immersed into liquid paraffin (at 90 °C) for 60 s.

3. Then, the paper is removed from the paraffin solution, allowed to solidify at room temperature and placed above sheet of paper without paraffin, thus forming a sandwiched paper piece.

4. A metal stamp is preheated at 150 °C on a hot plate for 2 min.

5. Then, the metal stamp is put in contact with sandwiched piece and ca. 0.1 MPa of pressure is applied to the metal stamp for 2 s. This process is necessary to enable the thermal transfer of paraffin from top paper to bottom paper, thus forming hydrophobic barriers.

6. The layout of the proposed μPADs (45 mm × 45 mm) consisted of eight circular detection zones for bioassays interconnected by microfluidic channels and one central zone as sample inlet.

7. All channels are nominally fabricated with a 10 mm length and 3 mm width. The diameter values for detection and central zones are 5 and 10 mm, respectively.

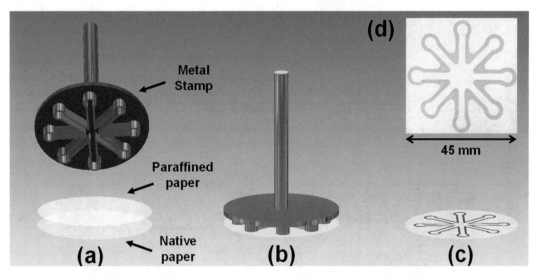

Fig. 1 Scheme showing the fabrication process of μPADs based on stamping process. In (**a**), a paraffinized paper is placed over the native paper surface; in (**b**), a metal stamp preheated at 150 °C is brought into contact with the layered paper pieces; in (**c**), it is presented a typical μPAD fabricated by the stamping method. The optical micrograph in (**d**) depicts a real image showing the stamped μPAD. Reproduced from ref. 4 with permission

1. Laser cutting is also selected out of the many techniques used to fabricate µPADs due to its simplicity of being a straightforward single step and quick process.

2. A schematic representation of the laser cutting step can be seen in Fig. 2.

3. The µPAD is first designed using CorelDraw™ X6 Software with dimensions of a 1.25 mm main channel that branches into three channels of the same width leading to three 5 mm detection circles on each end.

4. The design is sent to the laser engraver and cut at 600-dpi resolution using 30% power and 30% speed. A steady flow of N_2 gas impinged at the engraving point to minimize the risk of ignition.

5. The laser cuts the paper and define the geometry on the surface by the ablation process.

6. The laser cuts at a rate of approximately 5 s per µPAD.

3.2   Modification Steps

3.2.1   Oxidation Process with NaIO₄

1. Paper substrates are chemically modified by an oxidation process using $NaIO_4$ to allow the covalent coupling of enzymes on the cellulose surface.

2. Prior to stamping, the paper sheets are immersed in a petri dish containing 10 mL of 0.5 M $NaIO_4$ solution and allowed to react at room temperature in the dark for 30 min.

3. After oxidation, the paper sheets are washed three times with ultrapure water.

4. Then, µPADs are stamped according to procedure described on Subheading 3.1.2.

5. All detection zones are spotted with 0.75 µL of a mixed solution of 1-Ethyl-3-(3-dimethylaminopropyl)carbodiimide (EDC, 0.1 M) (Sigma-Aldrich; St. Louis, MO, USA) and N-Hydroxysuccinimide (NHS, 0.1 M) (Sigma-Aldrich; St. Louis, MO, USA) prepared in ultrapure water. Then, each detection zone is allowed to dry at room temperature for

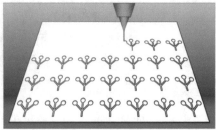

Fig. 2 Procedure for fabricatin µPADs by CO_2 laser process. (**a**) CO_2 laser machine. (**b**) Scheme showing the laser cut of a paper sheet to produce paper biosensor devices. All µPADs were cut using the vector mode, at 30% speed (of a maximum linear speed of 1.65 cm/s) and 30% power (of a maximum intensity of 30 W)

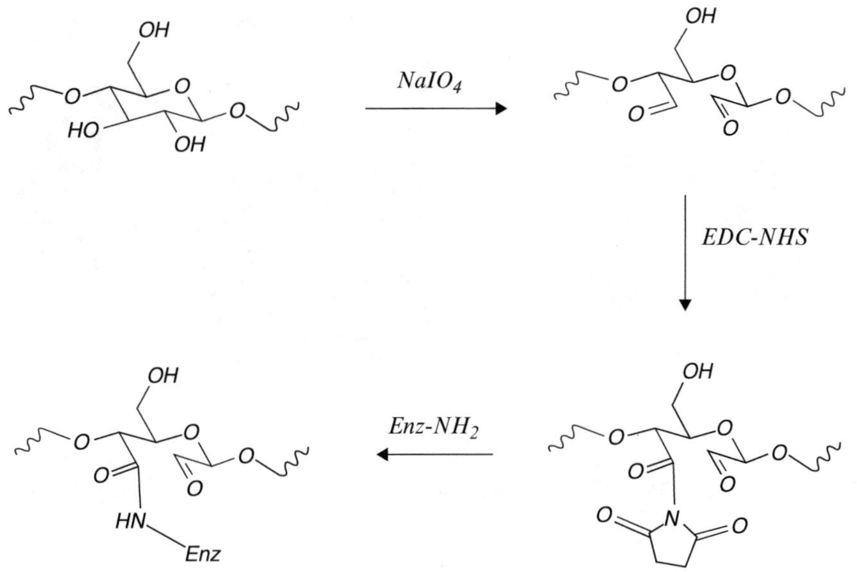

Fig. 3 Reactional scheme showing the modification process on paper surface with NalO$_4$ to obtain a paper biosensor. Adapted from ref. 17 with permission

20 min to convert the aldehyde groups of the cellulose to amine reactive esters.

6. Afterwards, enzymes are covalently conjugated on the modified paper surface by adding 0.75 µL of each enzyme solution to the detection zones.

7. All reactions involved in the modification process of the cellulose surface to obtain a biosensor in paper platform are shown in Fig. 3.

3.2.2   Incorporation of Silica Nanoparticle (SiO$_2$NP)

Silica nanoparticles (SiO$_2$NP) are also employed to modify the surface of the paper in order to accomplish a uniform distribution of the enzyme, therefore increasing the color homogeneity in the detection zones. The SiO$_2$NP incorporated on detection zone of µPAD can been seen in Fig. 4.

1. First, a silane-coupling reaction is employed to introduce amine groups onto the surface of the nanoparticles. 3.3 mg of 15 nm SiO$_2$NP are added to a 5% (v/v) solution of APTES in ethanol.

2. The length of time the nanoparticles remained in contact with the APTES solution is investigated from 1 to 24 h due to the variables that can affect the density of the amine groups, and ultimately the adsorption of the enzymes to the modified µPAD.

3. The incorporation of SiO$_2$NP on µPADs is realized by two different procedures: dispensing the nanoparticle suspension

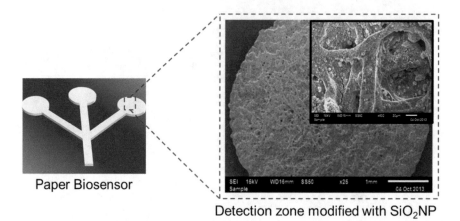

Fig. 4 Scheme of modification process with SiO$_2$NP. Scanning electron micrograph images show the silica incorporated on cellulose surface. The micrograph images was reproduced from ref. 5 with permission

using an automatic pipette and full immersion of the paper in the nanoparticle suspension.

4. The μPADs are carefully rinsed with 0.1 M PBS pH 6 to remove any loosely bound particles and finally allowed to dry at room temperature before use.

3.3 Enzymatic Procedures

3.3.1 μPADs—Oxidation Process

The enzymatic assays are performed by adding specific color reagents for each bioassay on the detection zones. The processes are described below.

1. For the glucose assay, each detection zone is spotted with 0.75 μL of solution 1 and dried at room temperature for 10 min. Then, 0.75 μL of solution 4 is spotted in each detection zone and allowed to dry at room temperature for 10 min.

2. For the uric acid assay, each detection zone is spotted with 0.75 μL of solution 2 and dried at room temperature for 10 min. Afterwards, 0.75 μL of solution 7 is added in each detection zone and allowed to dry at room temperature for the same time used previously.

3.3.2 μPADs—SiO$_2$NP Incorporation

1. For the lactate assay, each detection zone is spotted with 1 μL of solution 2 and dried at room temperature for 15 min. Afterwards, the zone is spotted with 1 μL of solution 5 and dried for the same duration as before.

2. For the glucose assay, the detection zone is spotted with 1 μL of solution 1 and dried at room temperature for 15 min. Afterwards, the zone is spotted with 1 μL of solution 4 and dried for additional 15 min.

3. For the glutamate assay, the detection zone is spotted with 1 μL of solution 3 and dried at room temperature for 15 min. Then

the zone is spotted with 1 μL of solution 6 and dried for 10 min.

4. In order to perform simultaneously all three bioassays on μPADs, a 10 μL solution containing the analytes is introduced at the bottom stem of the μPAD and allowed to wick up the three channels by capillarity to the detection zones.

3.4 Colorimetric Detection System

The scanner mode of a DeskJet multifunctional printer (Hewlett-Packard, model F4280) is used to capture the colorimetric information related to the semi-quantitation of glucose, lactate, glutamate, and uric acid.

1. Following the colorimetric enzymatic procedures described in Subheading 3.3, images of the μPADs are taken with a scanner using 600-dpi resolution.

2. The recorded images are first converted to a 24-bit color scale (RGB or CMYK dimension) and then analyzed in either Corel Photo-Paint™ or Adobe Photoshop software. Some free softwares, as for example ImageJ, can also be used to obtained the pixel information.

3. The arithmetic mean of the pixel intensity within each test zone is obtained with the histogram tool of graphic software and used to obtain the color intensity.

4. The color intensity achieved is used to achieve the analyte concentration and thus to perform quantitative measurements.

4 Paper-Based Biosensor: Characterization Step

The poor color uniformity and homogeneity associated with colorimetric measurements represent some the most important shortcomings on μPADs (*see* **Notes 1** and **2**). Aiming to solve this problem, the oxidation of cellulose and incorporation of SiO_2 nanoparticles are used as analytical strategies to support the enzymes required for the selected bioassays (*see* **Note 3**). The improvements achieved with both methodologies are described below.

4.1 Oxidation with $NaIO_4$

1. The oxidation process is used for the conversion of the hydroxyl groups on paper surface to aldehyde groups. Then, these groups are chemically activated with an EDC/NHS solution enabling the covalent coupling with different enzymes.

2. The conversion process is monitored by infrared spectroscopy. The technique confirmed the presence of aldehyde groups on paper surface after the oxidation step. As it can be observed in Fig. 5, the specific band at 1727 cm^{-1}, characteristic of aldehyde groups, is found only on the oxidized paper.

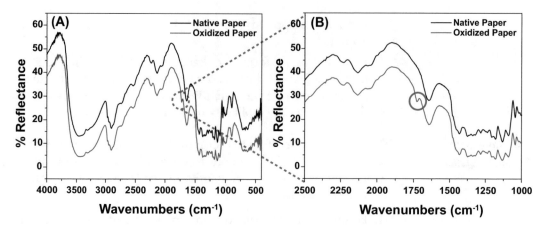

Fig. 5 Characterization of native and oxidized paper by Fourier transform infrared spectroscopy. Reproduced from ref. 4 with permission

3. Besides the characterization process, the feasibility of the μPADs is tested by performing the glucose and uric acid assays. The procedure is performed by spotting the specific color reagents on each detection zone and ca. 40 μL of standard solution on the central zone. After 15 min, the images are obtained by colorimetric detection as previously described in Subheading 3.4.

4. The chemical modification has improved the color uniformity for both assays, as observed in Fig. 6. The values found of color gradient is lower than 10%. The color gradient or uniformity values are obtained from the standard deviation provided by the histogram tools on the graphic software. The better gradient is expected once the oxidized process ensures the chemical bond between activated groups on the paper surface and amino group from each enzyme.

4.2 Incorporation of SiO₂NP

1. The performance of amino group incorporation on silica nanoparticle is evaluated on color intensity and gradient response. As it can be observed in Fig. 7, the modification time has significant effects on the color intensity but little effect on the color gradient. These behaviors can be interpreted by relation between the amount of protein adsorbed on paper modified with SiO$_2$NP and the strength of the interaction. In short time of reaction just few amino groups are present on silica surface to support the enzyme dispensed, and in long reaction times the modification can make the surface too hydrophobic. Considering these results, 3 h is selected as the optimum time for the silica nanoparticles to be modified with the amino groups.

2. The best method to incorporate the SiO$_2$NP on paper surface is also investigated. As described on Subheading 3.2, basically two different (dispense by pipette and full immersion) procedures are used. The better homogeneity distribution of

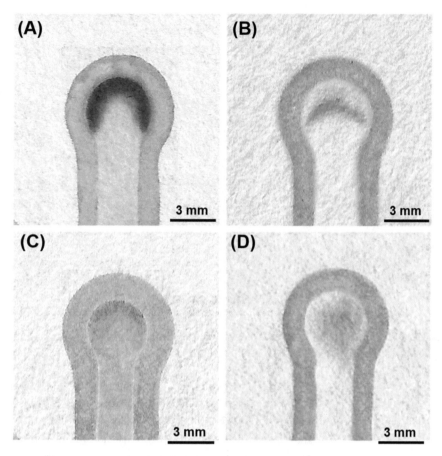

Fig. 6 Optical micrographs showing the improvements on the color uniformity for glucose (*left images*) and uric acid assays (*right images*) before (**a**, **b**) and after (**c**, **d**) chemical oxidation on paper surface. Reproduced from ref. 4 with permission

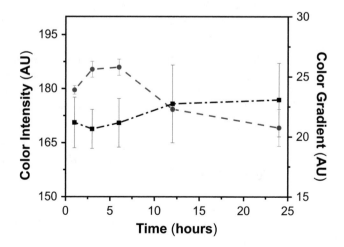

Fig. 7 Effect of modification time of silica nanoparticles with APTES on the color intensity and gradient. Reproduced from ref. 5 with permission

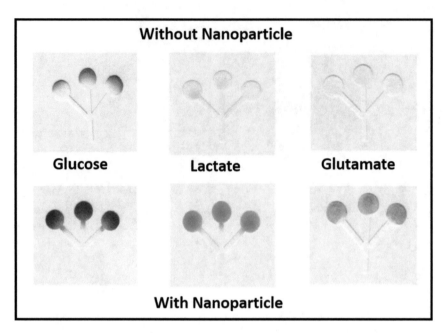

Fig. 8 Optical images showing the colorimetric assays for glucose, lactate and glutamate assay with and without nanoparticle. Reproduced from ref. 5 with permission

nanoparticle on paper surface is achieved when the paper is full immersed in the suspension solution (*see* Fig. 4 on SEM image). This system is used for the remaining experiments.

3. Once the optimized conditions for the incorporation process have been established, the feasibility of these devices are shown for lactate, glucose, and glutamate bioassays. In this case the μPADs are prepared as described in Subheading 3.3.2.

4. All three selected reactions are catalyzed by enzymes with well-known properties, and use a simple, oxygen-dependent reaction that produces H_2O_2. The H_2O_2 is then utilized to oxidize a chromogenic agent in a secondary reaction catalyzed by HRP.

5. The enzymatic tests are carried out on the paper devices with nanoparticles and the results are compared with those conducted on native μPADs. As can be observed on Fig. 8 significant improvements in the signal intensity and color homogeneity are obtained when silica nanoparticles are incorporated on paper device. In the glucose assay, for example, the increase in color intensity is 46% while the color gradient decreased by 75%.

4.3 Clinical Assay with Artificial Urine Sample

1. μPADs modified with SiO_2NP have been explored for the detection of glucose, lactate, and glutamate in artificial urine samples.

2. Once demonstrated individual assays on modified μPADs, all three assays (glutamate, lactate, and glucose) are simultaneously

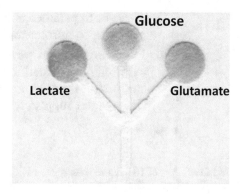

Fig. 9 Optical image showing the analysis of an artificial urine sample spiked with lactate, glucose, and glutamate on the proposed paper based biosensor. Reproduced from ref. 5 with permission

performed on the same devices using artificial urine as biological sample (Fig. 9). For these experiments, urine (prepared as described in the Subheading 2.5) is spiked with glucose (7.33 mM), lactate (2.14 mM), and glutamate (4.21 mM). These levels are chosen once they can be related to different biological disorders such as diabetes, liver disturbers, and risk of migraines [39–41].

3. The pixel intensity information is correlated with concentration using a calibration curve for each analyte. The linear equations for glucose, lactate, and glutamate are $y = 2 + 8.7$ [Glucose]; $y = 22 + 14$[Lactate] and $y = 16.5 + 7.5$[Glutamate], respectively. The concentration values obtained from glucose, lactate, and glutamate are 7.9 ± 0.6 mM, 2.2 ± 0.2 mM, and 5.5 ± 1.0 mM, respectively.

4. The modified μPAD has exhibited good analytical performance for simultaneous assays. The relative standard deviation (RSD) values achieved for colorimetric measurements ranged from 0.9% to 8%. The estimated error between found concentrations and known concentrations is ca. 20%.

5. Based on the results, μPADs have offered good accuracy and precision when used as colorimetric biosensors. They also have great potential for quantitative determination of clinically relevant analytes. In addition, these biosensors are fabricated using a low cost substrate allowing their use in places with limited resources.

5 Notes

1. The poor color uniformity and homogeneity associated with colorimetric measurements represent one the most important shortcomings on μPADs.

2. These both parameters have affected the reliability of low-cost platform, compromising its use in clinical assays.

3. In our experiment, we used two different strategies associated with cellulose oxidation and incorporation of SiO_2 nanoparticles aiming the chemical modification of paper surface to provide a better support for enzyme adsorption.

References

1. Parolo C, Merkoci A (2013) Paper-based nanobiosensors for diagnostics. Chem Soc Rev 42:450–457

2. Nery EW, Kubota LT (2013) Sensing approaches on paper-based devices: a review. Anal Bioanal Chem 405:7573–7595

3. Martinez AW, Phillips ST, Carrilho E, Thomas SW III, Sindi H, Whitesides GM (2008) Simple telemedicine for developing regions: camera phones and paper-based microfluidic devices for real-time, off-site diagnosis. Anal Chem 80:3699–3707

4. Garcia PT, Cardoso TMG, Garcia CD, Carrilho E, Coltro WKT (2014) A handheld stamping process to fabricate microfluidic paper-based analytical devices with chemically modified surface for clinical assays. RSC Adv 4:37637–37644

5. Evans E, Gabriel EFM, Benavidez TE, Coltro WKT, Garcia CD (2014) Modification of microfluidic paper-based devices with silica nanoparticles. Analyst 139:5560–5567

6. Gomes HI, Sales MGF (2015) Development of paper-based color test-strip for drug detection in aquatic environment: application to oxytetracycline. Biosens Bioelectron 65:54–61

7. Sicard C, Glen C, Aubie B, Wallace D, Jahanshahi-Anbuhi S, Pennings K, Daigger GT, Pelton R, Brennan JD, Filipe CD (2015) Tools for water quality monitoring and mapping using paper-based sensors and cell phones. Water Res 70:360–369

8. Zhang Y, Zuo P, Ye B-C (2015) A low-cost and simple paper-based microfluidic device for simultaneous multiplex determination of different types of chemical contaminants in food. Biosens Bioelectron 68:14–19

9. Chung J, Kim S, Bernhardt R, Pyun J (2005) Application of SPR biosensor for medical diagnostics of human hepatitis B virus (hHBV). Sens Actuators B 111:416–422

10. Lin Z, Chen L, Zhang G, Liu Q, Qiu B, Cai Z, Chen G (2012) Label-free aptamer-based electrochemical impedance biosensor for 17β-estradiol. Analyst 137:819–822

11. Coltro WKT, de Santis Neves R, de Jesus Motheo A, Da Silva JAF, Carrilho E (2014) Microfluidic devices with integrated dual-capacitively coupled contactless conductivity detection to monitor binding events in real time. Sens Actuators B 192:239–246

12. Delaney TL, Zimin D, Rahm M, Weiss D, Wolfbeis OS, Mirsky VM (2007) Capacitive detection in ultrathin chemosensors prepared by molecularly imprinted grafting photopolymerization. Anal Chem 79:3220–3225

13. Gamby J, Abid J-P, Abid M, Ansermet J-P, Girault H (2006) Nanowires network for biomolecular detection using contactless impedance tomoscopy technique. Anal Chem 78:5289–5295

14. Lima RS, Piazzetta MH, Gobbi AL, Rodrigues-Filho UP, Nascente PA, Coltro WK, Carrilho E (2012) Contactless conductivity biosensor in microchip containing folic acid as bioreceptor. Lab Chip 12:1963–1966

15. Martinez AW, Phillips ST, Butte MJ, Whitesides GM (2007) Patterned paper as a platform for inexpensive, low-volume, portable bioassays. Angew Chem Int Ed 46:1318–1320

16. Martinez AW, Phillips ST, Whitesides GM, Carrilho E (2009) Diagnostics for the developing world: microfluidic paper-based analytical devices. Anal Chem 82:3–10

17. Pelton R (2009) Bioactive paper provides a low-cost platform for diagnostics. Trends Anal Chem 28:925–942

18. Martinez AW, Phillips ST, Wiley BJ, Gupta M, Whitesides GM (2008) FLASH: a rapid method for prototyping paper-based microfluidic devices. Lab Chip 8:2146–2150

19. Klasner S, Price A, Hoeman K, Wilson R, Bell K, Culbertson C (2010) Paper-based microfluidic devices for analysis of clinically relevant analytes present in urine and saliva. Anal Bioanal Chem 397:1821–1829

20. Abe K, Suzuki K, Citterio D (2008) Inkjet-printed microfluidic multianalyte chemical sensing paper. Anal Chem 80:6928–6934

21. Wang J, Monton MRN, Zhang X, Filipe CDM, Pelton R, Brennan JD (2014) Hydrophobic

sol-gel channel patterning strategies for paper-based microfluidics. Lab Chip 14:691–695

22. Carrilho E, Martinez AW, Whitesides GM (2009) Understanding wax printing: a simple micropatterning process for paper-based microfluidics. Anal Chem 81:7091–7095

23. Nie J, Liang Y, Zhang Y, Le S, Li D, Zhang S (2013) One-step patterning of hollow microstructures in paper by laser cutting to create microfluidic analytical devices. Analyst 138:671–676

24. Evans E, Gabriel EFM, Coltro WKT, Garcia CD (2014) Rational selection of substrates to improve color intensity and uniformity on microfluidic paper-based analytical devices. Analyst 139:2127–2132

25. Zhang Y, Zhou C, Nie J, Le S, Qin Q, Liu F, Li Y, Li J (2014) Equipment-free quantitative measurement for microfluidic paper-based analytical devices fabricated using the principles of movable-type printing. Anal Chem 86:2005–2012

26. Dungchai W, Chailapakul O, Henry CS (2009) Electrochemical detection for paper-based microfluidics. Anal Chem 81:5821–5826

27. Noiphung J, Songjaroen T, Dungchai W, Henry CS, Chailapakul O, Laiwattanapaisal W (2013) Electrochemical detection of glucose from whole blood using paper-based microfluidic devices. Anal Chim Acta 788:39–45

28. Santhiago M, Wydallis JB, Kubota LT, Henry CS (2013) Construction and electrochemical characterization of microelectrodes for improved sensitivity in paper-based analytical devices. Anal Chem 85:5233–5239

29. Yamada K, Takaki S, Komuro N, Suzuki K, Citterio D (2014) An antibody-free microfluidic paper-based analytical device for the determination of tear fluid lactoferrin by fluorescence sensitization of Tb3+. Analyst 139:1637–1643

30. Delaney JL, Hogan CF, Tian J, Shen W (2011) Electrogenerated chemiluminescence detection in paper-based microfluidic sensors. Anal Chem 83:1300–1306

31. Yu J, Ge L, Huang J, Wang S, Ge S (2011) Microfluidic paper-based chemiluminescence biosensor for simultaneous determination of glucose and uric acid. Lab Chip 11:1286–1291

32. Wleklinski M, Li Y, Bag S, Sarkar D, Narayanan R, Pradeep T, Cooks RG (2015) Zero volt paper spray ionization and its mechanism. Anal Chem 87:6786–6793

33. Bag S, Hendricks PI, Reynolds JC, Cooks RG (2015) Biogenic aldehyde determination by reactive paper spray ionization mass spectrometry. Anal Chim Acta 860:37–42

34. Apilux A, Siangproh W, Praphairaksit N, Chailapakul O (2012) Simple and rapid colorimetric detection of Hg(II) by a paper-based device using silver nanoplates. Talanta 97:388–394

35. Wang H, Li Y-j, Wei J-f, Xu J-r, Wang Y-h, Zheng G-x (2014) Paper-based three-dimensional microfluidic device for monitoring of heavy metals with a camera cell phone. Anal Bioanal Chem 406:2799–2807

36. Salles M, Meloni G, de Araujo W, Paixão T (2014) Explosive colorimetric discrimination using a smartphone, paper device and chemometrical approach. Anal Methods 6:2047–2052

37. Souza FR, Duarte-Junior GF, Garcia PT, Coltro WKT (2014) Avaliação de dispositivos de captura de imagens digitais para detecção colorimétrica em microzonas impressas. Quim Nova 37:1171–1176

38. Gabriel EFM, Coltro WKT, Garcia CD (2014) Fast and versatile fabrication of PMMA microchip electrophoretic devices by laser engraving. Electrophoresis 35:2325–2332

39. Dungchai W, Chailapakul O, Henry CS (2010) Use of multiple colorimetric indicators for paper-based microfluidic devices. Anal Chim Acta 674:227–233

40. Lin C-C, Tseng C-C, Chuang T-K, Lee D-S, Lee G-B (2011) Urine analysis in microfluidic devices. Analyst 136:2669–2688

41. Ragginer C, Lechner A, Bernecker C, Horejsi R, Moller R, Wallner-Blazek M, Weiss S, Fazekas F, Schmidt R, Truschnig-Wilders M, Gruber HJ (2012) Reduced urinary glutamate levels are associated with the frequency of migraine attacks in females. Eur J Neurol 19:1146–1150

Chapter 21

A Smartphone-Based Colorimetric Reader for Human C-Reactive Protein Immunoassay

A.G. Venkatesh*, Thomas van Oordt, E. Marion Schneider, Roland Zengerle, Felix von Stetten, John H.T. Luong, and Sandeep Kumar Vashist*

Abstract

A smartphone-based colorimetric reader (SBCR), comprising a Samsung Galaxy SIII mini, a gadget (iPAD mini, iPAD4, or iPhone 5s) and a custom-made dark hood and base holder assembly, is used for human C-reactive protein (CRP) immunoassay. A 96-well microtiter plate (MTP) is positioned on the gadget's screensaver to provide white light-based bottom illumination only in the specific regions corresponding to the well's bottom. The images captured by the smartphone's back camera are analyzed by a novel image processing algorithm. Based on one-step kinetics-based human C-reactive protein immunoassay (IA), SBCR is evaluated and compared with a commercial MTP reader (MTPR). For analysis of CRP spiked in diluted human whole blood and plasma as well as CRP in clinical plasma samples, SBCR exhibits the same precision, dynamic range, detection limit, and sensitivity as MTPR for the developed IA (DIA). Considering its compactness, low cost, advanced features and a remarkable computing power, SBCR is an ideal point-of-care (POC) colorimetric detection device for the next-generation of cost-effective POC testing (POCT).

Key words Smartphone-based colorimetric reader, Human C-reactive protein, Graphene, One-step kinetics, Sandwich immunoassay, Monitoring inflammation

1 Introduction

Cost-effective point-of-care (POC) devices are of importance for the readout of in vitro diagnostic (IVD) assays in decentralized, remote, and personalized settings. Current smartphones with advanced features provide a promising digital platform for POC diagnostics and mobile healthcare [1]. Smartphones are equipped with a high resolution camera, a powerful processor with high storage capacity, wireless connectivity, real-time geo-tagging, secure data management, and cloud computing. The potential of

---

*These authors are contributed equally to this work.

Avraham Rasooly and Ben Prickril (eds.), *Biosensors and Biodetection: Methods and Protocols Volume 1: Optical-Based Detectors*, Methods in Molecular Biology, vol. 1571, DOI 10.1007/978-1-4939-6848-0_21,
© Springer Science+Business Media LLC 2017

smartphones in such applications is phenomenal considering above seven billion cellphone users with more than 70% subscription in developing countries [2]. Smartphones are constantly upgraded with more advanced features at low cost. Thus, many smartphone devices and smart applications have been commercialized recently for health and fitness, diabetes and weight management, and measurement of routine healthcare characteristics [3]. In addition, many specialized smartphone-based bioanalytical applications have been demonstrated in the last few years [1]. Consequently, smartphones will have a significant impact on healthcare monitoring and management to revolutionize the fields of IVD and mobile healthcare by enabling real-time on-site analysis and telemedicine opportunities in remote settings [4].

The microtiter plate (MTP)-based enzyme linked immunosorbent assay (ELISA), a subject of over 350,000 publications, is the gold standard of IVD in clinical and bioanalytical settings. Therefore, there is an immense need for a portable and cost-effective SBCR for the readout of IVD assays at decentralized and remote settings. Considering the extensive outreach of cellphone that covers more than 95% of the world population, the developed SBCR will be of immense utility for IVD and mobile healthcare.

This chapter describes an extremely low-cost SBCR (Figs. 1 and 2) that employs a commercial gadget (iPAD mini, iPAD4, or iPhones 5s), an inexpensive polyamide dark hood and a polyamide base holder [5]. It serves as the colorimetric readout of graphene-based CRP IA [6] (Fig. 3) and compared with commercial MTPR for its analytical performance. As an important biomarker for infection and inflammation [7], CRP analysis is needed for a wide range of diseases, disorders, and pathophysiological conditions, such as neonatal sepsis [8–12], inflammasome related diseases [13], depressive and posttraumatic stress [14–16], diabetes [17–19], and cardiovascular diseases [20–24]. It is a signal-enhanced IA format, where graphene nanoplatelets (GNPs) provide an increased surface area for the higher binding of capture CRP antibody (Ab). The Ab is covalently bound to MTP in a leach-proof technique by admixing EDC-activated Ab with GNPs in APTES inside the MTP wells (Fig. 3). The DIA detectes CRP with linearity of 0.3–81 ng/mL and a limit of detection (LOD), a limit of quantification (LOQ) of 0.07 and 0.9 ng/mL, respectively (Fig. 4). The DIA enables a precise and specific determination of CRP in diluted human whole blood and plasma (Fig. 4a), as it is unaffected by immunological reagents and nonspecific control proteins (Fig. 4b). There is the leach-proof binding of capture Ab as the Ab-prebound MTPs retained their activity at 4 °C in 0.1 M PBS, pH 7.4 for up to 6 weeks (Fig. 4c). Moreover, there is no batch-to-batch variability for various preparations of GNP-anti-CRP Ab (Fig. 4d). The bioanalytical performance of the DIA employing SBCR is comparable to that of the conventional sandwich ELISA using MTPR and the

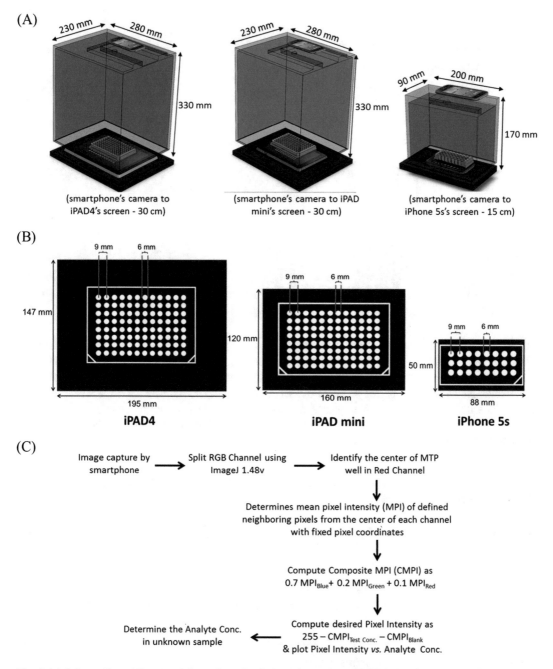

Fig. 1 (**a**) Schematics of the smartphone-based colorimetric readers (SBCRs) developing using the gadgets' (iPAD4, iPAD mini, or iPhone 5s) screen-based bottom illumination, Samsung Galaxy SIII mini's back camera (5 MP) based imaging, and a custom-made polyamide dark hood and polyamide base holder assembly. (**b**) Screensavers used for the screen-based bottom illumination of the 96-well microtiter plate (MTP) in gadgets. (**c**) Image processing algorithm employed for the analysis of smartphone-captured colorimetric images

clinically accredited analyzer-based IA. Similarly, the SBCR readout based DIA is identical in sensitivity to MTPR readout based DIA, as confirmed by the overlay plot of the normalized signals (Fig. 5).

(A)

(B)

(C)

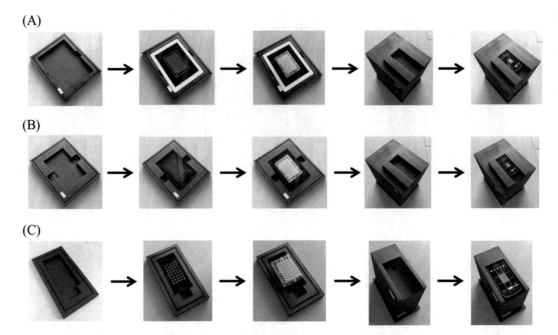

Fig. 2 Typical SBCR assemblies, demonstrating the steps involved for SBCR-based colorimetric readout of immunoassays, for (**a**) iPAD4, (**b**) iPAD mini, and (**c**) iPhone 5s. The process involves the placement of the gadget, with an illuminated screensaver, within the engraved cavity of the base holder. The MTP with colorimetric products is then placed on top of the screensaver after aligning with the respective illumination spots. Subsequently, the dark hood is placed on top of the base holder and the smartphone imaging is performed after placing the smartphone in the containment provided at the top of the dark hood and aligning the smartphone's back camera with the hole

2   Materials

2.1 Preparation of Capture Ab-Bound MTP

1. Nunc microwell 96-well polystyrene plates, flat bottom (non-treated), sterile (Catalog No. P7491, Sigma-Aldrich).

2. Eppendorf microtubes (1.5 mL; Catalog No. Z 606340, Sigma-Aldrich).

3. Deionized water (18.2 MΩ, DIW). (Direct-Q®3 Water Purification System, Millipore).

4. −70 °C freezer (operating range −60 to −86 °C) (New Brunswick).

5. 2–8 °C refrigerator (Future, UK).

6. Direct-Q®3 water purification system (Millipore, USA).

7. Mini incubator (Labnet Inc., UK).

8. PVC fume cupboard Chemflow range (CSC Ltd.).

9. KOH pellets (99.99%), semiconductor grade (Catalog No. 306568, Sigma-Aldrich).

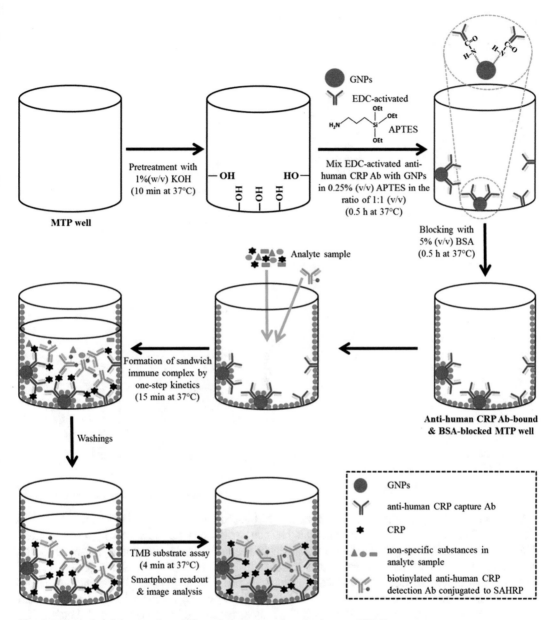

Fig. 3 Schematic of the developed bioanalytical procedure for human CRP IA

10. 3-aminopropyltriethoxysilane (3-APTES) (Catalog No. A3684, Sigma-Aldrich).

11. Graphene nanoplatelets (diameter 5 μm) (Catalog No. Grade 2 Graphene nanoplatelets, Cheap Tubes Inc): GNPs (1 mg) are mixed with 1 mL of 0.25% APTES and sonicated for 1 h.

12. 1-ethyl-3-(3-dimethylaminopropyl)-carbodiimide hydrochloride (EDC) (Catalog No. 22981, Thermo Scientific).

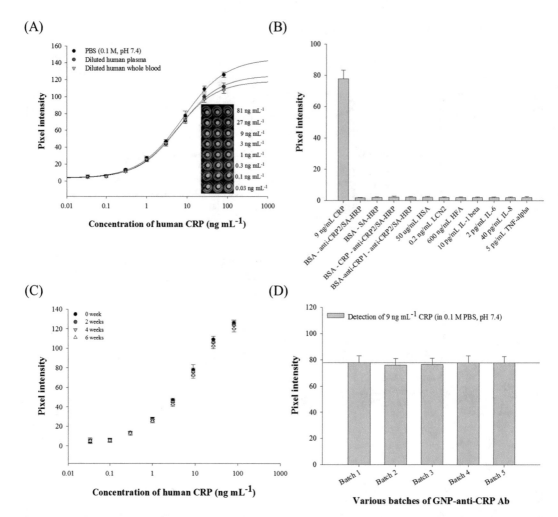

Fig. 4 The bioanalytical performance of the developed SBCR-based human CRP IA. (**a**) Detection of CRP in PBS (10 mM, pH 7.4), diluted human plasma and diluted human whole blood. (**b**) Experimental process controls. Anti-CRP1 and anti-CRP2 are capture and detection antibodies, respectively. The control proteins employed are human serum albumin (HSA), human lipocalin-2 (LCN2), human fetuin A (HFA), interleukin (IL)-1 beta, IL-6, IL-8, and tumor necrosis factor (TNF)-alpha. (**c**) Storage stability of anti-human CRP capture Ab-bound graphene-functionalized MTPs stored at 4 °C in PBS for 6 weeks. (**d**) Determination of batch-to-batch variability for various preparations of GNP-anti-CRP Ab, as demonstrated for the detection of 9 ng/mL CRP by the developed graphene-based IA. All experiments are performed in triplicate with the error bars representing the standard deviation. Reproduced with permission from Elsevier Inc. [6]

13. BupH phosphate buffered saline packs (0.1 M sodium phosphate, 0.15 M sodium chloride, pH 7.2) (Catalog No. 18372, Thermo Scientific).

14. BupH MES buffered saline packs (0.1 M MES [2-(N-morpholino)ethane sulfonic acid], 0.9% (w/v) sodium chloride, pH 4.7) (Catalog No. 28390, Thermo Scientific).

15. Human CRP Ab (Catalog No. MAB17071, R&D Systems).

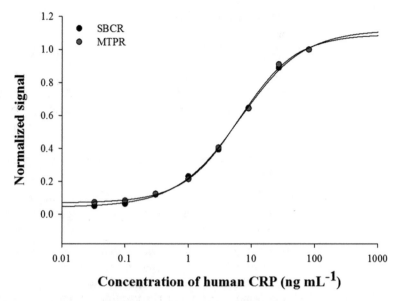

Fig. 5 An overlay plot of the developed human CRP IAs based on the use of SBCR and MTPR, and employing the normalized signals (the signal response obtained for a particular human CRP concentration divided by the signal response obtained for the highest human CRP concentration). Reproduced with permission from Elsevier Inc. [6]

16. Blocker BSA in PBS (10×), pH 7.4, 10% w/v (Catalog No. 37525, Thermo Scientific).

17. 0.1% BSA in PBS, pH 7.4 is used as the binding buffer.

18. 0.05% Tween® 20 in PBS, pH 7.4 (PBST) is used as the washing buffer.

2.2 Human CRP IA

1. Recombinant human CRP (Catalog No. 1707-CR, R&D Systems).

2. Biotinylated mouse anti-human CRP detection Ab (Catalog No. BAM17072, R&D Systems).

3. Streptavidin-conjugated horseradish peroxidase (SA-HRP, Catalog No. , R&D Systems).

4. Biotinylated anti-CRP detection Ab conjugated to SA-HRP is prepared by adding 1 µL of biotinylated anti-CRP detection Ab (0.5 mg/mL) to 1 µL of SA-HRP to 2998 µL of the binding buffer followed by 20 min of incubation at RT. The resulting concentration of biotinylated anti-CRP detection Ab is 0.17 µg/mL, while SA-HRP dilution is 1:3000.

5. 3,3′,5,5′-Tetramethylbenzidine (TMB) substrate kit (Catalog No. 34021, Thermo Scientific).

    (a) TMB solution (0.4 g/L).

    (b) Hydrogen peroxide solution (containing 0.02% v/v H_2O_2 in citrate buffer).

6. Human whole blood (HQ-Chex level 2) (Catalog No. 232754, Streck).

7. Human serum (CRP free) (Catalog No. 8CFS, HyTest Ltd.).

8. Recombinant human serum albumin (HSA, Catalog No., R&D Systems), human fetuin A (HFA, Catalog No., R&D Systems), human lipocalin 2 (LCN2, Catalog No., R&D Systems), interleukin (IL)-1β (Catalog No., R&D Systems), IL-6 (Catalog No., R&D Systems), IL-8 (Catalog No., R&D Systems) and tumor necrosis factor (TNF)-α (Catalog No., R&D Systems).

9. The CRP spiked diluted human whole blood or serum samples (0.3–81 ng/mL) are prepared by mixing the desired CRP concentrations in 1:100 diluted human whole blood/serum.

10. *Anonymized EDTA plasma samples from patients:* The leftover anonymized EDTA plasma samples of patients are provided by Dr. Eberhard Barth, University Hospital Ulm, Germany. They are diluted 1:1000 and 1:4000 in the binding buffer so that the CRP concentration in these samples falls within the linear range of the DIA, which enables the detection of an entire pathophysiological concentration range of human CRP.

11. Tecan Infinite M200 Pro microplate reader (Tecan, Austria GmbH).

12. Roche COBAS® 8000 modular analyzer.

2.3 SBCR Analysis

1. Samsung Galaxy SIII mini.

2. iPAD mini, iPAD4, iPhone 5s.

3. Custom-made polyamide dark hood and polyamide base holder assembly (Figs. 1 and 2).

4. Image J 1.48v (freely available to download and use at http://imagej.nih.gov/ij).

3 Methods

3.1 Preparation of Capture Ab-Bound MTP

1. *PBS:* A BupH PBS pack is dissolved in 100 mL of autoclaved DIW. The resulting volume is adjusted to 500 mL using autoclaved DIW. Each BupH PBS pack makes 500 mL of PBS at pH 7.2 (*see* Notes 1 and 2).

2. *MES:* A BupH MES pack is dissolved in autoclaved DIW with a final concentration of 500 mL. Each BupH MES pack makes 500 mL of MES at pH 4.7 (*see* Notes 1 and 3).

3. *APTES:* The commercial APTES solution (99% purity) is reconstituted in autoclaved DIW to make an effective 0.25% (v/v) solution (*see* Note 4).

4. *EDC*: An effective concentration of 8 mg/mL is made by reconstituting EDC in 0.1 M MES buffer, pH 4.7 (*see* **Notes 5** and **6**).

5. *Binding buffer*: 0.1% BSA in PBS, pH 7.4.

6. *Washing buffer*: 0.05% Tween® 20 in PBS, pH 7.4 (PBST).

7. *GNPs*: GNPs (1 mg) is sonicated in 1 mL of 0.25% APTES for 1 h.

8. *EDC-activated anti-human CRP Ab*: EDC (0.4 mg or 2 mM) is dissolved in 100 μL of 0.1 M MES, pH 4.7. Thereafter, 990 μL of the anti-human CRP Ab (4 μg/mL) is incubated with 10 μL of EDC (4 mg/mL) for 15 min at RT (*see* **Note 7**).

9. Each MTP well is incubated with 100 μL of 1% (w/v) KOH in DIW for 10 min and then washed five times with 300 μL DIW (*see* **Note 8**).

10. EDC-activated Ab is mixed with GNPs (1 mg/mL) in 0.25% APTES in the ratio of 1:1 (v/v). The resulting anti-human CRP Ab solution (100 μL, 2 μg/mL, 0.5 mg/mL GNPs and 0.125% APTES) is then added to the MTP wells and incubated for 30 min at RT. The Ab-bound and GNPs-functionalized MTP wells are washed five times with 300 μL of PBST.

11. The MTP wells are blocked by incubating with 300 μL of 5% (w/v) BSA for 30 min at 37 °C and washing subsequently with 300 μL of wash buffer five times. Washing can also be performed with an automatic plate washer. This step is essential to prevent nonspecific binding to the unbound sites available on the GNPs and the MTP (*see* **Notes 9** and **10**).

3.2 Human CRP IA

1. *CRP spiked diluted human whole blood or serum*: The CRP spiked samples (0.3–81 ng/mL) are prepared by mixing the desired CRP concentrations in 1:100 diluted human whole blood/serum.

2. *EDTA plasma samples from anonymized patients*: The EDTA plasma samples from anonymized patients are diluted 1:1000 and 1:4000 in the binding buffer, which makes the resulting CRP concentration in these samples to fall within the DIA linear range, thereby enabling the detection of an entire pathophysiological concentration range of human CRP.

3. Biotinylated anti-CRP detection Ab (100 μL, 0.17 μg/mL) pre-conjugated to SA-HRP and 100 μL of CRP (varying concentrations; 0.3–81 ng/mL) are dispensed sequentially to the Ab-bound and BSA-blocked MTP wells, and incubated for 15 min at 37 °C (*see* **Note 11**).

4. The sandwich immune complex-bound MTP is washed with 300 μL of PBST five times to remove nonspecifically bound substances and excess IA reagents.

5. A TMB-H_2O_2 mixture (100 μL) is added to each MTP well and incubated for 4 min at RT to allow the enzymatic reaction by horseradish peroxidase to develop color. TMB acts as a hydrogen donor for the reduction of hydrogen peroxide to water by the enzyme. The resulting diimine causes the solution blue and its absorbance can be read at 650 nm.

6. The enzymatic reaction is stopped by adding 50 μL of 2 N H_2SO_4 to each MTP well (*see* **Note 12**).

7. If there are problems such as no development of color or a high background, **Notes 13** and **14** should be followed.

3.3 **SBCR Analysis**

1. The colorimetric readout of the MTP is performed by smartphone imaging using our developed SBCR set up (Figs. 1 and 2). The gadget (iPAD 4, iPAD mini or iPhone 5s) is placed in the designated groove of the base holder followed by placing the 96-well MTP on the gadget's screensaver that provides white light-based bottom illumination only in the specific regions corresponding to the MTP well's bottom. The dark box is then placed on top of the base holder in the alignment groove. Subsequently, the Samsung SIII mini's back camera-based imaging of the colorimetric reaction in the MTP is performed by placing the smartphone inside the designated groove on top of the dark hood (*see* **Note 15**).

2. The desired pixel intensity (PI) of the captured smartphone image is then determined by our novel image analysis algorithm using Image J 1.48v (http://imagej.nih.gov/ij) (Fig. 1c). The color channel of the image is split and the pixel coordinate of an individual MTP well's center is identified using the red channel image. A mean of neighboring pixels (varies based on the resolution of the camera) from the center is calculated individually for all the channels. The composite mean PI (CMPI) is derived from the following color-weighted formula [CMPI = $0.7 \text{ MPI}_{Blue} + 0.2 \text{ MPI}_{Green} + 0.1 \text{ MPI}_{Red}$]. The desired PI is determined as $255 - \text{CMPI}_{Test\ Conc} - \text{CMPI}_{Blank}$. The resulting CMPI is plotted against their respective—CRP concentrations using the four-parameter logistic based standard curve analysis in SigmaPlot version 11.2 software. The determination of CMPI and PI is done in Microsoft Excel after taking the PI values of red, blue and green channels of the smartphone-captured image using image J. The concentration of unknown CRP samples is determined on the basis of their CMPI from the resulting calibration plot. As described in Subheading 3.2, a standard curve is plotted using various known concentrations of CRP (0.3–81 ng/mL). A schematic of the image analysis algorithm, comprising various steps, is provided in Fig. 6.

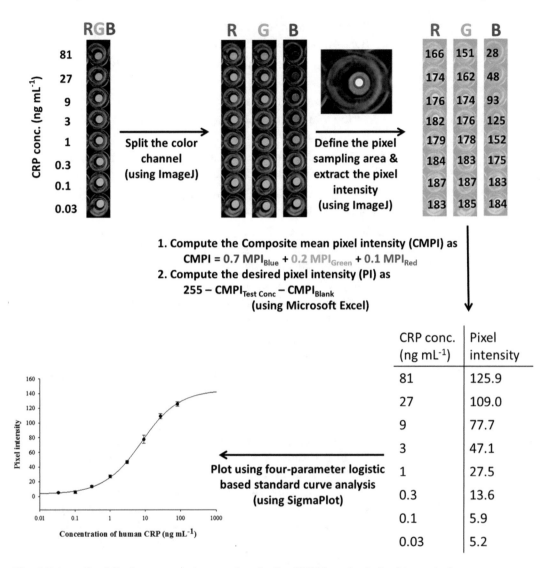

Fig. 6 Schematic of the image analysis procedure for the SBCR based colorimetric readout

1. Compute the Composite mean pixel intensity (CMPI) as
   $$CMPI = 0.7\ MPI_{Blue} + 0.2\ MPI_{Green} + 0.1\ MPI_{Red}$$
2. Compute the desired pixel intensity (PI) as
   $$255 - CMPI_{Test\ Conc} - CMPI_{Blank}$$
   (using Microsoft Excel)

CRP conc. (ng mL^{-1})	Pixel intensity
81	125.9
27	109.0
9	77.7
3	47.1
1	27.5
0.3	13.6
0.1	5.9
0.03	5.2

Plot using four-parameter logistic based standard curve analysis (using SigmaPlot)

4   Notes

1. Inhalation of BupH PBS and MES buffer packs must be avoided and the solutions should be prepared in autoclaved DIW (18.2 MΩ).

2. The reconstituted BupH PBS is good for a week at RT and up to 4 weeks at 4 °C.

3. The reconstituted BupH MES can be stored at RT for up to 2 weeks.

4. The APTES solution should be prepared fresh just prior to use.

5. The recommended concentration of EDC should be used for the optimal cross-linking of capture Ab.

6. The aliquots of EDC can be stored at $-20\ ^\circ$C for up to 6 months.

7. The reconstituted Ab and antigen, stored at 2–8 $^\circ$C, are good within a month. The aliquots stored at -20 to $-70\ ^\circ$C are good for up to 6 months.

8. KOH treatment over 10 min may cause strong surface aberration, which in turn modifies its properties.

9. The blocker BSA is filtered with 0.2 μm pore size filter paper prior to use to avoid contamination.

10. The human CRP concentrations must be prepared in BSA-preblocked sample vials to minimize the analyte loss due to its nonspecific surface binding.

11. SA-HRP should not be frozen and simply stored in the dark at 4 $^\circ$C as streptavidin is light-sensitive. In the case of its exposure to light, a freshly prepared SA-HRP solution should be used.

12. Personal protective equipment (PPE), such as chemical safety glasses, chemical-resistant shoes and lab coats, must be used for handling sulfuric acid, H_2O_2, KOH, APTES, and EDC, in a fume cabinet. Skin contact must be avoided with sulfuric acid, a strong corrosive agent and an irritant. In the case of skin contact, the exposed area should be washed immediately with acid neutralizers and medical advice should be sought urgently. KOH can cause severe burns. Its concentration must be 1% (w/v) in autoclaved DIW as higher concentrations may affect the surface properties, which leads to a decreased binding of capture Ab. APTES is a skin and eye irritant, and highly toxic to the kidney. Being hygroscopic, EDC absorbs moisture that leads to its activity loss, so it should be equilibrated to RT before opening the sample vial.

13. If there is no development of color in DIA, the possible reasons could be the degradation of KOH; loss of activity of APTES due to hydrolysis; loss of activity of EDC due to hydrolysis; inactivity of capture Ab, detection Ab or CRP; and/or degradation of TMB or H_2O_2. This problem could be solved by employing a new batch or stock of these chemicals or immunological reagents.

14. If there is a high background signal, this could be due to the loss of activity or contamination of BSA. Therefore, a fresh stock of BSA should be used to obviate this problem.

15. The illumination of the gadget's screen is set to maximal settings before its use in the developed SBCR.

Acknowledgments

We thank PD Dr. Eberhard Barth for providing the leftover anonymized EDTA plasma samples of patients treated by intensive care at University Hospital Ulm, Germany for validating the DIA.

References

1. Vashist SK, Mudanyali O, Schneider EM et al (2014) Cellphone-based devices for bioanalytical sciences. Anal Bioanal Chem 406: 3263–3277

2. http://www.itu.int/en/ITU-D/Statistics/ Documents/publications/mis2013/MIS2013_ without_Annex_4.pdf. Accessed 8 June 2015

3. Vashist SK, Schneider EM, Luong JHT (2014) Commercial smartphone-based devices and smart applications for personalized healthcare monitoring and management. Diagnostics 4:104–128

4. Ozcan A (2014) Mobile phones democratize and cultivate next-generation imaging, diagnostics and measurement tools. Lab Chip 14:3187–3194

5. Vashist SK, van Oordt T, Schneider EM et al (2015) A smartphone-based colorimetric reader for bioanalytical applications using the screen-based bottom illumination provided by gadgets. Biosens Bioelectron 67:248–255

6. Vashist SK, Schneider EM, Zengerle R et al (2015) Graphene-based rapid and highly-sensitive immunoassay for C-reactive protein using a smartphone-based colorimetric reader. Biosens Bioelectron 66:169–176

7. Marnell L, Mold C, Du Clos TW (2005) C-reactive protein: ligands, receptors and role in inflammation. Clin Immunol 117:104–111

8. Chiesa C, Natale F, Pascone R et al (2011) C reactive protein and procalcitonin: reference intervals for preterm and term newborns during the early neonatal period. Clin Chim Acta 412:1053–1059

9. Chiesa C, Panero A, Osborn JF et al (2004) Diagnosis of neonatal sepsis: a clinical and laboratory challenge. Clin Chem 50:279–287

10. Dollner H, Vatten L, Austgulen R (2001) Early diagnostic markers for neonatal sepsis: comparing C-reactive protein, interleukin-6, soluble tumour necrosis factor receptors and soluble adhesion molecules. J Clin Epidemiol 54:1251–1257

11. Nguyen-Vermillion A, Juul SE, McPherson RJ et al (2011) Time course of C-reactive protein and inflammatory mediators after neonatal surgery. J Pediatr 159:121–126

12. Tappero E, Johnson P (2010) Laboratory evaluation of neonatal sepsis. Newborn Infant Nurs Rev 10:209–217

13. http://www.nomidalliance.org/downloads/ comparative_chart_front.pdf. Accessed 8 June 2015

14. Danner M, Kasl SV, Abramson JL et al (2003) Association between depression and elevated C-reactive protein. Psychosom Med 65:347–356

15. Raison CL, Capuron L, Miller AH (2006) Cytokines sing the blues: inflammation and the pathogenesis of depression. Trends Immunol 27:24–31

16. Spitzer C, Barnow S, Volzke H et al (2010) Association of posttraumatic stress disorder with low-grade elevation of C-reactive protein: evidence from the general population. J Psychiatr Res 44:15–21

17. Lin MS, Shih SR, Li HY et al (2007) Serum C-reactive protein levels correlates better to metabolic syndrome defined by International Diabetes Federation than by NCEP ATP III in men. Diabetes Res Clin Pract 77:286–292

18. Morimoto H, Sakata K, Oishi M et al (2013) Effect of high-sensitivity C-reactive protein on the development of diabetes as demonstrated by pooled logistic-regression analysis of annual health-screening information from male Japanese workers. Diabetes Metab 39:27–33

19. Testa R, Bonfigli AR, Sirolla C et al (2008) C-reactive protein is directly related to plasminogen activator inhibitor type 1 (PAI-1) levels in diabetic subjects with the 4G allele at position −675 of the PAI-1 gene. Nutr Metab Cardiovasc Dis 18:220–226

20. Kuo HK, Yen CJ, Chen JH et al (2007) Association of cardiorespiratory fitness and levels of C-reactive protein: data from the National Health and Nutrition Examination Survey 1999–2002. Int J Cardiol 114:28–33

21. Park HE, Cho GY, Chun EJ et al (2012) Can C-reactive protein predict cardiovascular events in asymptomatic patients? Analysis based on plaque characterization. Atherosclerosis 224:201–207

22. Ridker PM (2003) High-sensitivity C-reactive protein and cardiovascular risk: rationale for screening and primary prevention. Am J Cardiol 92:17K–22K

23. Sicras-Mainar A, Rejas-Gutierrez J, Navarro-Artieda R et al (2013) C-reactive protein as a marker of cardiovascular disease in patients with a schizophrenia spectrum disorder treated in routine medical practice. Eur Psychiatry 28:161–167

24. Wilson AM, Ryan MC, Boyle AJ (2006) The novel role of C-reactive protein in cardiovascular disease: risk marker or pathogen. Int J Cardiol 106:291–297

Chapter 22

A Novel Colorimetric PCR-Based Biosensor for Detection and Quantification of Hepatitis B Virus

Li Yang, Mei Li, Feng Du, Gangyi Chen, Afshan Yasmeen, and Zhuo Tang

Abstract

Hepatitis B virus (HBV) can cause viral infection that attacks the liver and it is a major global health problem that puts people at a high risk of death from cirrhosis of the liver and liver cancer. HBV has infected one-third of the worldwide population, and 350 million people suffer from chronic HBV infection. For these reasons, development of an accurate, sensitive, and expedient detection method for diagnosing, monitoring, and assessing therapeutic response of HBV is very necessary and urgent for public health and disease control. Here we report a new strategy for detection of viral load quantitation of HBV based on colorimetric polymerase chain reaction (PCR) with DNAzyme-containing probe. The special DNAzyme adopting a G-quadruplex structure exhibited peroxidase-like activity in the presence of hemin to report colorimetric signal. This method has shown a broad range of linearity and high sensitivity. This study builds an important foundation to achieve the specific and accurate detection level of HBV DNA with a low-cost and effective method in helping diagnosing, preventing and protecting human health form HBV all over the world, and especially in developing countries.

Key words Hepatitis B virus, Molecular diagnosis, Biosensor, G-quadruplex, DNAzyme, Peroxidase-like activity, Colorimetric PCR

1 Introduction

Hepatitis B virus (HBV) can cause a potential life-threatening liver infection and pathology ranging from self-limited illness to chronic hepatitis, cirrhosis, and hepatocellular carcinoma [1]. Approximately 400 million people are infected globally by HBV and about one-third of the world population has been infected once in their lives [2]. It is an important health hazard for the worldwide population and the most serious type of viral hepatitis.

Presently, serologic immunity and nucleic acid testing are the most commonly used methods for HBV diagnosis, prevention, and treatment in clinical medicine [3]. The HBV serological tests such as ELISA assays are used to distinguish acute, self-limited infections from chronic HBV infections and monitor vaccine-induced

Avraham Rasooly and Ben Prickril (eds.), *Biosensors and Biodetection: Methods and Protocols Volume 1: Optical-Based Detectors*, Methods in Molecular Biology, vol. 1571, DOI 10.1007/978-1-4939-6848-0_22,
© Springer Science+Business Media LLC 2017

immunity by testing for a series of serological markers of HBV. But this method can only be used to qualitative diagnosis with low sensitivity and cause false-positive frequently. Besides these, nucleic acid detection of HBV DNA based on the PCR techniques is also used to quantify HBV viral load and track the effectiveness of therapeutic drugs [4]. With PCR amplification, specific segment of viral DNA can be magnified and a large number of copies of the target DNA sequence are produced across several orders of magnitude [5]. Compared to the traditional immunological assays, the nucleic acid assays show higher sensitivity, accuracy, and specificity, which get rid of the evident limit for the accurate detection of HBV DNA in the progression of disease as well as the therapeutic effect. Because the nucleic acid assays can solve these problems which are brought about by immunological test and offer several obvious advantages, it is the most reliable diagnosis technique for detecting and monitoring HBV infection as well as assessing therapeutic response.

TaqMan technique in nucleic acid assays is one of the most popular methods used for detection of HBV DNA. The said technique uses a TaqMan probe containing a fluorophore and a quencher at both the ends respectively. In the single-stranded form, florescence is not detected because of fluorescence resonance energy transfer (FRET). When the PCR extension progresses, a DNA polymerase with the activity of $5'$–$3'$ exonuclease leads to degrading the fluorophore-modified DNA probe that will anneal to the target strand [6]. But the main limitation of the TaqMan method is the high-cost modified flurogenic oligonucleotide probes and sophisticated equipment, which are restricted to well-equipped laboratories and less accessible to many ordinary users.

G-quadruplexes are higher-order DNA and RNA structures formed from G-rich sequences that are built around tetrads of hydrogen-bonded guanine bases. The criteria for a potential quadruplex sequence is restricted to: $G_{3-5}N_{L1}G_{3-5}N_{L2}G_{3-5}N_{L3}G_{3-5}$, where N_{L1-3} are loops of unknown length, within the limits $1 < N_{L1-3} < 7$ nt. DNAzymes are catalytic single-stranded deoxyribonucleic acids that are obtained through in vitro selection. The DNAzyme we use has been found to exhibit peroxidase-like activity by forming G-quadruplex structure and binding with hemin. In recent years, this DNAzyme has been used to design various colorimetric or chemiluminescent assays. Herein, we report a novel approach for both qualitative and quantitation of HBV DNA based on our previous work [7]. This method takes the key advantage of the TaqMan technology (i.e., the elegant use of the $5'$-exonuclease activity of Taq polymerase) [8], and applies a low-cost catalytic DNA molecular beacon as probe [9]. In the process of PCR amplification, Taq DNA polymerases cleave the probe and release a DNAzyme sequence [10–12], which has been embedded in the probe. After PCR amplification, the DNAzyme can form

G-quadruplex and bind with hemin, possessing a peroxidase-like activity which catalyzes the oxidation of different substrates by H_2O_2 to generate either colorimetric or fluorometric signals [13–15].

2 Materials

2.1 Preparation of ^{32}P Labeled Probe

1. T4 polynucleotide kinase (10 U/μL; TransGen Biotech, Beijing, China).

2. [γ-^{32}P] ATP (Furui Biological Engineering, Beijing, China).

3. Probe-HBV-2 (Sangon Biotech, Shanghai, China) is dissolved in sterile water.

2.2 Clinical Serum Sample Extraction

1. QIAamp DNA Blood Mini Kit (Qiagen).

2. HBV infection clinical serum samples (from HBV patients).

3. Normal clinical serum samples (from normal population).

2.3 Amplification

1. HBV DNA template (the DNA is extracted from the HBV infection clinical serum samples) is dissolved in sterile water, stored at −20 °C.

2. Primers (forward primer: 1.5 μM, reverse primer: 1 μM; Sangon Biotech, Shanghai, China) are dissolved in sterile water, stored at −20 °C.

3. Probe (^{32}P labeled, 0.6 μM; Sangon Biotech, Shanghai, China) is dissolved in sterile water, stored at −20 °C.

4. 1.2 mM dNTPs (600 μM dUTP, 200 μM dATP, 200 μM dCTP, 200 μM dGTP; TransGen Biotech, Beijing, China).

5. FastStart Taq DNA polymerase (5 U; Roche).

6. Taq DNA polymerase (5 U; TransGen Biotech, Beijing, China).

7. UNG (1 U; Takara).

8. $MgCl_2$ (0.5 mM).

9. 1× Taq polymerase buffer (15 mM Tris–HCl (pH 8.2), 30 mM KCl, 5 mM $(NH_4)_2SO_4$, 2.5 mM $MgCl_2$, 0.002% BSA; Trans-Gen Biotech, Beijing, China).

10. C1000 thermal cycler (Bio-Rad).

2.4 Colorimetric Detection

1. 400 mM NaCl solution is prepared by dissolving 0.2340 g NaCl in 10 ml of ultrapure water.

2. Hemin (Alfa Aesar) is dissolved in DMSO.

3. ABTS (2,2′-azinobis-(3-ethylbenzthiazoline-6-sulphonate)) (Wolsen, Xi'an, China) is dissolved in ultrapure water.

4. H_2O_2 (Bodi Chemical Holding Co., Ltd., Tianjin, China) is diluted in ultrapure water.

5. Varioskan Flash ($\lambda = 414$ nm, 40 readings with a 30 s interval are recorded; Thermo Scientific).

2.5 TaqMan Assay

1. Hepatitis B Viral DNA Quantitative Fluorescence Diagnostic Kit (48 Tests; Sansure Biotech, Changsha, China).

2. Real-Time PCR System (Thermo Scientific).

3 Methods

3.1 The Principle of the Colorimetric Detection

1. The principle of our detection strategy is depicted in Fig. 1. The DNA probe containing three parts of sequences (A, B, C) is designed to form a hairpin structure at room temperature. After denaturation step of PCR, the concentration of probes is much higher than that of HBV dsDNA templates, so the chance for probe to hybridize with the targeted HBV single strand DNA is much higher than that of its complementary single strand DNA. Then, in the annealing step of PCR, the loop domain (part C) of the probe hybridizes to the conserved region of HBV genome, which will be amplified by two primers. The stem part of the probe contains blocking sequence A (black) and a DNAzyme sequence B (blue) that could report the detection result through oxidation reaction [16–20]. The blocking sequence A is used to prohibit DNAzyme sequence B fold into catalytic G-quadruplex structure at room temperature to decrease the background of colorimetric assay [7]. As the

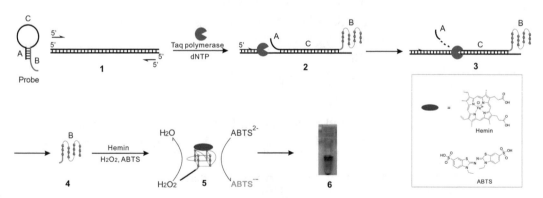

Fig. 1 Strategy of the colorimetric PCR-based detection of HBV. A part of the probe is the blocking sequence, B represents the sequence for DNAzyme forming G-quadruplex, loop C is the complementary sequence of the conserved region of HBV genome. (*1*) Melting of double-stranded DNA and hairpin loop during PCR denaturation. (*2*) Hybridization of probe and primers to targeted ssDNA respectively during PCR annealing step. (*3*) Cleaving of the hybridized probe by DNA polymerase during primers extension. (*4*) Release of G-quadruplex sequence after cleavage by DNA polymerase. (*5*) Horseradish peroxidase-mimicking DNAzymes formation with the addition of hemin, reacting with ABTS and H_2O_2. (*6*) Colorimetric test result

PCR proceeding, the probe formed a stable duplex with target sequence of HBV genome and Taq DNA polymerase with $5'-3'$ exonuclease activity can cleave the obstruction of part A, consequently released the part B. After PCR amplification, the cleaved fragment of the probe could form a stable G-quadruplex structure and bind with hemin to exhibits the peroxidase-like activity to oxidize ABTS in the presence of H_2O_2, leading to produce a colorimetric green product. With the exception of the HBV sequence, the blocking sequence and the sequence for DNAzyme forming G-quadruplex can be used for detection of other targets.

3.2 DNA Extraction from Clinical Serum Sample

1. Extraction of DNA from the HBV infection and normal clinical serum samples from HBV patients or normal population is carried out using QIAamp DNA Blood Mini Kit as described by the manufacturer (Qiagen).

3.3 The Serum Calibration Curve Preparation

1. 10^8 copies/mL clinical serum samples are chosen to dilute with negative serum by decuple to ten copies successively. These samples are extracted with QIAamp DNA Blood Mini Kit (Qiagen) to build the calibration curve.

3.4 Manual Design of Probe and Primers

1. According to our previous procedure, this method requires two consecutive PCR steps to achieve the successful detection of a target DNA, which include the amplification of target sequence and subsequently the cleavage of probe [7]. In our experiment, three different hairpin probes (probe-HBV-1, probe-HBV-2, and probe-HBV-3) and corresponding primers are designed against the different conserved regions of the HBV genome including overlapping genes encoding X-protein, S gene, and X gene region.

2. Under the optimized conditions, Fig. 2 shows that probe-HBV-2 for S gene region of HBV genome exhibits the best colorimetric result comparative to that of other probes. Furthermore, this region of HBV genome is rather conserved in different genotypes of HBV strains and can be extensively and accurately used to detect all HBV genotypes. Therefore, the probe-HBV-2 has been proved to be the best fitted probe for HBV detection based on our method. The selected sequences are:

Primer-HBV-Forward-2:CCTGGTTATCGCTGGATGTGT,
Primer-HBV-Reverse-2:GGACAAACGGGCAACATACCTT,
Probe-HBV-2: CCCTACCCA TTCATCCTGCTGCTATG CCTCATCTTCTT **TGGGTAGGG** CGGGTTGGGAAA - NH_2 (*see* **Note 1**).

The sequence marked with box represents the stem part of Probe-HBV-2. Sequence in the second box would be closed

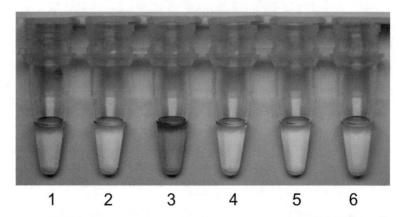

Fig. 2 Colorimetric PCR reaction with three different probes. *Tube 1*: probe-HBV-1 with HBV DNA targets; *tube 2*: probe-HBV-1 without targets; *tube 3*: probe-HBV-2 with targets; *tube 4*: probe-HBV-2 without targets; *tube 5*: probe-HBV-3 with targets; *tube 6*: probe-HBV-3 without targets

at room temperature. The underlined sequence could anneal to S gene region of HBV genome and is the loop domain of the probe. Sequence for DNAzyme forming G-quadruplex is shown in boldface.

3.5 One-Step PCR

1. Thaw $10\times$ Taq polymerase buffer, dNTPs, and primers.

2. The PCR is performed in 50 µL volume containing 2 µL HBV DNA template, 1.5 µM Primer-HBV-Forward-2, 1 µM Primer-HBV-Reverse-2, 0.6 µM Probe-HBV-2, 1.2 mM dNTPs (600 µM dUTP, 200 µM dATP, 200 µM dCTP, 200 µM dGTP), 5 U of FastStart Taq DNA polymerase, 1 U of UNG (*see* **Note 2**), 0.5 mM $MgCl_2$, $1\times$ Taq polymerase buffer (15 mM Tris–HCl (pH 8.2), 30 mM KCl, 5 mM $(NH_4)_2SO_4$, 2.5 mM $MgCl_2$, 0.002% BSA).

3. The negative control is prepared without HBV DNA.

4. The amplification conditions are as follows: 20 °C for 10 min; 95 °C for 2 min followed by 40 cycles of 94 °C for 30 s and 60 °C for 90 s.

3.6 ^{32}P Labeled Selected Probe PCR Reaction

1. To verify the probe had been cleaved by DNA polymerase successfully, Probe-HBV-2 and mark are 5′ end-labeled with $[\gamma\text{-}^{32}P]$ ATP. The reaction mixture containing oligonucleotides with 10 µCi $[\gamma\text{-}^{32}P]$ and 10 U of T4 polynucleotide kinase is incubated for 1 h at 37 °C for DNA phosphorylation.

2. The labeled product is purified by 10% denaturing PAGE.

3. The 5′ ^{32}P labeled probe-HBV-2 is added into the PCR reaction.

4. The PCR product is used for PAGE analysis. As shown in Fig. 3, in the positive control containing HBV genome, a

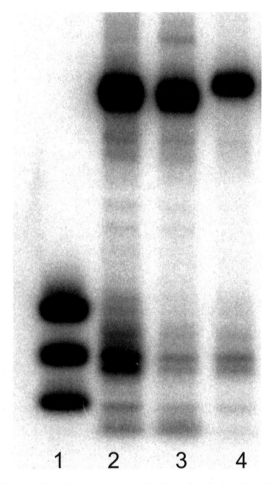

Fig. 3 PAGE analysis of the cleavage of isotope-labeled probe-HBV-2 in PCR amplification. *Lane 1*, from top to bottom 17-nt, 13-nt, 9-nt DNA marker; *lane 2*, PCR reaction containing isotope-labeled probe-HBV-2 and HBV DNA targets; *lane 3*, negative control: PCR reaction without HBV DNA targets; *lane 4*, ^{32}P-labeled probe-HBV-2 as marker

13-nt fragment is released from probe-HBV-2 during the PCR reaction (lane 2, Fig. 3). Comparatively, in the negative control, without HBV target, the probe displayed no cleavage (lane 3, Fig. 3). When Taq DNA polymerase encounters the hybridized complementary duplex between the target site of HBV genome and the probe-HBV-2, it cleaves the probe at the second nucleotide of stable double-stranded domain. Then, a 13-nt ^{32}P-labeled fragment from probe-HBV-2 is released. The result is consistent with our previous research about 5′–3′ exonuclease activity of DNA polymerase. So these results prove that the final colorimetric result of positive control is caused by the cleavage of probe to release enzymes as reporter in the process of PCR amplification.

**3.7 Colorimetric
Detection**

1. NaCl at the concentration of 400 mM (*see* **Note 3**) is added to the PCR products.

2. The reaction mixture is heated at 94 °C for 5 min.

3. The reaction mixture is incubated at room temperature for 30 min.

4. Hemin (2 μM), ABTS (2.4 mM), and H_2O_2 (2 mM) (*see* **Note 4**) are added successively.

5. Colorimetric signals of different concentrations of HBV DNA are recorded (Fig. 4a). Samples containing different concentrations of HBV DNA (from 10 to 10^7 copies) has been detected, the colorimetric results reveal a gradual increase of color intensity from light to dark green, while the negative control containing no HBV target remained colorless. As few as ten copies of HBV DNA could be qualitatively detected with naked eyes in 50 $μL^{-1}$ reaction mixture. The method represents a simply visual way for HBV DNA detection, which refrains from the advanced equipment and harsh detection conditions.

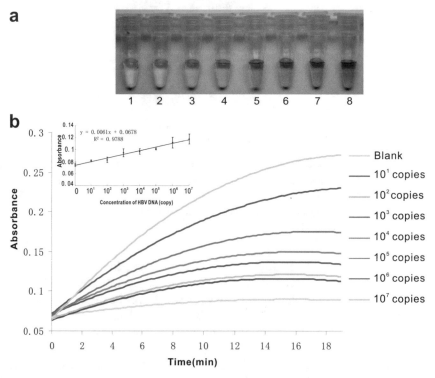

Fig. 4 Colorimetric detection of HBV DNA. (**a**) Photograph of the colorimetric detection of different concentration of HBV DNA: *tube 1*, negative control (without HBV DNA); *tubes 2–8* containing 10, 10^2, 10^3, 10^4, 10^5, 10^6, 10^7 copies of HBV DNA, successively. (**b**) Time-dependent colorimetric detection of different concentrations of HBV DNA. The inset is a calibrated curve of the average absorbance (414 nm) at 3 min plotted against the number of HBV DNA. The *solid line* indicates linear least squares fitting between 10 and 10^7 copies of HBV DNA, and their formulation is FI = 0.0061 lg(C_{HBVDNA}) + 0.0678 (R^2 = 0.9788). The correlation coefficient is 0.9788. The error bars were determined by standard deviation (SD) of the triplicate data

6. Absorbance detection of ABTS⁻ is performed at 414 nm. About 40 readings with a 30 s interval are recorded. With the aid of simple UV-spectrometer of microplate-reader, more accurate quantitative detection can be realized. A time-dependent optical absorption change (414 nm) is recorded (experiments are conducted in triplicate), and the relationship between different concentrations of HBV DNA and absorbance is studied (Fig. 4b). The optical density is proportional to the concentration of HBV DNA over the range of 10–10^7 copies. The inset is a calibration curve with colorimetric intensity data at 414 nm obtained from the spectrum, corresponding to different concentrations of HBV DNA (the average values of triplicate data are used for plotting at 3 min). The calibration curve reveals that there is a good linear relationship between absorbance and the concentration of HBV DNA with a correlation coefficient of 0.9788, which shows linear dynamic ranges from 10 to 10^7 copies. We test the detection limit from 10^7 to 10 copies by serial dilutions, and the ten copies is the lowest concentration in samples that we can dilute and detect reproducibly. So a minimum detection for ten copies of HBV DNA is achieved based on our method.

3.8 Comparation with TaqMan Assay

1. TaqMan assay is the most advanced technique for HBV detection, which is widely used in clinical diagnosis, so we compare our colorimetric assay with the quantitative TaqMan PCR assay. Seven different concentrations of clinical HBV serum samples and seven negative clinical serum samples are measured with the TaqMan method.

2. Prepared reagents according to the instruction of Hepatitis B Viral DNA Quantitative Fluorescence Diagnostic Kit.

3. 200 μL of negative control, positive control, quantitative reference A–D, and the sample under test is respectively added to seven clean 1.5 ml microcentrifuge tubes, each containing 300 μL of DNA extraction solution. Mix completely by vibrating for about 10 min.

4. Add 100 μL of DNA extraction solution 2-mix to each tube. Mix for 10 s, then quiescence at room temperature for 10 min.

5. Place the microcentrifuge tubes into the magnetic bead separator for 3 min after a short centrifugation, then sucked out the solution slowly.

6. 600 μL DNA extraction solution 3 and 200 μL DNA extraction solution 4 are added to each tube. Mix for 5 s.

7. Place the microcentrifuge tubes into the magnetic bead separator again after a short centrifugation. 3 min later, there are two layers of the solution. Discard the lower layer and transfer all of the microcentrifuge tubes to tube rack.

8. Add 50 µL of PCR-mix to each tube. Mix completely. The mixed solution is transferred to clean 0.2 ml PCR tube respectively.

9. Apply the mixed solution for Real-Time PCR under the instruction of Hepatitis B Viral DNA Quantitative Fluorescence Diagnostic Kit.

10. Apply colorimetric assay to detect the same clinical sample in triplicate; meanwhile consecutive serum samples ranging from 10^1 to 10^8 copies/mL are determined in triplicate to record the time-dependent optical absorption changes (Fig. 5a) and

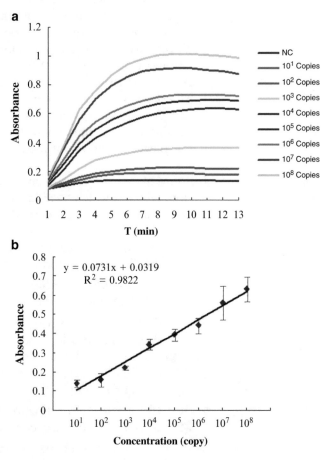

Fig. 5 Calibration curve establishment of HBV serum samples. (**a**) Time-dependent colorimetric detection of different concentrations of HBV DNA (414 nm). (**b**) Calibrated curve of the average absorbance (414 nm) at 3 min plotted against the number of HBV DNA. The *solid line* indicates linear least squares fitting between 10 and 10^8 copies of HBV DNA, and their formulation is FI = 0.0731 lg(C_{HBVDNA}) + 0.0319 (R^2 = 0.9822). The correlation coefficient is 0.9822. The error bars were determined by standard deviation (SD) of the triplicate data

the calibration curve which used the average values of triplicate data is obtained plotting at 3 min (Fig. 5b). It shows a good linear relationship between absorbance and the concentration of HBV DNA. The correlation coefficient reached 0.9822 and their formulation is $FI = 0.0731 \lg(C_{HBVDNA}) + 0.0319$. The results reveal the excellent proximity of estimated values in curve showed the correlation between the two assays is very high, whose correlation coefficient attained 0.913 (Fig. 6). Besides, the other negative clinical samples determined by TaqMan assay are also negative measured in triplicate by our colorimetric method, which failed to produce a detectable PCR product.

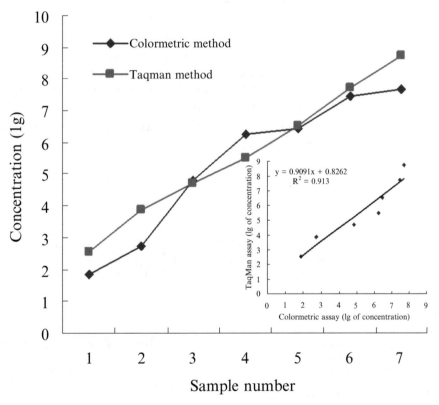

Fig. 6 Comparison of colorimetric and TaqMan assay. Serum samples of consecutive concentration were measured by colorimetric and TaqMan assay. *Red points* represent the values of TaqMan assay and *blue points* represent the values of colorimetric assay. The inset is the comparison of the values by colorimetric and TaqMan assay. The concentration of target used for experiment was copies/ml. "lg" represents the denary logarithm value of the concentration of target. *X*-axis represents the lg value of concentration by colorimetric assay and *Y*-axis represents the lg value of concentration by TaqMan assay. The correlation coefficient is 0.913

4   Notes

1. The catalytic beacons probe could be extended from the 3′ end by DNA polymerase unexpectedly, which could interfere the forming of right structure of DNAzyme and the following colorimetric reaction as well. The probe with amino modifier at the 3′ end has been introduced into our PCR amplification because the modification on the probe could prohibit the undesired extension caused by the DNA polymerase.

2. Uracil N-glycosylase (UNG) treatment system is most commonly used method to prevent the contamination caused by PCR amplicon. To overcome the shortage of low-efficiency PCR amplification caused by dUTP, we optimized the PCR conditions, containing the PCR temperature, cycles, times, buffer constitution, the quantity of enzyme, concentration of sensor and primer, and concentration of ion.

3. NaCl solution can help the original probe form hairpin structure, which keep solution colorless. So the concentration of the NaCl solution should be tested to hold the background of original probe solution.

4. To ensure the result of colorimetric reaction, Hemin, ABTS, and H_2O_2 are freshly prepared.

Acknowledgments

This study was supported by the financial supports from the CAS (Hundreds of Talents Program), National Science Foundation of China (Grant No. 21172215 and No. 21102140), Innovation Program of the CAS (Grant No. KSCX2-EW-J-22).

References

1. Mahoney FJ (1999) Update on diagnosis, management, and prevention of hepatitis B virus infection. Clin Microbiol Rev 12: $351–366

2. Loeb KR, Jerome KR, Goddard J et al (2000) High-throughput quantitative analysis of hepatitis B virus DNA in serum using the TaqMan fluorogenic detection system. Hepatology 32:626–629

3. Gish RG, Locarnini SA (2006) Chronic hepatitis B: current testing strategies, clinical gastroenterology and hepatology. Off Clin Pract J Am Gastroenterol Assoc 4:666–676

4. Krajden M, McNabb G, Petric M (2005) The laboratory diagnosis of hepatitis B virus. Can J Infect Dis Med Microbiol J Can Mal Infect Microbiol Med/AMMI Can. 16:65–72

5. Mullis KB, Faloona FA (1987) Specific synthesis of DNA in vitro via a polymerase-catalyzed chain reaction. Methods Enzymol 155:335–350

6. Lee LG, Connell CR, Bloch W (1993) Allelic discrimination by nick-translation PCR with fluorogenic probes. Nucleic Acids Res 21:3761–3766

7. Du F, Tang Z (2011) Colorimetric detection of PCR product with DNAzymes induced by 5′-nuclease activity of DNA polymerases. Chembiochem Eur J Chem Biol 12:43–46

8. Holland PM, Abramson RD, Watson R et al (1991) Detection of specific polymerase chain reaction product by utilizing the 5′–3′ exonuclease activity of Thermus aquaticus DNA polymerase. Proc Natl Acad Sci U S A 88:7276–7280

9. Xiao Y, Pavlov V, Niazov T et al (2004) Catalytic beacons for the detection of DNA and telomerase activity. J Am Chem Soc 126:7430–7431

10. Li Y, Sen D (1996) A catalytic DNA for porphyrin metallation. Nat Struct Biol 3:743–747

11. Travascio P, Li Y, Sen D (1998) DNA-enhanced peroxidase activity of a DNA-aptamer-hemin complex. Chem Biol 5:505–517

12. Travascio P, Witting PK, Mauk AG et al (2001) The peroxidase activity of a hemin–DNA oligonucleotide complex: free radical damage to specific guanine bases of the DNA. J Am Chem Soc 123:1337–1348

13. Deng M, Zhang D, Zhou Y et al (2008) Highly effective colorimetric and visual detection of nucleic acids using an asymmetrically split peroxidase DNAzyme. J Am Chem Soc 130:13095–13102

14. Majhi PR, Shafer RH (2006) Characterization of an unusual folding pattern in a catalytically active guanine quadruplex structure. Biopolymers 82:558–569

15. Kong DM, Wu J, Wang N et al (2009) Peroxidase activity-structure relationship of the intermolecular four-stranded G-quadruplex-hemin complexes and their application in Hg^{2+} ion detection. Talanta 80:459–465

16. Emilsson GM, Breaker RR (2002) Deoxyribozymes: new activities and new applications. Cell Mol Life Sci 59:596–607

17. Baum DA, Silverman SK (2008) Deoxyribozymes: useful DNA catalysts in vitro and in vivo. Cell Mol Life Sci 65:2156–2174

18. Navani NK, Li Y (2006) Nucleic acid aptamers and enzymes as sensors. Curr Opin Chem Biol 10:272–281

19. Breaker RR (1997) DNA enzymes. Nat Biotechnol 15:427–431

20. Liu J, Cao Z, Lu Y (2009) Functional nucleic acid sensors. Chem Rev 109:1948–1998

Chapter 23

CCD Camera Detection of HIV Infection

John R. Day

Abstract

Rapid and precise quantification of the infectivity of HIV is important for molecular virologic studies, as well as for measuring the activities of antiviral drugs and neutralizing antibodies. An indicator cell line, a CCD camera, and image-analysis software are used to quantify HIV infectivity. The cells of the P4R5 line, which express the receptors for HIV infection as well as β-galactosidase under the control of the HIV-1 long terminal repeat, are infected with HIV and then incubated 2 days later with X-gal to stain the infected cells blue. Digital images of monolayers of the infected cells are captured using a high resolution CCD video camera and a macro video zoom lens. A software program is developed to process the images and to count the blue-stained foci of infection. The described method allows for the rapid quantification of the infected cells over a wide range of viral inocula with reproducibility, accuracy and at relatively low cost.

Key words Charge-coupled device (CCD), Imaging, HIV, Infectivity, Quantify, Rapid

1 Introduction

The quantification of a biological process can sometimes be cumbersome or labor-intensive. In the field of virology, several assays have been developed to measure the infectivity of the human immunodeficiency virus (HIV) during a single round of replication: plaque assays [1], syncytium formation assays [2, 3], focal immunoassays [4], assays using indicator cell lines expressing β-galactosidase or luciferase [5–12], and assays using fluorescently labeled reporter viruses [13, 14]. The quantification of infection in several of these assays requires counting the number of plaques, syncytia, foci, or cells within an infected population of cells. High throughput assays may utilize plate readers or flow cytometers. The use of

Portions reprinted from the Journal of Virological Methods, Vol. 137, J.R. Day, L.E. Martínez, R. Šášik, D.L. Hitchin, M.E. Dueck, D.D. Richman and J.C. Guatelli, A computer-based, image-analysis method to quantify HIV-1 infection in a single-cycle infectious center assay, pp. 125–133, Copyright 2006, with permission from Elsevier.

Avraham Rasooly and Ben Prickril (eds.), *Biosensors and Biodetection: Methods and Protocols Volume 1: Optical-Based Detectors*, Methods in Molecular Biology, vol. 1571, DOI 10.1007/978-1-4939-6848-0_23, © Springer Science+Business Media LLC 2017

indicator cell lines, such as those containing an integrated β-galactosidase gene under the control of the HIV long terminal repeat (LTR), has enabled the straightforward determination of viral infectivity by visualizing infected cells that are stained blue with 5-bromo-4-chloro-3-indolyl-β-D-galactopyranoside (X-gal). The Magi-CCR5 [6] and P4R5 [7] indicator cell lines are frequently used as targets of infection. These are HeLa-based adherent cells that have been engineered to express CD4, the HIV receptor, as well as CCR5, one of the two predominantly used coreceptors; HeLa cells naturally express CXCR4, the other major coreceptor used by HIV.

To quantify the number of infected cells by using these β-galactosidase-based indicator cell lines, one approach is to manually count blue cells by eye through a microscope. This laborious method is subject to observer error and is difficult when the density of infected cells is high, potentially limiting the dynamic range of the assay. To facilitate the counting of infected cells with speed, accuracy, and optimal dynamic range, a charge-coupled device (CCD) camera is used and software is developed to analyze digital images of infected cell monolayers [15]. Accurate and reproducible cell counts over a wide range of viral inocula are obtained with this method. CCD camera technology has been available for years, but the explosion in the use of digital cameras for personal use has accelerated technology development and facilitated the incorporation of affordable CCD detection in the laboratory.

2 Materials

2.1 Imaging Apparatus

1. *Five megapixel CCD color camera* with real-time viewing, C-mount optical lens interface, and FireWire IEEE 1394 digital interface: MicroPublisher 5.0 RTV (Model # MP5.0-RTV-CLR-10, QImaging, Burnaby, BC, Canada) (*see* **Note 1**).

2. *Macro video zoom lens*: Optem 18–108 mm, f/2.5, C-mount (Qioptiq Imaging Solutions, Rochester, NY, formerly Thales Optem, Inc.) (*see* **Note 2**).

3. *Copy stand* with a 1/4″-20 threaded mounting screw (model # CS-3, Testrite Instrument Company, Inc.). Something similar would suffice.

4. *Fluorescent light box* (typically used for viewing X-ray films) (*see* **Note 3**).

5. *Camera mount adapter*, custom made by our university machine shop, consisting of a tripod mounting screw (1/4″-20 thread) and a rectangular piece of aluminum (2.25 × 1 × 0.25 in.) with two drilled holes (*see* **Note 4**).

Use of the adapter requires that the copy stand have a mounting rod that can extend away from the stand.

6. *Personal computer for capturing and analyzing images*: Windows operating system, 1 available FireWire port (IEEE 1394) or a FireWire expansion card.

2.2 Image Acquisition and Analysis Software

1. *Image acquisition software*, QCapture Pro, v.5.0.0.16, with USB key (dongle) and drivers for PC (QImaging, Burnaby, BC, Canada). The Micropublisher 5.0 camera is bundled with QCapture Suite software, however, the upgrade to QCapture Pro is purchased to increase control over the image capture process.

2. *Image analysis software*: application (5MGL.exe) and parameter file (5MGL.par), collectively termed the "Romanizer." The software is developed in-house in the Fortran 95 programming language for the specific purpose of counting HIV-infected cells. Use of the software for other purposes may not be suitable. Alternative image-analysis applications exist, both freeware and for purchase (*see* **Note 5**).

2.3 HIV Infectivity Assay

1. *Indicator cells*: P4R5 HeLa cells (P4.R5 MAGI cells, AIDS Research and Reference Reagent Program, Division of AIDS, NIAID, NIH contributed by Dr. Nathaniel Landau). The P4R5 cell line contains the β-galactosidase gene under the control of the HIV-1 promoter region (LTR). The derivation of the parental P4 line has been reported [7]. P4R5 cells express on their surface the HIV entry receptors, CD4, CXCR4, and CCR5.

2. *Complete media*: Dulbecco's Modified Eagle Medium (DMEM; Gibco, Grand Island, NY) supplemented with 10% fetal bovine serum (FBS; GemCell, Woodland, CA), 100 U/mL penicillin, 100 µg/mL streptomycin (pen/strep; Gibco, Grand Island, NY), 2 mM L-glutamine (Gibco, Grand Island, NY) and 1 µg/mL puromycin. Store at 4 °C.

3. *48-well tissue culture plate* (Cat. # 353078, BD Falcon, San Jose, CA).

4. *Phosphate buffered saline*, 1× (PBS; Invitrogen, Carlsbad, CA).

5. *Fix solution*: 1% formaldehyde, 0.2% glutaraldehyde in PBS. Can be made in advance and stored in the dark at 4 °C for 1–2 months.

6. *X-gal stain solution made fresh*: For each 1.0 mL of solution, combine 949 µL PBS, 20 µL 0.2 M potassium ferrocyanide, 20 µL 0.2 M potassium ferricyanide, 1.0 µL 2.0 M Mg_2CL, and 10 µL 40 mg/mL X-gal (5-bromo-4-chloro-3-indolyl-β-D-galactopyranoside). Stock solutions of potassium ferrocyanide, potassium ferricyanide, Mg2Cl, and X-gal can be prepared and

stored in aliquots at −20 °C. X-gal is dissolved in dimethyl sulfoxide (DMSO) and should be stored in the dark. The X-gal solution will turn yellow over time, but this does not affect the activity. Discard the stock if it becomes green/brown.

7. The HIV virus is either isolated and cultured from blood, or produced in vitro by transient transfection of proviral plasmid DNA into cultured cells. Viruses do not need a complete HIV genome, but they must be competent for entry into P4R5 cells and they must be capable of expressing the HIV Tat protein. Tat transactivates the HIV LTR promoter region and will enable expression of β-galactosidase in the P4R5 indicator cells.

3   Methods

In order to facilitate the counting of blue-stained HIV-infected cells, a CCD camera, macro lens, and a computer program are set up to capture and process images of plated cells. The setup does not require the exact equipment listed in Subheading 2. A camera, lens, light source, computer, and a method to mount the camera are the basic required elements. The ideas described here can be adapted to different configurations to suit the needs of the application. The method for analyzing the images will vary depending on the final goal of the analysis. Although the software we developed to count HIV-infected cells may not be suitable for other applications, alternative software programs may provide the necessary tools for your particular application.

3.1   Setup of Imaging Apparatus

1. Place the copy stand on a table and the light box (or light source of choice) on the base of the stand.

2. Assemble the camera by screwing the lens onto the camera.

3. Mount the camera to the stand using the 1/4″-20 threaded mount. The most straightforward way to mount the camera is illustrated in Fig. 1a. An alternative method to mount the camera requires a custom-built adapter (*see* **Note 6**). The alternative method is illustrated in Fig. 1b and a photograph of the setup is shown in Fig. 2. Screw the mount adapter onto the copy stand, then attach the camera to the adapter using the tripod mount screw (Fig. 3).

4. The height of the camera above the light box is adjustable and depends on the zoom setting of the lens. In our setup, the distance between the surface of the light box and the mounting screw hole of the camera is 14.25 in. (36.2 cm). The tip of the lens is 6.25 in. (15.9 cm) above the light box. Level the light box and camera using a standard bubble level.

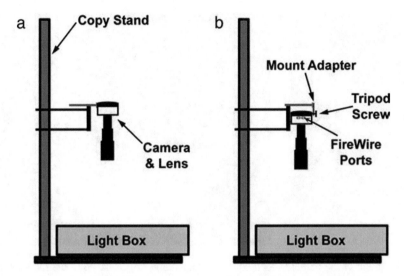

Fig. 1 Schematic diagrams of the CCD camera setup. (**a**) The simplest setup using direct attachment of the camera to the copy stand. The FireWire ports on the camera are on the backside in the perspective shown. (**b**) Use of an aluminum mount adapter and tripod screw to mount the camera rotated at 180° (*see* **Notes 4** and **6**)

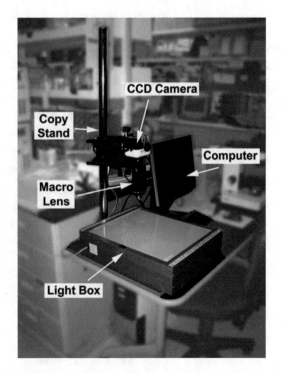

Fig. 2. Photograph of the actual camera setup showing the light box, copy stand, CCD camera, macro lens, and computer

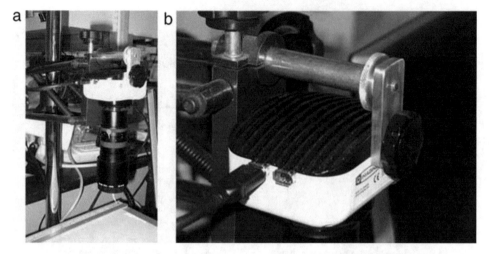

Fig. 3. (**a**) Photograph of the camera mounted to the copy stand. (**b**) Close-up of the mounted camera with mount adapter

5. Attach a FireWire cable to either one of the ports on the camera. Do not connect the cable to the computer until after loading the software and drivers.

3.2 Capture and Analysis of Images

3.2.1 Capturing Images

1. These instructions assume the use of the Micropublisher 5.0 RTV camera, Optem 18–108 mm macro video zoom lens, and QCapture Pro software. Other cameras, capture software, and analysis software may be used. Please follow the directions included with your specific components.

2. Follow the setup instructions from QImaging to load the camera drivers and QCapture Pro software. You will connect the FireWire cable to the computer during this process.

3. To define the capture settings, launch QCapture Pro and click the camera icon on the toolbar (or select Acquire/Video-Digital from the menu) to open the acquisition dialog box. Use the Basic Dialog (rather than the Advanced Dialog) by clicking the "Basic Dialog" button. If the button in the lower left says "Advanced" then you are already looking at the Basic Dialog.

4. Check the settings by clicking the "More >>" button at the bottom of the window. This expands the window. Using version 5.0.0.16 of QCapture Pro and with the specific light source we used, the settings are as follows (Fig. 4) (*see* **Note** 7 for possible variations):
   - *Exp Acq*: 00.300.00, Adjust Exp for Binning (checked).
   - *Binning*: Pvw (2 × 2 for focusing, 4 × 4 for moving the plate); Acq (1 × 1).
   - *Capture area dimensions*: Left (320), Top (0), Right (2239), Bottom (1919) [final resolution, 1920 × 1920 pixels].

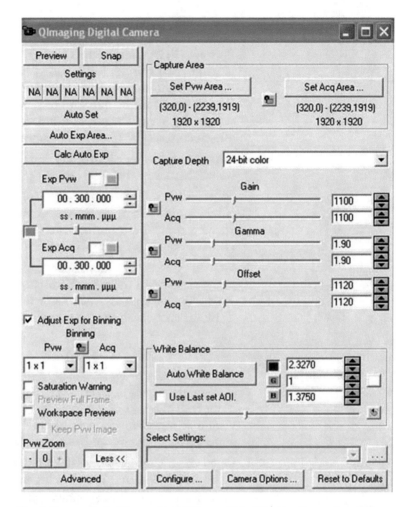

Fig. 4. Screen shot of the capture settings within QCapture Pro capture software

- *Capture depth*: 24-bit color.
- Gain: 1100; Gamma: 1.90; Offset: 1120.
- *White balance*: R (2.327), G (1.0), B (1.375).

5. Click "Less <<" to close the settings side of the window, then click "Preview" to see a live image.

6. Turn on the light source and place a 48-well plate (or object to be imaged) under the camera.

7. Adjust the aperture (top ring on lens) to change the amount of light passing through the lens. The amount of this adjustment will depend on the desired result. We adjusted the aperture so that a slightly gray image appeared in the preview window (*see* **Note 8**).

8. Adjust zoom (middle ring) and focus (bottom collar) to fill the preview window with a single well of the 48-well plate. An

initial height adjustment of the camera above the light box may be necessary as well, to achieve the desired result. Use 2 × 2 Preview (Pvw) binning to focus, then switch to 4 × 4 for moving the plate. As long as Acq binning remains at 1 × 1 the full resolution image will be acquired. The binning status of Pvw does not affect the acquired image, but only the image seen in the preview window. Binning at 4 × 4 allows for faster on-screen response of plate movement (*see* **Note 9**).

9. Capture images by clicking the "Snap" button. Move the 48-well plate to view the next well, click "Snap," and repeat for the entire plate.

10. After acquiring all the images, close the live preview window by clicking the "Stop" button, then the "X" in the upper right corner of the acquisition window.

11. Save the images to a new folder. A quick way to save all the images is to select the menu Window/Close All, then click "Yes All." When prompted, type a name for each file. Save the files as TIFF for analysis by the Romanizer software.

3.2.2 Analyzing Images

1. The Romanizer program will automatically count blue-stained HIV-infected cells in 1920 × 1920 pixel, 24-bit color images. Note that in addition to counting blue-stained cells, the software may also count overly clumped cells, dust, or fibers that appear dark in the image (*see* **Note 10**).

2. Open the file folder where the Romanizer program is stored. Copy 5MGL.exe and 5MGL.PAR into the folder where you saved your images. The program must reside in the same folder as your TIFF images, and it will analyze all images within that folder.

3. Double-click 5MGL.exe to launch the counting program. You should see one of your images with a red circle superimposed upon it (Fig. 5). The circle defines the region to be analyzed. Using the computer mouse, click and drag the circle to the center of your well to check its size. The circle should be a little smaller than the size of the well to prevent analyzing the edge of the well. If you need to adjust the size follow these steps:

   (a) Close the program by clicking Exit on the menu.

   (b) Open the file 5MGL.PAR file (*see* **Note 11**) using a simple text editor such as the Notepad.

   (c) Increase or decrease the number next to "circle radius in tiff pixels" depending on the desired change in the circle radius.

   (d) Save your changes and close the file.

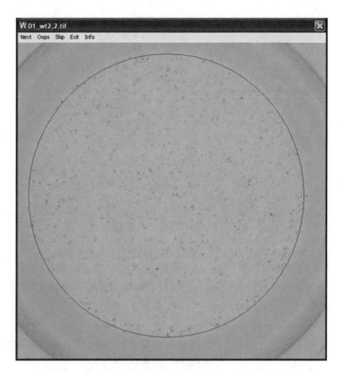

Fig. 5. Screen shot of the Romanizer program. The *red circle* indicates the area that will be analyzed by the program. The position of the analysis circle was defined for each image because of potential variations in the placement of the well within the digital image

   (e) Launch 5MGL.exe again to see the result of the size change.

   (f) Repeat **steps a–e** if needed to find the appropriate size.

4. Position the circle in the center of the well, then click "Next." The next image in the folder will appear. Repeat the procedure for the remaining images. Due to variation in the placement of the well by hand within the frame of the digital photograph, the location of the analysis circle needs to be defined separately for each image analyzed.

5. When all circles have been defined, the program will process the images and start counting. Press "OK" when it has finished. Figure 6a shows a digital photograph taken with the Micropublisher 5.0. The resulting image after being processed by the software is shown in Fig. 6b.

6. After background subtraction and processing, the images are placed in the file folder with an "f" preceding each filename. The results of the analysis are exported to a Microsoft Excel file called "results.xls." "Simple count" is a raw count of spots found. "Smart count" takes into consideration that some

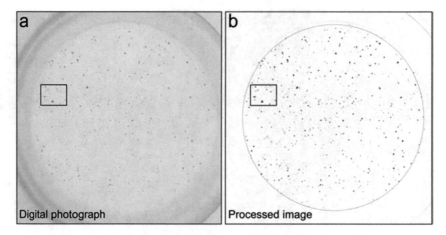

Fig. 6. Image-analysis of HIV-infected P4R5 cells. (**a**) Digital photograph of a monolayer of P4R5 cells infected with HIV and stained with X-gal within one well of a 48-well tissue culture plate using a MicroPublisher 5.0 camera and macro video zoom lens. (**b**) The same well after processing by the analysis software. The *boxed region* is magnified in Fig. 7

spots may in fact be several adjacent cells, as it counts the adjacent cells separately. In most cases, "Smart count" is the preferred result to use (*see* **Note 12**).

3.3 HIV Infectivity Assay

1. These instructions describe an assay to quantify the infectivity of live HIV virus. Sterile cell culture technique and safe laboratory practices for handling live HIV are beyond the scope of this description. We performed these assays in a Biosafety Level 3 (BL3) facility approved by the Environmental Health and Safety department of our institution.

2. The P4R5 cell line is an adherent cell line derived from HeLa cells. The cells are engineered to express β-galactosidase after infection with HIV. Maintain them according to standard cell culture procedures for adherent cell lines.

3. Plate cells at a density of 2×10^4 cells per well in a 48-well tissue culture plate and place them in a humidified, 37 °C, 5% CO_2 incubator overnight.

4. The next day, carefully aspirate the media, avoiding contact with the cell monolayer. Infect the cells in duplicate wells with 100 µL of serial dilutions of virus in complete media. After a 2 h incubation at 37 °C, add 400 µL of complete media. Incubate for 2 days at 37 °C.

5. Carefully aspirate the media from the wells and add 1 mL of fix solution to each well. Incubate for 5 min at room temperature. The fix solution renders the virus non-infectious.

6. Aspirate the fix solution and wash the wells twice with PBS.

7. Add 250 μL of X-gal stain solution and incubate overnight at 37 °C in a non-CO_2 incubator.

8. Aspirate the stain solution and wash once with PBS. Wash again with deionized water. Aspirate the water from the well.

   After wiping down the outside of the plate with 70% ethanol, the plate can be removed from the BL3 and handled at the lab bench.

9. Invert the plate over a paper towel and allow the wells to dry.

10. The plate is now ready for imaging and analysis. Figure 7 illustrates the final result in a magnified region of Fig. 6. Syncytia (fused neighboring HIV-infected cells) of various sizes and single cells are accurately distinguished within the monolayer of P4R5 cells. The magnified image illustrates the amount of detail seen in the original digital photograph (Fig. 7a), and for comparison, the same region is photographed using the 20× objective of a Leitz Labovert microscope (Fig. 7b). In the processed image, every object that is counted is marked with a

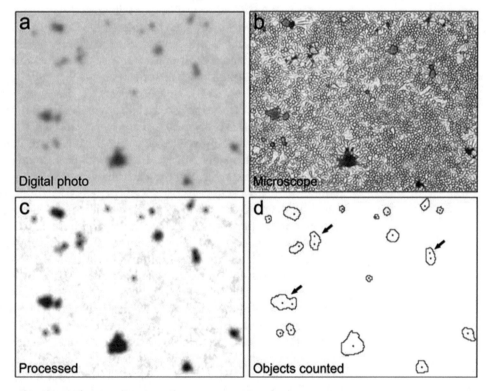

Fig. 7 Magnified view of HIV-infected P4R5 cells. Magnified images of the *boxed region* of Fig. 6 depicting (**a**) the original digital photograph, (**b**) the same region viewed through the 20× objective of a microscope, (**c**) the processed image including small *red dots* as indicators of the objects counted and (**d**) outlines of infected cells surrounding *red dot markers*. The *arrows* indicate cell groups counted singly using Simple count and as two cells using Smart count

small red dot to allow the user to examine more closely how the software is performing and exactly what objects in the image are counted (Fig. 7c). To illustrate this more clearly, the outlines of infected foci are shown encircling each red dot (Fig. 7d). Two or more red dots within a group of cells that come in contact in the image are counted as one infection, using "Simple" count and as multiple foci using "Smart" count (*see* **Note 12**).

4   Notes

1. With today's advances in digital camera technology, there are many types of affordable options. The first parameter to consider when choosing a camera is the spatial resolution that the camera provides. The resolution corresponds to the ability to discriminate between points that are closely spaced in the image [16]. We initially used an analog CCD camera with a standard VGA resolution of 640×480 pixels (approximately 0.3 megapixels). The camera is able to resolve the larger blue-stained foci, but smaller ones remained undetected. Furthermore, the analysis software is not able to distinguish between small foci in close proximity. The resolution of the camera chosen depends largely on the object to be imaged. Smaller objects require larger arrays of pixels (higher resolution) so that they can be discriminated from one another. We upgraded to a 5 megapixel digital CCD camera (2560×1920 pixels) when they became more readily available and reasonably priced. In addition to resolution, a second consideration in choosing a camera is the sensitivity requirement. The sensitivity of a camera refers to the lowest signal that can be detected [17]. If the object to be imaged is fluorescent or chemiluminescent, a camera with increased sensitivity may be required. For example, QImaging makes a camera that will detect a single photon of light (Rolera-MGi); however the resolution of the camera is only 512×512. An increase in camera sensitivity is often accomplished at the expense of resolution. Higher sensitivity applications can also be subjected to an increase in noise; therefore, the signal-to-noise ratio becomes important, as well. Because heat generated in a camera can cause an increase in noise (dark noise), cooled cameras, such as the cooled version of the Micropublisher (MP5.0-RTV-CLR-10-C), can maintain a lower temperature and reduce the amount of noise in the system [18]. Choosing the right camera for your application requires careful consideration of these factors. Consult camera manufacturers for more information and advice.

2. Like the choice of camera, the choice of lens depends on the specific object to be imaged. Macro lenses are designed for close-up photography and are capable of focusing on objects within a short distance. Using a lens with zoom capability enables greater control of the size of the object in the frame. The Optem lens we used is recommended to us and it meets our needs well.

3. The choice of a light source also depends on what is to be imaged. Illumination from above or from below depends on the transparency of the subject. For our purposes, a simple fluorescent light box provided ample light to illuminate the wells of a 48-well plate from below. The light source does not necessarily need to emit a specific intensity of light because there are camera (aperture) and software (exposure) settings that can be adjusted to change the brightness of the resulting image. The color spectra emitted by the light source is also not a large factor because white balance settings can be adjusted within the capture software. A commonly used light box for viewing X-ray films is more than adequate.

4. A tripod mounting screw and a rectangular piece of aluminum are the materials used for the mount adapter (Fig. 8). The dimensions of the aluminum rectangle are 2.25 × 1 × 0.25 in. (5.7 × 2.5 × 0.6 cm). Two holes are made in the aluminum, one large enough for the tripod mounting screw to pass through easily, and the other threaded so that it would screw onto the copy stand mount (1/4″-20 thread) (Fig. 8). The holes are 1.25 in. (3.2 cm) apart. The adapter works only if the copy stand mounting rod can extend far enough and away from the stand to leave at least 3.2 in. (8.1 cm) of space

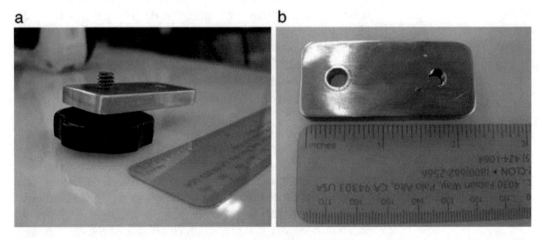

Fig. 8 (**a**) Photograph of the mount adapter with tripod mounting screw. (**b**) Photograph of mount adapter holes. One hole was drilled wide enough for the tripod mounting screw to pass through easily. The second hole was drilled with a 1/4″-20 thread for mounting onto the copy stand

for the camera body (Fig. 3). This custom-built adapter allowed the camera to be rotated 180° and mounted on the copy stand.

5. The Romanizer program is freely available for use in the type of analysis described here. Its use for counting other kinds of objects of different sizes may not produce acceptable results. Processed images should be scrutinized carefully to determine the program's performance. You may download a copy of the software and setup instructions here: http://cfar.ucsd.edu/links/downloads/software. QCapture Pro performs some basic analyses and may be one alternative. Another freely available program that performs image analysis is the NIH Image/ImageJ program developed by the National Institutes of Health [19].

6. When the camera is mounted directly on the camera stand, the image on-screen is inverted, i.e., moving a microplate to the left caused the on-screen live preview image to move to the right, and moving the plate up caused the preview to move down. One can become accustomed to this type of motion, but it is not ideal. The capture software did not have the option to reverse the orientation of the image, but future versions may. Our solution to the problem is to rotate the camera 180°. Unfortunately there is no threaded mounting hole on the opposite side of the camera. Instead, we designed a mount adapter and asked our university machine shop to construct it (Fig. 8). This custom-built adapter allowed the camera to be rotated and mounted on the copy stand.

7. For consistency between experiments, the same settings are used each time. Some changes have been made in newer versions of QCapture Pro such that the parameters may not be exactly the same as shown here. We used the default settings for gain, gamma and offset, and since version 5.0.0.16 they have changed the value ranges for gain and offset. The default settings are recommended; alternately, consult QImaging for advice. Different light boxes will most likely require different settings for red, green, and blue than those shown here. White balance should be set using the Auto White Balance feature of the software. The exposure of 00.300.000 (0.3 s) can be adjusted to change the brightness of the image. Differences in light box intensity may require a change to the exposure time and/or the aperture (*see* **Note 8**). Finally, setting the capture dimensions to a square area (1920 × 1920 pixels) rather than using the full rectangular capture area (2560 × 1920) reduced TIFF file size and eliminated unnecessary image data on either side of the circular well.

8. One way to adjust the brightness of the image is to adjust the aperture. Turning the aperture ring on the lens changes the

diameter of a circular opening. A larger aperture will allow more light to reach the CCD array. For consistency in plate analysis, the same settings should be used for all experiments, including lens aperture. In our experiments, the aperture of the lens is empirically adjusted so that the images are not too bright to obscure faint blue cells, and not too dark to increase background counts of uninfected wells. Once an aperture setting is selected, the aperture ring is marked so that the same setting is used consistently.

9. Binning is a function in which the charges from adjacent pixels on the CCD array are combined. This is accomplished at the expense of spatial resolution [17]. For example, 2×2 binning combines the charge from a 2×2 grid of pixels (4 pixels total) into one signal, effectively reducing the resolution two times in both the x and y axes. Similarly, 4×4 binning combines the charge from 16 adjacent pixels into one signal. The greatest benefit to binning is an increase in signal output and dynamic range [17]. The increase in signal-to-noise ratio allows for detection of lower light levels. In QCapture Pro we checked the box "Adjust Exp for Binning." When binning is increased from 1×1 (no binning) to 2×2 or 4×4, the exposure time is reduced to compensate for the increase in signal. Another benefit of binning is that the resolution of the image is reduced and it is more easily processed by the computer, effectively speeding up the on-screen response time when moving the object. We used binning solely for that purpose. Increased signal is not necessary because there is ample light emitted from the fluorescent light box.

10. An evenly plated monolayer of target cells is critical to preventing cells from overgrowing and forming dark clumps that may be detected by the software. Using clean reagents that will not leave dust or fibers in the wells is also important.

11. The parameter file, 5MGL.PAR, consists of the following default settings:
    - Circle radius in tiff pixels: 860.
    - Window size in screen pixels: 640.
    - Noise cutoff: 45.
    - Spot area cutoff in pixels: 10.
    - Minimum distance of spot maxima: 3.
    - Size of the structuring element: 10.

Other than the circle radius, these settings should not be changed. The window size may need adjustment depending on the screen resolution of the computer. If you feel you need to adjust some of the other parameters, do so in a careful manner, comparing the results before and after changes to assess the modified performance of the program.

12. To count HIV infection, blue-stained cells are identified as dark spots in all three channels of the RGB spectrum even though they appeared predominantly blue to the naked eye. 5MGL.exe therefore analyzed a gray-scale image created by adding all three channels of the original 24-bit color TIFF image. To remove the impulse noise from the image, gray scale thinning and thickening morphological filters are applied with a point structuring element [20]. Because the illumination of the image is not perfectly uniform, straightforward thresholding of the image cannnot be used to count the spots. Instead, using a square structuring element whose size is larger than the typical spot size, the software calculated the morphological opening of the image and then subtracted it from the image. The resulting image has a uniform, fluctuating background. Signal and noise are then separated by thresholding. The remaining islands (defined as unions of all side-to-side adjacent pixels with non-zero intensity) represented the actual blue-stained cells, plus a few small ghost spots that emerge by random agglomeration of noise after thresholding. The latter are typically much smaller than the islands representing actual blue cells and are removed by applying an area opening to the image (ibid). The remaining islands are considered as signals and are counted; this analysis is termed "Simple count." However, sometimes a single island represented several adjacent cells that should have been counted separately; this is especially the case under conditions of high inoculum. Because a stained cell is darker on the inside than near the edge, adjacent cells have separate dark intensity maxima, which could be counted individually. This analysis is termed "Smart count." Using Smart count, in terms of the intensity landscape, the hills on the islands rather than the islands themselves are counted. Both Simple and Smart counts are reported in the exported results file. For user verification of software performance, a processed version of each image containing a small red dot on every object counted is saved into the file folder containing the original images.

Acknowledgments

We thank the coauthors of the original publication, Laura Martí-nez, Roman Šášik, Douglas Hitchin, Megan Dueck, Douglas Richman, and John Guatelli for their contributions to this work. The system we have developed was conceived with ideas from Chris Aiken. We are also grateful to Prentice Higley, Sherry Rostami and Nanette Van Damme for assistance during setup and evaluation. This work was supported by grants AI27670, AI043638,

AI038201, the UCSD Center for AIDS Research (AI 36214), AI29164, AI047745, from the National Institutes of Health and the Research Center for AIDS and HIV Infection of the San Diego Veterans Affairs Healthcare System.

References

1. Harada S, Koyanagi Y, Yamamoto N (1985) Infection of HTLV-III/LAV in HTLV-I-carrying cells MT-2 and MT-4 and application in a plaque assay. Science 229:563–566

2. McKeating JA, McKnight A, McIntosh K, Clapham PR, Mulder C, Weiss RA (1989) Evaluation of human and simian immunodeficiency virus plaque and neutralization assays. J Gen Virol 70:3327–3333

3. Nara PL, Fischinger PJ (1988) Quantitative infectivity assay for HIV-1 and -2. Nature 332:469–470

4. Chesebro B, Wehrly K (1988) Development of a sensitive quantitative focal assay for human immunodeficiency virus infectivity. J Virol 62:3779–3788

5. Akrigg A, Wilkinson GW, Angliss S, Greenaway PJ (1991) HIV-1 indicator cell lines. AIDS 5:153–158

6. Chackerian B, Long EM, Luciw PA, Overbaugh J (1997) Human immunodeficiency virus type 1 coreceptors participate in postentry stages in the virus replication cycle and function in simian immunodeficiency virus infection. J Virol 71:3932–3939

7. Charneau P, Mirambeau G, Roux P, Paulous S, Buc H, Clavel F (1994) HIV-1 reverse transcription. A termination step at the center of the genome. J Mol Biol 241:651–662

8. Deng H, Liu R, Ellmeier W, Choe S, Unutmaz D, Burkhart M, Di Marzio P, Marmon S, Sutton RE, Hill CM, Davis CB, Peiper SC, Schall TJ, Littman DR, Landau NR (1996) Identification of a major co-receptor for primary isolates of HIV-1. Nature 381:661–666

9. Kimpton J, Emerman M (1992) Detection of replication-competent and pseudo-typed human immunodeficiency virus with a sensitive cell line on the basis of activation of an integrated beta-galactosidase gene. J Virol 66:2232–2239

10. Vodicka MA, Goh WC, Wu LI, Rogel ME, Bartz SR, Schweickart VL, Raport CJ, Emerman M (1997) Indicator cell lines for detection of primary strains of human and simian immunodeficiency viruses. Virology 233:193–198

11. Wei X, Decker JM, Wang S, Hui H, Kappes JC, Wu X, Salazar-Gonzalez JF, Salazar MG, Kilby JM, Saag MS, Komarova NL, Nowak MA, Hahn BH, Kwong PD, Shaw GM (2003) Antibody neutralization and escape by HIV-1. Nature 422:307–312

12. Montefiori DC (2009) Measuring HIV neutralization in a luciferase reporter gene assay. Methods Mol Biol 485:395–405

13. Tilton CA, Tabler CO, Lucera MB, Marek SL, Haqqani AA, Tilton JC (2014) A combination HIV reporter virus system for measuring post-entry event efficiency and viral outcome in primary CD4+ T cell subsets. J Virol Methods 195:164–169

14. Xu R, El-Hage N, Dever SM (2015) Fluorescently-labeled RNA packaging into HIV-1 particles: direct examination of infectivity across central nervous system cell types. J Virol Methods 224:20–29

15. Day JR, Martínez LE, Šášik R, Hitchin DL, Dueck ME, Richman DD, Guatelli JC (2006) A computer-based, image-analysis method to quantify HIV-1 infection in a single-cycle infectious center assay. J Virol Methods 137:125–133

16. Beyon JDE, Lamb DR (eds) (1980) Charge-coupled devices and their applications. McGraw-Hill, UK

17. Holst GC (1998) CCD arrays, cameras, and displays, 2nd edn. JCD Publishing/SPIE Optical Engineering Press, Winter Park, FL/Bellingham, WA

18. Howes MJ, Morgan DV (eds) (1979) Charge-couple devices and systems. John Wiley & Sons, New York, NY

19. NIH. (2005) NIH Image/ImageJ, a public domain program developed at the US. National Institutes of Health. Available at http://rsb.info.nih.gov/nih-image/

20. Soille P (2003) Morphological image analysis: principles and applications. Springer-Verlag, New York, NY

Chapter 24

"Dipstick" Colorimetric Detection of Metal Ions Based on Immobilization of DNAzyme and Gold Nanoparticles onto a Lateral Flow Device

Debapriya Mazumdar, Tian Lan, and Yi Lu

Abstract

Real-time, on-site detection and quantification of different trace analytes is a challenge that requires both searching a general class of molecules to recognize a broad range of contaminants and translating this recognition to easily detectable signals. Functional nucleic acids, which include DNAzymes (DNA with catalytic activity) and aptamers (nucleic acids that bind an analyte), are ideal candidates for the target recognition. These nucleic acids can be selected by a combinatorial biology method called in vitro selection to interact with a particular analyte with high specificity and sensitivity. Furthermore, they can be incorporated into sensors by attaching signaling molecules. Due to the high extinction coefficients and distance-dependent optical properties, metallic nanoparticles such as the commonly used gold nanoparticles have been shown to be very attractive in converting analyte-specific functional DNA into colorimetric sensors. DNAzyme directed assembly of gold nanoparticles has been used to make colorimetric sensors for metal ions such as lead, uranium, and copper. To make the operation even easier and less vulnerable to operator's errors, dipstick tests have been constructed. Here, we describe protocols for the preparation of DNAzyme-linked gold nanoparticles (AuNP) that are then immobilized on to lateral flow devices to make easy-to-use dipstick tests for metal ions.

Key words Dipstick, Nanoparticle, Sensor, Colorimetric, Lateral flow, DNAzyme, Metal ions

1 Introduction

1.1 DNAzymes or Deoxyribozymes as Sensors for Metal Ions

Nucleic acids have recently emerged as an important platform for selective molecular recognition, one major requirement for sensors. Long considered as passive molecules for the storage of genetic information, RNA and DNA molecules with catalytic function similar to protein enzymes were discovered in the early 1980s and 1990s, respectively [1–3]. These enzymes are called ribozymes (catalytic RNA) and deoxyribozymes or DNAzymes (catalytic DNA). The nucleic acid enzymes usually require a metal ion cofactor to perform their catalytic function and can be tailored to be specific for a particular metal ion. In addition, nucleic acids that

Avraham Rasooly and Ben Prickril (eds.), *Biosensors and Biodetection: Methods and Protocols Volume 1:*
Optical-Based Detectors, Methods in Molecular Biology, vol. 1571, DOI 10.1007/978-1-4939-6848-0_24,
© Springer Science+Business Media LLC 2017

bind to a target material with high specificity and affinity (analogous to protein antibodies) have also been obtained, and these are called aptamers [4–7]. Nucleic acid enzymes and aptamers have also been fused to form a new class of allosteric enzymes called aptazymes [8, 9]. Collectively, the nucleic acid enzymes, aptamers and aptazymes are termed functional nucleic acids. In this chapter, our focus is on DNAzymes that catalyze cleavage reactions in the presence of a metal ion.

Although a number of naturally occurring ribozymes have been discovered in nature [1, 2], DNAzymes have not been found in nature. DNAzymes and aptamers are obtained by a combinatorial biology technique called in vitro selection (also called *systematic evolution of ligands by exponential enrichment* (SELEX for aptamer selections)) in test tubes [4–6, 10]. Figure 1a is a schematic representation of the selection process. A large pool of nucleic acid sequences represented by the colored objects ($\sim 10^{14}$–10^{16} different sequences) is incubated with a target of interest in each round of selection. The 'winner sequences' which bind to the target analyte (in the case of aptamer selection) or catalyze a reaction in the presence of target (in the case of nucleic acid enzyme selection) are separated by various techniques such as gel electrophoresis, column separation and capillary electrophoresis. These "winners" are amplified using polymerase chain reaction to be used for the next round of selection. During each round of selection, the stringency can be increased by decreasing the interaction time between the target and the nucleic acid or by decreasing the concentration of the target. Iterative rounds of selection are continued until the pool is sufficiently enriched with sequences of desired sensitivity and specificity (represented by blue cubes in Fig. 1a). A review of in vitro selection of DNAzymes, including a detailed protocol has been published in a previous *Methods in Molecular Biology* series book [11]. This technique can be used to obtain nucleic acid sequences that recognize a target contaminant with sensitivity and specificity. In vitro selection has been used to obtain a number of metal specific-DNAzymes dependent on Pb^{2+} [12, 13], Zn^{2+} [14], Co^{2+} [15], Cu^{2+} [16], UO_2^{2+} [17], Hg^{2+} [18], and As^{5+} [18], some of which have been converted into fluorescent and colorimetric sensors as described in the following sections. A number of these metal ions, notably Pb^{2+}, Hg^{2+}, and As^{5+}, are heavy metals that are particularly toxic; UO_2^{2+} is a radionuclide. The maximum level of these metal ions in drinking water is strictly regulated by the US Environmental Protection Agency (EPA). Only a few sensors available today can detect metal ions below EPA regulatory levels, and therefore the selectivity of sensors should be improved in order to be practically applicable. For this purpose, the utility of DNAzymes as toxic metal ion sensors is of great importance. The predicted secondary structures of a few DNAzymes are shown in Fig. 1b. The strands in green represent the enzyme and the strands in black are

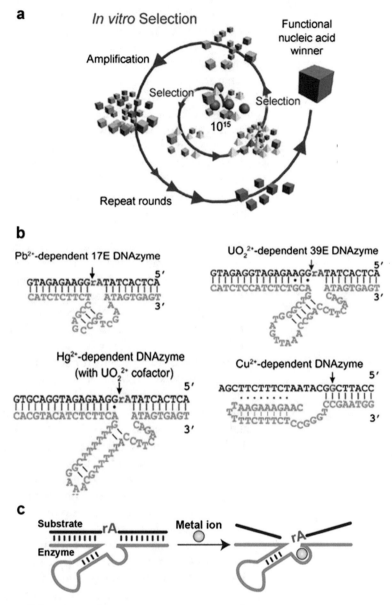

Fig. 1 In vitro selection of functional nucleic acids. (**a**) Schematic depiction of general in vitro selection scheme for obtaining functional nucleic acid that interacts with a specific target (contaminant). (**b**) Predicted secondary structures of metal specific RNA cleaving DNAzymes obtained by in vitro selection. The *black strands* represent the substrate and the *green strands* are the enzyme. The cleavage site is depicted by the *black arrow* and there is a single RNA base at the cleavage site of the Pb^{2+}, Hg^{2+} and UO_2^{2+} DNAzyme denoted by rA. (**c**) General schematic representation of the catalytic activity of metal ion dependent RNA cleaving DNAzyme. Strand in *black* represents the substrate strand with an embedded riboadenosine base (rA in *red*); strand in *green* represents the enzyme strand which is capable of recognizing a specific metal ion and catalyze the cleavage of a phosphodiester bond at the rA site

the nucleic acid substrate. All these DNAzymes (except the Cu^{2+} dependent DNAzyme) are RNA cleaving enzymes that catalyze the cleavage of a single phosphodiester bond within the rA (RNA) embedded in the DNA substrate. They share a similar secondary structure containing a substrate strand and an enzyme strand hybridizing to each other through the two arms on each side of the cleavage site (Fig. 1c). It has been determined that, unlike the DNA sequences in the single stranded loop, the sequence identity of two arms do not contribute significantly to either metal binding or activity, as long as they can form double stranded DNA to provide stability [12]. Therefore the sequence length and GC content can be designed so that, in the absence of the target metal ion, the enzyme strand and substrate strand have a melting temperature above the ambient temperature, making them hybridize to each other. In the presence of target metal ion, however, the cleavage of the substrate strand changed the melting temperature between the two halves of the cleaved products and enzyme strand to below the ambient temperature so that they can be released from the enzyme strand. Some of the fastest DNAzymes have a catalytic efficiency (k_{cat}/K_m) of 10^9 M^{-1} min^{-1} [19], rivaling that of protein enzymes and thus they are ideal for fast sensing.

As a major component for sensors, nucleic acids possess many advantages. First, DNA/RNA targeting essentially any molecule of choice can be obtained through combinatorial selections [4–6, 10], which provide a unique opportunity to construct a general sensing platform for a broad range of analytes. Second, nucleic acids, particularly DNA, are very stable and can be denatured and renatured many times without losing their binding abilities, allowing a long shelf life. Third, nucleic acids have predictable base pairing interactions, which have been proven to be very useful for rational sensor design that are difficult to attain when developing protein or organic molecule-based sensors. Finally, a broad range of DNA modifications can be chemically synthesized with relatively low cost, allowing convenient and precise conjugation. These properties make DNA and RNA ideal candidates to create sensors.

Since natural nucleic acids do not possess functional groups that can generate absorption in the visible region, external signaling labels need to be applied to convert them into sensors. To achieve this goal, many inorganic nanoparticles have been employed. The next few sections describe the properties of these nanoparticles and preparation of a DNAzyme based lateral flow device using gold nanoparticles.

1.2 Physical Properties of Nanoparticles

Depending on the composition, size, and shape of inorganic nanoparticles, a wide range of properties can be obtained. Inorganic metallic [20, 21], semiconductor [22], and magnetic nanoparticles [23] can all be used to make functional sensors with different detection modes. In this protocol, metallic nanoparticles are used.

Dispersed AuNPs (diameter from several nanometers to about 100 nm) display red colors resulting from their surface plasmons. In addition to such distance-dependent optical properties, AuNPs also possess very high extinction coefficients, which are usually 3–5 orders of magnitude higher than the brightest organic chromophores. Thiol-modified DNA can be attached to the surface of AuNPs and these functionalized nanoparticles can be crosslinked by complementary DNA to form blue-colored aggregates [24]. This process has been applied by Mirkin and coworkers to design highly sensitive and selective colorimetric sensors for DNA detection [25]. By using functional DNA (aptamers [20, 26], catalytic DNA or DNAzymes [27, 28], and aptazymes [29]) that can recognize a diverse range of analytes, AuNP-based colorimetric detection method can be applied to detect many analytes beyond DNA.

1.3 Nanoparticle Conjugated DNAzymes for Sensing Application

Many metal specific DNAzymes have been successfully converted into fluorescent sensors using a catalytic beacon technology [30] where the DNA enzyme and substrate are tagged with a combination of organic fluorophore and quencher. Sensors for metal ions such as Pb^{2+} [13], UO_2^{2+} [17], Hg^{2+} [31], Cu^{2+} [32] are reported in literature and these have been commercialized by ANDalyze Inc. using a platform that consists of consumable fluorescent sensor cartridges and a handheld fluorimeter.

Although fluorescent sensors are very sensitive and provide a method of precise quantitative detection, colorimetric sensors have been developed to make the detection possible by the naked eye. In 1996, Mirkin and coworkers utilized the DNA induced assembly of AuNPs to make a colorimetric sensor for nucleic acids [25]. Lu and coworkers expanded the scope of sensing to analytes beyond nucleic acids, by combining AuNPs with DNAzymes [27, 28, 33, 34]. The sensing method is depicted using the Pb^{2+} dependent 17E DNAzyme as a representative example (Fig. 2). AuNPs functionalized with short thiol modified DNA are assembled on the arms of the extended substrate which is in turn hybridized to the enzyme. Since each AuNP is functionalized with many DNA strands, blue aggregates are formed. In the presence of Pb^{2+}, the enzyme catalyzed cleavage of the substrate will disassemble the aggregate producing red color. The color can be spotted on a TLC plate and one such representative test is shown in the inset of Fig. 2. Red color of increasing intensity is seen with Pb^{2+}, whereas the sensors containing other metals are blue. The reaction is fast with color change occurring within 10 min under optimized conditions and the detection limit is ~100 nM.

Although not described in detail in this chapter, AuNP-based aptamer sensors for various analytes, including several small organic molecules are reported in literature [20, 33, 35–37]. The AuNPs can also be replaced by other metallic nanoparticles, so that different color changes can be achieved in the presence of different

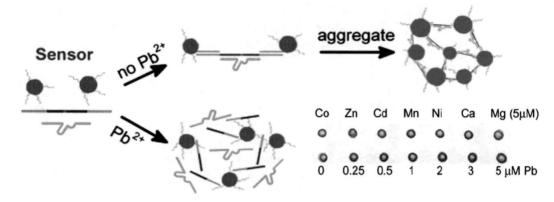

Fig. 2 DNAzyme based colorimetric sensor for Pb^{2+} detection. In the absence of Pb^{2+}, the oligonucleotide functionalized AuNPs are assembled on the substrate to form *blue* aggregates. When Pb^{2+} is present, the substrate is cleaved and the aggregate is disassembled to yield *red* colored dispersed AuNPs. *Inset*: Picture of sensor incubated with increasing concentrations of Pb^{2+} and with other metal ions (that are all 5 μM in concentration)

analytes. In addition to metallic nanoparticles, other nanomaterials including magnetic nanoparticles, quantum dots, nanotubes, or polymer beads can also be used to suit various applications.

1.4 Dipstick Tests Based on Immobilization of DNAzyme-Gold Nanoparticles onto a Lateral Flow Device

While these colorimetric biosensors are an important step toward real-time sensing as the signal is detectable by the naked eye, without the need for expensive instrumentation, they still require laboratory type operations, such as precise transfer and mixing of multiple solutions. In addition, although the sensitivity is high when the absorbance is recorded using a UV–Vis spectrophotometer, it is often difficult to distinguish the red color of dispersed nanoparticles against a blue background from the aggregates, particularly at low metal-ion concentrations. Furthermore, AuNPs are not very stable in solution state and are vulnerable to aggregation under a variety of conditions thereby making it difficult to store the sensors for a long period of time.

Lateral flow devices are an ideal platform for making dipstick type tests to further improve the performance of DNAzyme-AuNP colorimetric sensors. In addition to eliminating precise solution transfer and allowing separation of AuNPs to make it easier to distinguish colors, the reagents can be prepared in a dry or nearly dry state, making the device stable at ambient conditions for a long period of time. The home pregnancy test is the most commonly used lateral flow devices and several other antibody-based tests are also well-known; however, DNA-based dipstick tests are not common. Glynou et al. reported a lateral flow device for the detection of DNA [38]. To expand on the range of analytes detected, we previously reported dipstick tests for the detection of adenosine and cocaine using AuNPs conjugated to aptamers [39]. The protcol to make these aptamer-AuNP based dipsticks has been reported previously [40]. A paper-based bioassay using aptamers and the protein enzyme DNase I, which also involved the disassembly of nanoparticle aggregates that were dried onto paper substrates was also reported by Yingfu Li and coworkers [41]. Even though our previously reported methodology can be applied to almost any target for which aptamers can be obtained, because aptamers with high affinity for metal ions have been difficult to select, this methodology has not been applied to dipstick tests for metal ions. To extend the applicability of this test for metal ions, metal-specific DNAzymes are an excellent choice. However, it is not trivial to adopt the aptamer-based dipstick methodology by simply replacing aptamers with DNAzymes because DNAzymes undergo not only binding as aptamers do but also catalytic activity and product release, making the design more complicated. The Lu group has previously reported a method to convert DNAzyme-AuNPs into dipstick tests for metal ions, specifically for Pb^{2+} [42]. We showed that the cross-link based method used in previous aptamer-AuNP systems cannot be used for DNAzyme-AuNP system. Instead, we succeeded in developing a non-cross-linked DNAzyme-AuNP system for detection of Pb^{2+}. These dipstick tests are ideal for detection of lead in household paints, in accordance with the EPA defined threshold of 1 mg/cm^2 Pb^{2+} for paint to be classified as lead-based.

The 8–17 DNAzyme is used to construct the dipstick tests for Pb^{2+} because of its very high activity as shown by a fast cleavage rate (estimated $k_{obs} \sim 50$ min^{-1} at pH 7.0) [12]. Figure 3a shows its reaction scheme. In the presence of Pb^{2+}, the enzyme strand (called 17E) catalyzes the cleavage of the chimeric substrate (called 17S) at the single ribo-adenosine (rA) base. Unlike aptamer-based colorimetric tests, the presence of target does not cause immediate disassembly of the aggregates in the colorimetric tests. Disassembly in the case of DNAzymes has been shown to require heating [27] or the use of invasive DNA strands [33] to release the cleaved product trapped in the nanoparticle aggregates. Either of these methods leads to added complexity and are not highly feasible for lateral flow devices. Therefore, we decided to use an alternate approach that does not involve formation of nanoparticle aggregates and thus the detection is not based on a change in the optical properties of AuNPs.

In our scheme, the 8–17 DNAzyme is modified to form the construct shown in Fig. 3b. The 17S substrate is modified on the 3′ end to have a biotin moiety. On the 5′ side, 18 additional bases $(AAG)_6$ are added to act as a free site for DNA hybridization required for the capture of cleaved product. In addition, the 5′ end is functionalized with a thiol group in order to conjugate the substrate to 13 nm AuNPs. When the original 8–17 DNAzyme construct with 9 symmetric base pairs on either substrate binding

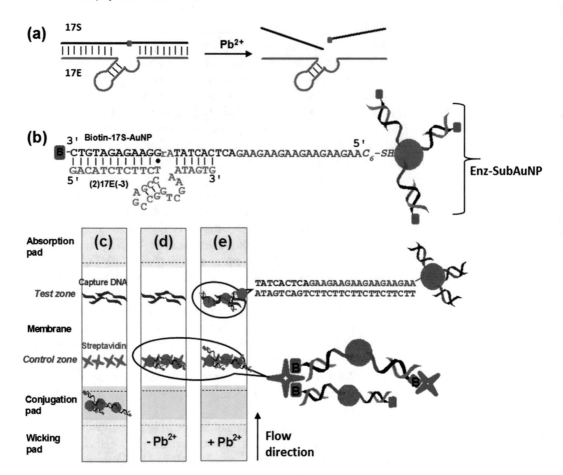

Fig. 3 (**a**) 8–17 DNAzyme reaction: In the presence of Pb^{2+}, the 17E enzyme catalyzes the cleavage of the chimeric substrate, 17S at the single ribo-linkage (rA). (**b**) Modified 8–17 construct conjugated to AuNPs (called Enz-SubAuNP) used for the dipstick tests. B inside the *purple box* denotes biotin. (**c**) Assembled lateral flow device. (**d**) Negative control: In the absence of Pb^{2+}, AuNP-uncleaved substrate is captured at the control zone via streptavidin-biotin interaction, producing a single *red line*. (**e**) Positive test: Substrate is cleaved in the presence of Pb^{2+} and the AuNP-cleaved product migrates beyond the control zone to be captured at the test zone by hybridization to complementary DNA. Two lines are produced

arm is tested, this construct failed to give a positive signal as designed. To facilitate the release of the cleavage product without compromising the binding of the substrate, three bases are deleted from the 5′ end of the enzyme and two bases are added to the 3′ end of the 17E enzyme. The shortened arm facilitates the release of the cleaved product on the 5′ end after cleavage reaction, while the overall stability of the construct before cleavage is maintained due to the extra base-pairing on the opposite arms. The enzyme–substrate complex is prepared by hybridizing the enzyme to the substrate conjugated to AuNPs and this was referred to as Enz-SubAuNP.

The lateral flow device is constructed using a Millipore Assembly kit by placing four overlapping pads on a backing. Streptavidin and capture DNA are applied on the capture zone and test zone of the membrane, respectively and the complex, Enz-SubAuNP is spotted on the conjugation pad and allowed to dry (Fig. 3c). We hypothized that if the dipstick is dipped in a flow buffer the Enz-SubAuNP complex will be rehydrated. In the absence of Pb^{2+}, the substrate would remain uncleaved and Enz-SubAuNP would migrate on the membrane till it reached the control zone, where the biotin-containing Enz-SubAuNP could be captured by streptavidin, thus producing a single red line at the control zone (Fig. 3d). In the presence of Pb^{2+}, the substrate would be cleaved, and the cleaved product would migrate past the control zone to be captured at the test zone by a 27 base long DNA sequence complementary to the cleaved substrate piece (called capture DNA), producing a red line at the test zone. Since the cleavage reaction may not be 100% complete, the positive tests would normally result in two red lines, one being the cleaved product and the other being the uncleaved enzyme–substrate (Fig. 3e).

2   Materials

1. Oligonucleotides: All oligonucleotides are purchased from a commercial source (e.g., Integrated DNA Technologies Inc., (Coralville, IA)). The oligonucleotides are purified by HPLC or polyacrylamide gel electrophoresis (PAGE) to ensure high purity.

2. Other chemicals: Hydrogen tetrachloroaurate (III) ($HAuCl_4$), trisodium citrate dihydrate, tris-(2-carboxyethyl)phosphine hydrochloride (TCEP), tris-(hydroxymethyl) aminomethane (Tris), lead chloride, sodium hydroxide, concentrated hydrochloric acid (HCl), and concentrated nitric acid (HNO_3) are purchased from Sigma-Aldrich (St. Louis, MO). Streptavidin is purchased from Promega (Madison, WI).

3. Buffers: Tris–HCl buffers are used in the experiments. 500 mM of Tris–HCl buffer stock at pH 8.0 is prepared by adding hydrochloric to 500 mM Tris solution until the desired pH value is achieved. The buffer stock solutions are incubated with metal chelating resin (iminodiacetic acid, sodium form, Aldrich) overnight to eliminate trace divalent metal ions. Finally, the buffer stock solutions are filtered through 0.2 μm syringe filters (Nalgene, Rochester, NY) and stored in a $-20\,°C$ freezer.

4. Equipment: A two-neck flask (100 ml), a condenser, and a stopper; hot plate with magnetic stirring and a stir bar; disposable scintillation vials (20 ml), polypropylene microcentrifuge

tubes (1.7 ml; catalog no. MCT-175-C; Axygen Scientific), temperature-controlled UV–visible spectrophotometer (Hewlett-Packard 8453), quartz UV–visible cell (Hellma), 0.2-μm syringe filter (Nalgene), and Sep-Pak desalting column (Waters).

5. Lateral flow devices: Lateral flow devices are constructed from membranes and pads obtained from Millipore Corporation (Bedford, MA). Hi-Flow Plus Cellulose Ester Membrane with a nominal capillary flow time of 240 s/4 cm and a nominal membrane thickness of 135 μm direct cast onto 2 mil polyester backing is used. Cellulose fiber sample pads and glass fiber conjugate pads are also purchased from Millipore. Avery Self-Adhesive Laminating Sheets, 9 × 12 in., are obtained from Amazon.com.

3 Methods

3.1 Preparation of AuNPs

Preparation of high quality AuNPs ensures the success of subsequent steps of the experiment. For current applications, we chose to synthesize ~13 nm diameter AuNPs for the following reasons. First, the protocol for such synthesis is well-established and requires only a simple mixing step [43]. Second, the resulting AuNPs can be readily used for conjugation of thiol-modified DNA [25], and the conjugates are usually highly stable against aggregation. In our experience, if nanoparticles of 40 nm diameter are used, it is more difficult to obtain DNA conjugates with comparable stability.

1. Prepare 500 ml of aqua regia by mixing 3:1 concentrated HCl: HNO_3 in a large beaker in a fume hood. The color of the mixture changes to deep orange/red in several minutes. Be extremely careful when preparing and working with aqua regia. Wear goggles and gloves, and perform the experiment in a fume hood.

2. Soak a two-neck flask, magnetic stir bar, stopper, and condenser in the aqua regia solution for at least 15 min (*see* **Note 1**). The volume of the flask can vary depending on the scale of synthesis, and usually 100–500 ml is used. Rinse the glassware with copious amount of deionized water and then Millipore water.

3. Prepare 50 mM $HAuCl_4$ solution by dissolving the solid in Millipore water. Do not use metal spatula while weighing out the $HAuCl_4$. Filter the solution with a 0.2 μm pore size syringe filter. Prepare 38.8 mM trisodium citrate solution by dissolving the salt in Millipore water and filter the solution.

4. To prepare about 100 ml of AuNPs, add 98 ml of Millipore water into the two-neck flask. Add 2 ml of 50 mM $HAuCl_4$ solution so that the final $HAuCl_4$ concentration is 1 mM. Connect the water condenser to one neck of the flask, and place the stopper in the other neck. Put the flask on a hot plate to reflux while stirring.

5. When the solution begins to reflux, remove the stopper. Quickly add 10 ml of 38.8 mM sodium citrate, and replace the stopper. The color should change from pale yellow to grayish blue to deep red in 1 min. Allow the system to reflux for another 20 min.

6. Turn off the heating and allow the system to cool to room temperature (23–25 °C) under stirring. The diameter of such prepared AuNPs is ~13 nm. The extinction value of the 520 nm plasmon peak is ~2.4, and the nanoparticle concentration is ~13 nM. The color of the solution should be burgundy red, and the AuNP shape should be spherical under transmission electron microscopy (TEM). The prepared nanoparticles are stable for months when stored in a clean container (glass or plastic) at room temperature. Do not freeze the nanoparticles.

3.2 Functionalization of AuNPs with Thiol-Modified DNA

1. Soak disposable scintillation vials (20 ml volume) in 12 M NaOH for 1 h at room temperature (*see* **Note 2**). Rinse the vials with copious amounts of deionized water and then Millipore water. Be extremely careful when preparing and working with concentrated NaOH. Wear goggles and gloves. When preparing 12 M NaOH solution, the temperature of the system increases significantly. Occasional stirring is needed to avoid the condensation of solid NaOH on the bottom of the container. The concentrated NaOH solution can be reused many times for soaking glass vials.

2. Prepare 10 mM fresh TCEP solution by dissolving a tiny crystal of TCEP in Millipore water.

3. Pipette 9 μl of 1 mM thiol modified substrate DNA into a microcentrifuge tube.

4. Add 1 μl of 500 mM acetate buffer (pH 5.2) and 1.5 μl of 10 mM TCEP to the tube to activate the thiol-modified DNA. Incubate the sample at room temperature for 1 h. This activation step is necessary because the thiol-modified DNA from IDT is shipped in the oxidized form with a disulfide bond.

5. Transfer 3 ml of the already prepared AuNPs into the NaOH-treated glass vials, and then add the TCEP-treated thiol DNA with gentle shaking by hand.

6. Cap the vial and store in a drawer at room temperature for at least 16 h. Although all the operations described in this

protocol can be carried out under light, it is advised to keep nanoparticles in the dark for long-term storage.

7. After the initial incubation, add 30 μl of 500 mM Tris–HCl (pH 8.0) buffer dropwise to the vial with gentle hand shaking. The final Tris acetate concentration is 5 mM.

8. Add 300 μl of 1 M NaCl dropwise to the vial with gentle hand shaking. Cap the vial tightly and store them in a drawer for at least another day before use. When sealed tightly, the functionalized AuNPs can be stored at room temperature for several months. However, slow degradation of the DNA on AuNPs may happen to change the properties of the AuNPs. *See* **Note 3** for discussion on the storage of DNA-functionalized AuNPs.

3.3 Preparation of DNAzyme-Linked AuNP Aggregates

1. Transfer 500 μl of functionalized particles into two 1.7-ml microcentrifuge tubes.

2. Centrifuge the two tubes at $16{,}110 \times g$ at room temperature (23–25 °C) on a benchtop centrifuge for 15 min.

3. Gently remove the two tubes from the centrifuge. The supernatant should be clear, and the AuNPs should be at the bottom of the tubes. If a red color can still be observed in the supernatant, centrifuge for another 5 min.

4. Gently pipette off as much supernatant as possible to remove free DNA. Again, disperse the AuNPs in 200 μl of buffer containing 100 mM NaCl, 25 mM Tris–HCl, pH 8.0.

5. Centrifuge again for 10 min at $16{,}110 \times g$ at room temperature.

6. Remove the supernatant. Again, disperse the nanoparticles in 200 μl of buffer containing 100 mM NaCl, 25 mM Tris–HCl, pH 8.0. We found that most of the free DNA can be removed by two centrifugations. If desired, repeat **steps 2–5** to remove more of the free DNA.

7. To each tube containing the substrate DNA functionalized AuNP, add 25 μl buffer containing 25 mM Tris–HCl pH 8.0, 100 mM NaCl, 8% sucrose and mix the contents of two tubes together to a total volume of ∼ 50 μl. This constitutes the AuNP functionalized with the thiol modified substrate called SubAuNP. (*See* **Note 4** about optimizing sucrose concentration and salt concentration.)

8. To prepare the hybridized construct of substrate functionalized with gold nanoparticles and enzyme (called Enz-SubAuNP), add 5 μl of 1 mM enzyme to the 50 μl of SubAuNP.

9. Anneal by heating at 70 °C for 2 min and slowly cooling to room temperature over ∼1 h, to give the hybridized construct of substrate functionalized with gold nanoparticles and enzyme

(called Enz-SubAuNP). *See* **Note 5** for discussion on the design of DNAzymes for dipstick application.

3.4 Dipstick Tests

1. Preparation of Lateral Flow Devices
   Lateral flow devices are constructed from membranes and pads obtained from Millipore Corporation. Hi-Flow Plus Cellulose Ester Membrane with a nominal capillary flow time of 240 s/ 4 cm and a nominal membrane thickness of 135 μm direct cast onto 2 mil polyester backing is used. The flow time of 240 s/ 4 cm gives the highest sensitivity. Using a membrane which has faster flow times (such as 90 s/4 cm) will increase the speed, but reduce the sensitivity. The absorption pad and wicking pad are cut from Millipore cellulose fiber sample pads, and the conjugation pad is cut from the Millipore glass fiber conjugate pad.

   - Cut out the absorption pad (15 mm × 300 mm) and the wicking pad (15 mm × 300 mm) from Millipore cellulose fiber sample pads, and the conjugation pad (13 mm × 300 mm) from Millipore glass fiber conjugate pad.

   - Cut the membrane (50 mm × 300 mm) from the membrane sheet obtained from Millipore. Touch the membranes from the edges only.

   - Attach the absorption pad, wicking pad, and conjugation pad to a plastic backing which is at least 300 mm wide. The overlap for each pad should be ∼2 mm. The assembled components are cut using a paper cutter into individual lateral flow devices with a width ∼8 mm. Discard the lateral flow devices that are cut from the edges as those membranes may have been damaged during handling.

3.5 Application of Reagents on Lateral Flow Device and Detection

1. Apply 5 μl of the construct, Enz-SubAuNP on each conjugation pad. Apply 1.5 μl of 10 mg/ml streptavidin on the control zone and 1.5 μl of 1 mM capture DNA on the test zone of the membrane using a 2 μl pipet (as shown in Fig. 3c). Allow the devices to dry for 6–8 h.

2. In order to test the dipsticks, prepare solutions of various concentrations of Pb^{2+} dissolved in a flow buffer containing 25 mM Tris (pH 8.0), 30 mM NaCl, 1% glycerol. The glycerol is used to slow down the flow.

3. Dip the wicking pad into the buffer solution containing lead (Pb^{2+}) for 10 min or until the liquid has migrated to the adsorption pad.

4. Remove the device from the buffer and lay it down on a flat plastic surface for flow to continue for about 5 min. Scan the dipsticks when they are dry.

5. If Pb^{2+} is not present, a single red line is observed at the control zone (Fig. 4a). In the presence of Pb^{2+}, a second red line is

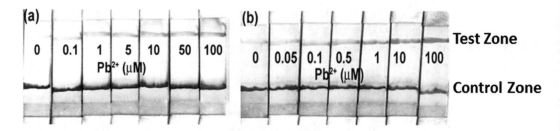

Fig. 4 Results of dipstick test for lead: (**a**) performance when the Enz-SubAuNP was pre-immobilized on the conjugate pad and the Pb^{2+} reaction occurred on the surface of the device; (**b**) performance when Enz-SubAuNP was allowed to react with Pb^{2+} in solution and then placed on the conjugate pad

observed at the test zone and its intensity increases with increasing Pb^{2+} concentration (Fig. 4a). This test can be qualitative or semiquantitative because a color chart can be used to estimate the Pb^{2+} concentration, like a pH paper. The detection limit for this test is determined to be ~5 μM. This detection limit is higher than ~0.1 μM reported for the DNAzyme-AuNP sensor in solution [2]. This is because the reaction is slowed down by diffusion limitation of Pb^{2+} to the DNAzyme active site on the surface, resulting in long reaction time (up to 1 h) [10]. Since the reaction on the lateral flow device takes place within 10 min, it is difficult to complete such a slow reaction on surface.

6. When the Pb^{2+}-induced cleavage reaction is done in solution and the lateral flow device as a medium to visualize the results, the sensitivity is improved. For the tests in which the Pb^{2+} reaction is performed in solution; add 1 μl of 6X concentration Pb^{2+} solution to 5 μl of Enz-SubAuNP construct and let the reaction proceed for 15 min. Apply this reaction mixture on the conjugation pad and then dip into the flow buffer for 10 min. Remove the device and lay it flat for 5 min and scan when dry (Fig. 4b).

7. A clear red line at the test zone can be seen in Fig. 4b at 0.5 μM Pb^{2+}, and thus the sensitivity is ~10 times better than the system shown in Fig. 4a. The sensitivity of this lateral flow device shown in Fig. 4b is also good in comparison with our previously reported colorimetric lead sensor that is based on the assembly of AuNPs and produces a blue to red color change with Pb^{2+} [2]. Although the detection limit for the solution-based method is 0.1 μM when the absorbance is measured using a UV–Vis instrument, it is difficult to visualize the red color below 5 μM Pb^{2+}, because of the large blue background from the aggregates. Thus, this lateral flow device provides a ~10-fold improvement in sensitivity for visual detection over the previously reported colorimetric sensor [2]. Further, performing the reaction in solution only requires one additional step of placing the reacted DNAzyme construct on the conjugation pad, which can be carried out using a dropper.

4   Notes

1. Care should be taken to make sure that no contamination is introduced during AuNP synthesis. Low quality AuNPs can result in poor DNA conjugation, which will induce nanoparticle precipitation during later processing steps such as centrifugation. Obtaining high-quality nanoparticles is the first important step toward the success of the experiment.

2. If the glass vials are not treated with concentrated NaOH, nanoparticles tend to stick to the surface of the vials, especially after addition of NaCl to the particles. If this problem occurs, the effective concentration of nanoparticles decreases.

3. The functionalized nanoparticles can still be used to form aggregates even after storing at room temperature for months, although their properties may change slightly with the passage of time as a result of events such as degradation of DNA. These changes could affect the properties of the prepared aggregates. To be sure that the results are consistent, use freshly functionalized AuNPs.

4. For lateral flow devices it is important to optimize the sucrose and NaCl concentration for drying the constructs on the conjugation pad. Sucrose is important for preserving the DNA-linked aggregates and aiding their rehydration. If sucrose is not added, then the aggregates do not disassemble. We tested sucrose concentration in the range of 0–30% and chose 8% sucrose as the optimum condition for drying our constructs. Optimizing the NaCl concentration (typically 50–500 mM) is important to obtain a balance between stabilizing the DNA base pairing interactions and achieving fast disassembly in the presence of analyte. The optimum concentration will depend upon the DNAzyme construct that is chosen for making dipsticks and therefore we recommend testing at least 3–5 separate NaCl concentrations when optimizing the tests.

5. It is important to optimize the sequence of the DNAzyme construct used for the dipstick tests, so that the cleavage efficiency is maintained but the release of the cleaved substrate is quick. For the system described here one arm is elongated to 11 base pairs, instead of the 9 base pairs in the original 17E DNAzyme (these bases have not been shown in the figure for clarity). The number of base pairs on the other arm (which in linked to the AuNP)is varied in order to facilitate release. For the construct shown in Fig. 5a, there are 9 base pairs, and the dipstick stick test with Pb^{2+} does not show any red line at the test zone. For the construct shown in Fig. 5b, there are 6 base pairs, and the dipstick stick test with Pb^{2+} shows a faint red line

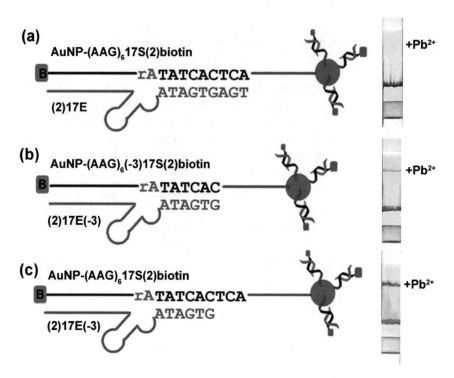

Fig. 5 Constructs used for the optimization of dipstick test. Construct (**c**) had the best performance in the presence of Pb^{2+} and was used for all other tests reported in this chapter. Construct (**a**) and (**b**) have different arm lengths and worse performance comparing to construct (**c**)

at the test zone. This is because the release of the cleaved substrate piece after Pb^{2+} reaction is easier as only 6 base pairs are holding it. For the construct shown in Fig. 5c, there are 6 base pairs; however, in this case only the enzyme arm is reduced in length to 6 bases whereas the substrate arm has 9 bases. The dipstick stick test with Pb^{2+} shows a dark red line at the test zone. By keeping the substrate intact, it is possible that the activity of the DNAzyme is higher as the structural perturbation is less. In addition, the capture of the cleaved piece at the test zone is better due to 3 additional base pairs with the capture DNA. Construct (c) is used in all other tests reported in this method, because its performance has been the best among the constructs tested.

Acknowledgment

This material is based on work from the following funding agencies—National Institute of Health (Grant no. ES016865), Department of Energy (DE- FG02-08ER64568), the National Science Foundation (Grant no. CTS-0120978 and DMI-0328162), and Department of

House and Urban Development (ILLHT0112-06). Yi Lu is a cofounder of both ANDalyze and GlucoSentient, Inc.

References

1. Kruger K, Grabowski PJ, Zaug AJ, Sands J, Gottschling DE, Cech TR (1982) Self-splicing RNA: autoexcision and autocyclization of the ribosomal RNA intervening sequence of Tetrahymena. Cell 31:147–157

2. Guerrier-Takada C, Gardiner K, Marsh T, Pace N, Altman S (1983) The RNA moiety of ribonuclease P is the catalytic subunit of the enzyme. Cell 35:849–857

3. Breaker RR, Joyce GF (1994) A DNA enzyme that cleaves RNA. Chem Biol 1:223–229

4. Tuerk C, Gold L (1990) Systematic evolution of ligands by exponential enrichment: RNA ligands to bacteriophage T4 DNA polymerase. Science 249:505–510

5. Ellington AD, Szostak JW (1990) In vitro selection of RNA molecules that bind specific ligands. Nature 346:818–822

6. Wilson DS, Szostak JW (1999) In vitro selection of functional nucleic acids. Annu Rev Biochem 68:611–647

7. Gold L, Polisky B, Uhlenbeck O, Yarus M (1995) Diversity of oligonucleotide functions. Annu Rev Biochem 64:763–797

8. Breaker RR (2002) Engineered allosteric ribozymes as biosensor components. Curr Opin Biotechnol 13:31–39

9. Hesselberth J, Robertson MP, Jhaveri S, Ellington AD (2000) In vitro selection of nucleic acids for diagnostic applications. Rev Mol Biotechnol 74:15–25

10. Famulok M (1999) Oligonucleotide aptamers that recognize small molecules. Curr Opin Struct Biol 9:324–329

11. Ihms HE, Lu Y (2012) In vitro selection of metal ion-selective DNAzymes. Methods Mol Biol 848:297–316

12. Brown AK, Li J, Pavot CMB, Lu Y (2003) A lead-dependent DNAzyme with a two-step mechanism. Biochemistry 42:7152–7161

13. Li J, Lu Y (2000) A highly sensitive and selective catalytic DNA biosensor for lead ions. J Am Chem Soc 122:10466–10467

14. Li J, Zheng W, Kwon AH, Lu Y (2000) In vitro selection and characterization of a highly efficient Zn(II)-dependent RNA-cleaving deoxyribozyme. Nucleic Acids Res 28:481–488

15. Brueshoff PJ, Li J, Augustine AJ, Lu Y (2002) Improving metal ion specificity during in vitro selection of catalytic DNA. Comb Chem High Throughput Screen 5:327–335

16. Carmi N, Shultz LA, Breaker RR (1996) In vitro selection of self-cleaving DNAs. Chem Biol 3:1039–1046

17. Liu J, Brown AK, Meng X, Cropek DM, Istok JD, Watson DB, Lu Y (2007) A catalytic beacon sensor for uranium with parts-per-trillion sensitivity and millionfold selectivity. Proc Natl Acad Sci U S A 104:2056

18. Vannela R, Adriaens P (2007) In vitro selection of Hg (II) and As (V)-dependent RNA-cleaving DNAzymes. Environ Eng Sci 24:73–84

19. Santoro SW, Joyce GF (1997) A general purpose RNA-cleaving DNA enzyme. Proc Natl Acad Sci U S A 94:4262–4266

20. Liu J, Lu Y (2006) Fast colorimetric sensing of adenosine and cocaine based on a general sensor design involving aptamers and nanoparticles. Angew Chem Int Ed Engl 45:90–94

21. Liu J, Lu Y (2006) Preparation of aptamer-linked gold nanoparticle purple aggregates for colorimetric sensing of analytes. Nat Protoc 1:246–252

22. Liu J, Lee JH, Lu Y (2007) Quantum dot encoding of aptamer-linked nanostructures for one pot simultaneous detection of multiple analytes. Anal Chem 79:4120–4125

23. Yigit MV, Mazumdar D, Kim H-K, Lee JH, Odintsov B, Lu Y (2007) Smart "turn-on" magnetic resonance contrast agents based on aptamer-functionalized superparamagnetic iron oxide nanoparticles. Chembiochem 8:1675–1678

24. Mirkin CA, Letsinger RL, Mucic RC, Storhoff JJ (1996) A DNA-based method for rationally assembling nanoparticles into macroscopic materials. Nature 382:607–609

25. Storhoff JJ, Elghanian R, Mucic RC, Mirkin CA, Letsinger RL (1998) One-pot colorimetric differentiation of polynucleotides with single base imperfections using gold nanoparticle probes. J Am Chem Soc 120:1959–1964

26. Liu J, Lu Y (2006) Smart nanomaterials responsive to multiple chemical stimuli with controllable cooperativity. Adv Mater 18:1667–1671

27. Liu J, Lu Y (2003) A colorimetric lead biosensor using DNAzyme-directed assembly of gold

nanoparticles. J Am Chem Soc 125:6642–6643

28. Liu J, Lu Y (2004) Accelerated color change of gold nanoparticles assembled by DNAzymes for simple and fast colorimetric Pb2+ detection. J Am Chem Soc 126:12298–12305

29. Liu J, Lu Y (2004) Adenosine-dependent assembly of aptazyme-functionalized gold nanoparticles and its application as a colorimetric biosensor. Anal Chem 76:1627–1632

30. Liu J, Lu Y (2006) Fluorescent DNAzyme biosensors for metal ions based on catalytic molecular beacons. Methods Mol Biol 335:275–288

31. Liu J, Lu Y (2007) Rational design of "Turn-On" allosteric DNAzyme catalytic beacons for aqueous mercury ions with ultrahigh sensitivity and selectivity. Angew Chem Int Ed Engl 46:7587–7590

32. Liu J, Lu Y (2007) A DNAzyme catalytic beacon sensor for paramagnetic Cu2+ ions in aqueous solution with high sensitivity and selectivity. J Am Chem Soc 129:9838–9839

33. Liu J, Lu Y (2005) Stimuli-responsive disassembly of nanoparticle aggregates for light-up colorimetric sensing. J Am Chem Soc 127:12677–12683

34. Liu J, Lu Y (2006) Design of asymmetric DNAzymes for dynamic control of nanoparticle aggregation states in response to chemical stimuli. Org Biomol Chem 4:3435–3441

35. Torabi SF, Lu Y (2014) Functional DNA nanomaterials for sensing and imaging in living cells. Curr Opin Biotechnol 28:88–95

36. Li L, Lu Y (2013) Functional DNA-integrated nanomaterials for biosensing. In: Fan C (ed) DNA nanotechnology. Springer, Berlin, Heidelberg, pp 277–305

37. Zhang J, Liu B, Liu H, Zhang X, Tan W (2013) Aptamer-conjugated gold nanoparticles for bioanalysis. Nanomedicine 8:983–993

38. Glynou K, Ioannou PC, Christopoulos TK, Syriopoulou V (2003) Oligonucleotide-functionalized gold nanoparticles as probes in a dry-reagent strip biosensor for DNA analysis by hybridization. Anal Chem 75:4155–4160

39. Liu J, Mazumdar D, Lu Y (2006) A simple and sensitive "dipstick" test in serum based on lateral flow separation of aptamer-linked nanostructures. Angew Chem Int Ed Engl 45:7955–7959

40. Lu Y, Liu J, Mazumdar D (2009) Nanoparticles/dip stick. In: Mayer G (ed) Nucleic acid and peptide aptamers, vol 535. Humana Press, New York, pp 223–239

41. Zhao W, Ali MM, Aguirre SD, Brook MA, Li Y (2008) Paper-based bioassays using gold nanoparticle colorimetric probes. Anal Chem 80:8431–8437

42. Mazumdar D, Liu J, Lu G, Zhou J, Lu Y (2010) Easy-to-use dipstick tests for detection of lead in paints using non-cross-linked gold nanoparticle-DNAzyme conjugates. Chem Commun 46:1416–1418

43. Handley DA (1989) Methods for synthesis of colloidal gold. In: Hayat MA (ed) Colloidal gold principles, methods, and applications, vol 1, 1st edn. Academic Press, San Diego, pp 13–32

Chapter 25

Liposome-Enhanced Lateral-Flow Assays for Clinical Analyses

Katie A. Edwards, Ricki Korff, and Antje J. Baeumner

Abstract

Clinical and environmental analyses frequently necessitate rapid, simple, and inexpensive point-of-care or field tests. These semiquantitative tests may be later followed up by confirmatory laboratory-based assays, but provide an initial scenario assessment important for resource mobilization and threat confinement. Lateral-flow assays (LFAs) and dip-stick assays, which are typically antibody-based and yield a visually detectable signal, provide an assay format suiting these applications extremely well. Signal generation is commonly obtained through the use of colloidal gold or latex beads, which yield a colored band either directly proportional or inversely proportional to the concentration of the analyte of interest. Here, dye-encapsulating liposomes as a highly visible alternative are discussed. The semiquantitative LFA biosensor described in this chapter relies on a sandwich immunoassay for the detection of myoglobin in whole blood. After an acute myocardial infarction (AMI) event, several cardiac markers are released into the blood, the most common of which are troponin, creatine kinase MB, C-reactive protein, and myoglobin. Due to its early release, myoglobin has value as an indicator of a recent heart attack amongst conditions which present with similar symptoms and its lack of elevation can effectively rule out a heart attack (Brogan et al., Ann Emerg Med 24:665–671, 1994). The assay described within relies on sandwich complex formation between a membrane immobilized capture monoclonal antibody against myoglobin, a detector biotinylated monoclonal antibody against a different epitope on myoglobin, and streptavidin-conjugated visible dye (sulforhodamine B)-encapsulating liposomes to allow for signal generation.

Key words Lateral-flow assay, Liposome, Whole blood, Myoglobin

1 Introduction

Clinical and environmental analyses frequently necessitate rapid, simple, durable, and inexpensive point-of-care or field tests. Lateral-flow assays (LFAs), such as those used in pregnancy tests, often fulfill these requirements. These assays rely on the appearance of a colored band, the intensity of which corresponds to the concentration of the target analyte in the sample. The goal for such assays is typically a qualitative "yes" or "no" answer, though some assays can provide semiquantitative results. LFAs

Avraham Rasooly and Ben Prickril (eds.), *Biosensors and Biodetection: Methods and Protocols Volume 1: Optical-Based Detectors*, Methods in Molecular Biology, vol. 1571, DOI 10.1007/978-1-4939-6848-0_25, © Springer Science+Business Media LLC 2017

are commercially available for the detection of a wide variety of analytes. These targets include pathogenic organisms such as *Legionella pneumophilia* [1, 2], *Leptospira* [3], *Cryptosporidium parvum*, and *Giardia lamblia* [4]; viruses such as HIV [5], influenza [6], and respiratory syncytial virus (RSV) [7, 8]; food allergens [9] and drugs of abuse [10]; health states such as pregnancy, ovulation, or menopause [11, 12]; diseases such as prostate cancer [13, 14]; or biological toxins such as botulinum toxin [15, 16], and verotoxins [17, 18]. Some lateral flow assays have proven to be equivalent or better than enzyme-linked immunosorbent assays (ELISAs), direct fluorescent antibody testing, and viral cultures [5, 6, 8, 19]. Lateral-flow assays used in a clinical setting can obviate the need for patients to return to the clinician's office for results [20] and can provide guidance for the administration of chemoprophylaxis after potential occupational exposure to viruses [21]. In environmental or occupational settings, LFAs provide an initial assessment for mobilization of resources and containment of potential threats. LFAs for research and more fundamental applications such as probe selection, PCR-product identification, genomic sequence search, and identification of single-nucleotide polymorphisms have been suggested [22–24].

Here we discuss in detail the use and preparation of LFAs for analytes present in whole blood, using the 16.7 kDa cardiac marker myoglobin as an example analyte. After an acute myocardial infarction (AMI) event, several cardiac markers are released into the blood, the most common of which are cardiac troponin I (cTnI), cardiac troponin T (cTnT), creatine kinase MB subform (CK-MB), C-reactive protein (CRP), and myoglobin. cTnI, cTnT, and CK-MB are initially elevated 4–6 h after an AMI event; C-reactive protein is initially elevated up to 6 h after the event, and myoglobin is initially elevated after 1–3 h [25, 26]. The relative increase in these markers over time following an incident of chest pain is shown in Fig. 1.

Due to the rapid and reliable release of myoglobin from damaged myocardium, tests that use serum myoglobin levels to determine the risk of a recent myocardial infarction have high sensitivities and high negative predictive values within 90 min of symptom onset [27]. Especially when combined with detection of other cardiac markers like troponin I, the absence of elevated serum myoglobin is useful in quickly ruling out acute myocardial infarction in patients with symptoms such as chest discomfort, dyspnea, and syncope [28, 29]. Since approximately 80% of patients presenting to emergency rooms with AMI-associated symptoms do not develop heart attacks [30], a test that can rapidly rule out AMI may allow hospitals to prevent unnecessary admission of patients to coronary care units. While a serum myoglobin test cannot confirm AMI since myoglobin release is not specific to cardiac muscle damage, detecting elevated levels of myoglobin in the blood

Hours After Chest Pain

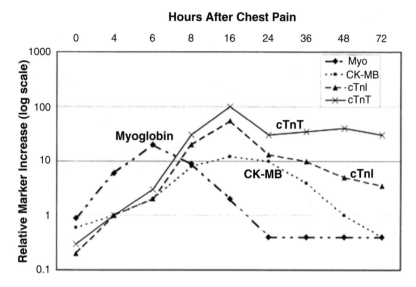

Fig. 1 Temporal release of myoglobin, CK-MB, and cTnT and cTnl. Reprinted with permission from Christenson RH, Azzazy HME. Biomarkers of Myocardial Necrosis: Past, Present and Future. In Morrow DA, ed. Cardiovascular Biomarkers: Pathophysiology and Clinical Management. Totowa, NJ: Humana Press, 2006

regardless of its source is important in preventing kidney damage. In rhabdomyolysis (severe muscle breakdown), serum myoglobin levels may rise to levels that lead to acute renal failure [31]. The approach described within allows for the analysis of myoglobin levels in whole blood using plasma separators incorporated into a LFA device and the use of dye-encapsulating liposomes as a highly visible signaling species in contrast to commonly used latex particles and colloidal gold.

1.1 Sandwich Immunoassay Format

While simple for the end-user to operate, from an engineering standpoint, the sandwich lateral flow immunoassay format is a relatively complicated multicomponent design encompassing a sample pad, conjugate pad, analytical membrane, absorbent pad, signaling species, blocking constituents, and multiple antibodies. When these components are assembled appropriately, the end-user simply needs to add a liquid sample and wait typically 5–15 min before a visible signal is assessed. The flow of the fluid in such an assay is detailed in Fig. 2.

A liquid sample ideally with minimal preparation requirements is applied to the sample pad via an opening in the housing. Depending on the sample matrix, the sample pad can serve as a particulate filter, site to retain interfering substances, and provide pH buffering capacity with the overall aim to prepare the fluid for subsequent analysis and dispense it evenly. Once the fluid passes through the sample pad, it then passes through a conjugate pad where it rehydrates a previously dehydrated signaling species, which is typically

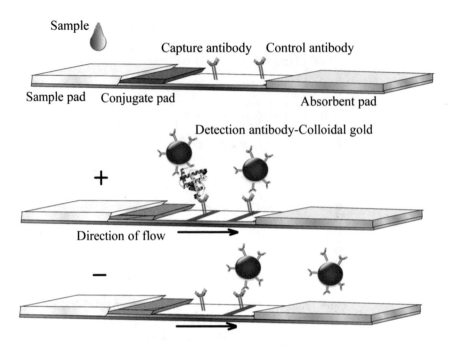

Fig. 2 Lateral flow assay fluid path. (*Top*) A liquid sample is applied to a sample pad through a port in the plastic housing (not shown). The liquid sample then passes through the conjugate pad where the signaling species (most commonly colloidal gold) with an attached detection antibody becomes rehydrated. The solution then passes onto the analytical membrane were capture and control antibodies are immobilized, then lastly is wicked by the absorbent pad at the opposite end of the assembly. (*Middle*) If the analyte is present, it can form sandwich complexes with the detection antibody on the colloidal gold and the capture antibody immobilized on the analytical membrane. A visible signal where the capture antibody is immobilized is observed as well as a visible signal where the control antibody is immobilized to indicate successful fluid flow and conjugate release. (*Bottom*) If the analyte is not present, a signal at only the control line is observed

colloidal gold or visibly dyed latex beads. The surface of the signaling species is functionalized with "detection" antibodies against the analyte of interest and can form antibody-antigen complexes with analyte in the sample solution if it is present. The fluid then passes onto the analytical membrane where a second antibody is immobilized at the test line. This "capture" antibody, most commonly a monoclonal antibody (mAb), can recognize a different epitope on the target analyte than that recognized by the detection antibody and serves to retain the passing analyte-signaling particle complexes. The sandwich complex formed between the immobilized capture antibody, analyte, and detection antibody-tagged particles allows a visible signal to be observed in the presence of the target analyte due to the attached signaling species. When the analyte is not present, no sandwich complexes are formed and thus no signal is generated. The intensity of the signal is proportional to the analyte concentration, which can be used for semiquantitative determinations. At a location downstream of the test line is a control line. This control line typically consists of a secondary antibody, which can bind to the primary antibody immobilized on the surface of the signaling species (e.g., a goat anti-mouse secondary antibody could be used as a control line if the primary antibody on the colloidal gold particles was a mouse anti-myoglobin antibody). This control line captures passing signaling species regardless of the presence or absence of the target analyte and allows the end user to confirm that sufficient fluid was applied to yield a valid assay and that the conjugate successfully released. At the distal end of the membrane lies an absorbent pad, which serves to draw fluid through the assay and retain excess fluid for the duration of the assay.

The inherent format of the sandwich immunoassay places certain constraints on the types of analytes that can be detected. In order to accommodate the binding of two separate antibodies which are ~150 kDa each, the analyte must be relatively large. As such, it is predominantly used for the quantification of large proteins or whole cells. While this format precludes the detection of small molecules, it has been used for quantification of oligopeptides including, for example, insulin C-peptide (31 amino acids (AA), 3 kDa) [32] provided an appropriate pair of antibodies is available. To satisfy the criteria for appropriate antibodies, either two antibodies that can bind to distinct, nonoverlapping epitopes on the target molecule are required, or if a single antibody type is used, the target molecule must have multiple copies of a single epitope. In practice, the method developer typically needs to determine which antibodies can successfully form a sandwich complex with a given analyte through in-house experimentation. With the price of commercially available primary antibodies typically ranging from ~$350 to $700/mg, this can be a costly trial and error process. Thus, it is a great benefit when "paired" antibodies that are known to work together in sandwich assays are available from antibody vendors. Other considerations to the choice of antibody include its level of purification and concentration. Antibodies used as capture antibodies in LFAs should not contain any stabilizing proteins or other species that may compete for adsorption to the nitrocellulose. Similarly, if the detection antibody is covalently attached to the signaling species, it should not contain stabilizing proteins or buffer constituents that will interfere with the conjugation chemistry. Unless the capture antibody has a very strong affinity, a minimum concentration of 1 mg/mL is recommended. The sensitivity of lateral flow assays stems in part from the basic design where fluid must pass the test line where it has the opportunity to contact the immobilized capture antibody with negligible mass transport limitations. This is in contrast to enzyme-linked immunosorbent assays (ELISAs) where diffusion from the bulk sample volume must occur to the plate surface where capture antibodies are immobilized [33]. While the latter can often achieve low limits of detection using much lower concentrations of antibodies (typically 5–10 µg/mL), extended incubation times (typically 1–2 h) are required. Given that there is a limited amount of time in which an analyte is in contact with antibody immobilized on the test line which is typically 0.5–1.0 mm wide, binding kinetics are key in LFAs. Aside from high affinity, the association rate is a critical parameter [33, 34].

1.1.1 Membrane Types

Most commonly, nitrocellulose membranes are used as analytical membranes for LFAs due to their high binding capacity, low cost, and wide availability [35]. The mechanism behind protein binding to nitrocellulose is not fully understood, but is believed to be noncovalent through hydrophobic, hydrogen bonding, and electrostatic interactions [36, 37]. The physical adsorption of nucleic acid probes through drying to nitrocellulose is believed to result in attachment through hydrophobic interactions, an effect which may be enhanced through modification with a poly-T tail [38]. We have also had success using polyethersulfone membranes for nucleic acid sandwich hybridization assays [22, 23, 39].

More recently, membranes based on electrospun nanofibers have been applied for LFAs [40]. Electrospun nanofibers are prepared by using a simple set-up consisting of a power voltage supply, a syringe with metal needle, a syringe pump and a ground plate. A polymer solution (dope) is filled into the syringe and pumped out at a steady flow rate. A high voltage (7–15 kV) is applied between the tip of the needle and the ground plate. A taylor cone is formed and a nanofiber in the size range of a few tens of nanometers to a few micrometers are collected on the grounded plate, depending on the spinning dope and electrospinning conditions chosen. Nitrocellulose and nylon via electrospinning have been shown for protein slot blot and Western blot applications [41]. By doping the composite with poly-L-lysine (PLL) or polylactic acid (PLA), specific functional groups including amino and carboxylic acid groups, respectively, may be incorporated [42]. Additionally, through control over polymer dope composition, voltage, distance between needle tip and ground plate control over the mat thickness and porosity can be achieved and offers significant surface area for immobilization.

1.1.2 Membrane Properties

The ideal properties of the membrane include particle and pore size consistency, hydrophilicity, high protein binding, and durability. As nitrocellulose membranes are a fibrous polymeric matrix rather than uniform particles, the membranes are characterized by their capillary rise rate (or capillary flow time), rather than by a specific particle or pore size. This rate is the amount of time required for fluid to flow 4 cm, and based on a survey of current commercial offerings from major manufacturers, is generally between 60 and 280 s. Since fluid flow speed in lateral flow membranes decreases with the distance traveled, the capillary rise rate is a more consistent measure of the speed of flow [43]. This parameter is important to consider during assay development since a fast wicking rate allows for more rapid results and lower background, but may adversely affect assay sensitivity since less time for interaction between the target and the antibodies immobilized at the test line is available [44].

Commercially available nitrocellulose membranes vary in physical properties based on the effective "pore size," the amount of trapped air, and the thickness of their nitrocellulose layer. These properties affect the surface area available for immobilization of capture reagents. The binding capacity of nitrocellulose itself is inherently high, with estimates ranging from 50 to 200 μg IgG/cm² [45]. The thickness of commercially available membranes typically ranges from 100 to 150 μm. Thicker membranes can allow more sample fluid to be accommodated by the membrane during the assay and more capture reagent to be applied. However, the latter does not necessarily yield greater sensitivity in lateral flow assays with optical detection since due to the opacity of nitrocellulose, only the top ~10 μm of membrane is visible to the user [45]. One advantage to thicker membranes is improved tensile strength, however, for ease of handling, durability, and avoidance of adhesive interactions, a supported membrane should be chosen. These membranes are manufactured by directly casting the nitrocellulose membrane material onto a plastic (polyester or cellulose acetate) backing. The backing thickness typically ranges from 50 to 225 μm [43]. Alternatively, non-supported membranes can be laminated manually prior to use. LFAs may be further made more durable and user-friendly by encasing them in an injected molded plastic housing such as that shown in Fig. 3.

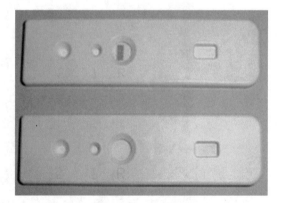

Fig. 3 Lateral flow assay membranes assembled in injection molded housings. In the presence of the target analyte, a sandwich complex forms with dye-encapsulating liposomes resulting in a magenta-colored band at the capture zone that is proportional to target concentration (*top*). In the absence of target analyte, no band is visible as no sandwich complex has formed (bottom). Though not necessary, these LFA membranes were housed in more user friendly packaging using a plastic cassette made using injection molding and designed for addition of liposome/target mixture in hole #1 and addition of running buffer in hole #2. Readings can be taken in the "R" hole

In addition, a sample pad, conjugate release pad, and wicking material are often utilized for LFAs to allow for sample pretreatment such as filtration or pH adjustment; storage of dehydrated signaling species; and accommodation of larger fluid volumes, respectively. These are critical components for any assay developed in complicated sample matrices and also those intended for consumer usage where the number of steps needs to be minimized. For the assays described in this chapter, the conjugate pad is omitted since the liposomes used to provide the signal are maintained in the solution phase. The sample pad is selected based on the type of sample matrix that is employed. For the assays described within, a sample pad that is capable of retaining white and red blood cells from whole blood while allowing the plasma to pass is required. The ideal sample pad minimizes hemolysis from red blood cells, which can contribute to an undesired colored background signal on the nitrocellulose membranes (Fig. 4).

The capture antibody can be applied to the membrane either manually using a pipettor, with a thin-layer chromatography (TLC) applicator, or similar approaches. Different binding characteristics, buffers (pH, salt), application rate, and biorecognition element concentration lead to differences in thickness in the capture zone line. Once the biorecognition element is applied, the membrane material is typically immersed in a blocking agent, which is used to

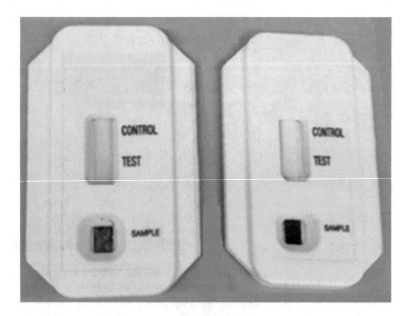

Fig. 4 Whole blood applied to lateral flow assay membranes assembled in injection molded housings to test sample pad types. No test or control lines were immobilized. Significant hemolysis was observed in the left assembly which yielded a *pink colored background* which could potentially obscure weak signals, whereas a clear background was available in the right assembly

reduce nonspecific binding of the assay components. Common agents include proteins such as bovine serum albumin (BSA), casein, and gelatin; non-ionic surfactants such as Tween 20, Triton X-100, or Brij; or synthetic polymers such as polyvinylpyrrolidone (PVP), polyvinyl alcohol (PVA), and polyethylene glycol (PEG). These components compete for protein binding sites, interfere with hydrophobic interactions, and protein binding, respectively [46, 47]. In addition to serving as a blocking agent, Tween 20 has also been reported to have a renaturing effect on immobilized antigens, which can improve assay sensitivity through enhanced antigen–antibody binding [48, 49]

1.2 Detection Mechanisms

LFAs typically utilize antibody-labeled colloidal gold [5, 19, 50] as a visualization method. Colloidal gold ranges in size 2–250 nm, though 30–80 nm particles are preferred for LFAs [24]. A diameter of 40 nm has been reported to be optimal, allowing for clear visualization, dense packing of these small particles at the capture zone, and reduced steric hindrance for protein binding [51]. These particles have a characteristic red color, allowing for visual detection or semiquantitative measurements with portable reflectometers or scanners [52–54]. Proteins are associated with gold particles ionically through the particle's negative charge with positively charged amino acids such as lysine; through hydrophobic interactions of amino acids such as tyrosine and tryptophan; and through sharing of electrons between gold and sulfur atoms of cysteine [51]. When labeled with anti-biotin or streptavidin, such species allow for recognition of biotinylated biorecognition elements, such as DNA oligonucleotides or antibodies. The sensitivity of LFAs using such particles can be increased with silver enhancement [55]. Alternatively, antibody-tagged latex particles [56, 57], up-converting phosphors [58], superparamagnetic particles [59], or visible dye-encapsulating liposomes [60] have been used as a signal enhancement means. The latter species are the focus of this chapter.

Liposomes are vesicles formed through the association of phospholipid molecules in an aqueous environment, yielding a structure with the hydrocarbon tails forming a lipid bilayer and hydrophilic headgroups directed at both the aqueous core and external aqueous medium. One of the common methods for liposome formation is known as the reverse-phase evaporation method [61, 62]. Here, phospholipids are dissolved in organic solvent; the mixture is introduced to an aqueous medium containing high concentrations of visible dye; the solvent is removed under vacuum; the resulting liposomes are passed through defined pore-size membranes to improve homogeneity and reduce the number of lipid bilayers (lamellarity); then the unencapsulated material is removed through size-exclusion chromatography and dialysis. We commonly use sulforhodamine B (SRB) dye, which is relatively inexpensive, highly soluble in water, has a high molar extinction coefficient, and has good photostability. However, the most appropriate dye color can be chosen to ensure good visibility against any background signal in the LFA that is not removed by the sample pad. For example, for whole blood, blue liposomes made using Patent Blue dye can improve optical detection against a potentially red or brownish background.

The advantages of dye-encapsulating liposomes as a label for immunoassays include long-term stability of the encapsulated signaling molecules; the substantial signal enhancement afforded by the encapsulation of hundreds of thousands of dye molecules within the large interior liposomal volume; the instantaneous signal provided through either visual detection of intact liposomes or the release of hydrophilic dye molecules from their aqueous cores upon surfactant-induced liposome lysis; and the ease of labeling through the covalent modification of lipid headgroups or the direct incorporation of hydrophobic biorecognition elements into their lipid bilayers (Fig. 5) [60, 63–65].

The stability of the liposomes is a function of their lipid composition, buffer composition, and storage conditions. In the formulation that we most commonly employ (described within), we have found that streptavidin tagged sulforhodamine B-encapsulating liposomes have retained their functionality in LFAs for at least 400 days when stored at either 4 or 21 °C, while loss of encapsulated dye and signal occurs at elevated temperatures [66].

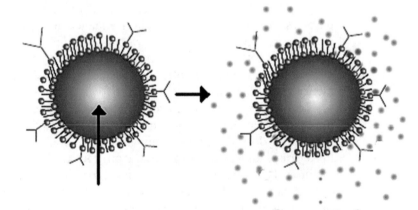

Fig. 5 Liposome structure. (**a**) Biorecognition elements can be covalently conjugated to or inserted into lipid bilayers through hydrophobic interactions (not to scale). (**b**) The large internal volume of unilamellar vesicles can encapsulate hundreds of thousands of hydrophilic signaling molecules and provide for their stability. (**c**) Surfactant introduction can provide for instantaneous signal enhancement through release of encapsulants. Fluorophores encapsulated within liposomes at high concentrations undergo self-quenching with is overcome upon release into the surrounding medium

1.3 Universal Assays   Both immunoassays and nucleic acid based lateral flow assays can also be made in a universal format, obviating the need for the generation of specifically labeled membranes and detection elements. This can be done through membrane immobilization of streptavidin [23] or anti-fluorescein antibodies [22] and subsequent addition of biotinylated or fluorescein-labeled capture antibodies, respectively, with the remaining assay components. Protein A/G [67], streptavidin/avidin [22, 68], or generic oligonucleotide [23] labels can be conjugated to the liposomes to form generic species capable of facile recognition of the Fc′ region of antibodies, biotinylated biorecognition elements, or complementary generic oligonucleotides, respectively. Anti-fluorescein, anti-biotin, or anti-digoxigenin are also options for common universal liposome tags [69]. These assays are of particular interest in research laboratories where a variety of antibodies or probes need to be screened or where different analytes need to be tested. From a commercial and manufacturing standpoint, the universal format is of interest since only one type of membrane and liposome need to be prepared which simplifies packaging of tests for any analyte.

2 Materials

Materials listed here are those that have been used successfully in our laboratory, though substitutions may be made. Unless otherwise specified, reagents are molecular biology grade and are purchased from VWR (Bridgeport, NJ). Table 1 lists the biomolecules used for the detection of myoglobin as an example.

2.1 Liposome Preparation

1. Bath sonicator (Aquasonic Model 150D, VWR, Bridgeport, NJ).

2. Rotary evaporator (Model R-114, Buchi, New Castle, DE).

3. 50-mL round bottom flask (Catalog # 80068-756, VWR).

4. Mini-extruder (Catalog # 610000, Avanti Polar Lipids, Alabaster, AL), including two 1 mL syringes, Teflon supports and o-rings, and extruder holder/heating block. When purchasing initially, the membranes and supports listed in #5 of this section are included with this catalog number as of the date that this chapter was prepared.

5. Extrusion membranes (0.4 and 1.0 μm polycarbonate membranes*, 19 mm) and filter supports (Catalog # 610007, 610010, and 610014, respectively, Avanti Polar Lipids).

6. 1,2-dipalmitoyl-*sn*-glycero-3-phosphocholine (DPPC), cholesterol, 1,2-dipalmitoyl-*sn*-glycero-3-[phospho-rac-(1-glycerol)], sodium salt (DPPG), 1,2-dipalmitoyl-*sn*-glycero-3-phosphoethanolamine-N-glutaryl, sodium salt (N-glutaryl DPPE) (Catalog #: 850355, 700000, 840455, and 870245, respectively, Avanti Polar Lipids).

Table 1
Reagents needed for the development of a LFA for human myoglobin

Function	Information	Vendor	Catalog number
Capture antibody	Anti-myoglobin (mouse, mAb, IgG$_1$)	Meridian Life Science, Inc.	H86703M
Detection antibody	Biotinylated anti-myoglobin (mouse, mAb, IgG$_1$)	Meridian Life Science, Inc.	H86142B
Control antibody	Anti-streptavidin (goat, pAb)	Vector Labs	SP-4000
Liposome tag	Streptavidin	Thermo Fisher Scientific	S888
Test matrix	Myoglobin-free serum	Meridian Life Science, Inc.	N86580H
Test analyte	Recombinant full-length myoglobin	Meridian Life Science, Inc.	A01428H
Function	Information	Vendor	Catalog Number
Capture antibody	Anti-myoglobin (mouse, mAb, IgG$_1$)	Meridian Life Science, Inc.	H86703M
Detection antibody	Biotinylated anti-myoglobin (mouse, mAb, IgG$_1$)	Meridian Life Science, Inc.	H86142B
Control antibody	Anti-streptavidin (goat, pAb)	Vector Labs	SP-4000
Liposome tag	Streptavidin	Thermo Fisher Scientific	S888
Test matrix	Myoglobin-free serum	Meridian Life Science, Inc.	N86580H
Test analyte	Recombinant full-length myoglobin	Meridian Life Science, Inc.	A01428H

7. Chloroform, methanol, and isopropyl ether (HPLC grade).

8. 10×HEPES-saline buffer is composed of 0.1 M HEPES, 2.0 M sodium chloride, and 0.1% (w/v) sodium azide, adjusted to pH 7.5 with NaOH.

9. 1×HEPES–saline–sucrose buffer (1×HSS) is prepared by dissolving 205.4 g sucrose (0.2 M) in 300 mL 10× HEPES-saline and 900 mL water, then bringing the final volume to 3 L with deionized water. However, we usually prepare a 2 M stock solution of sucrose instead of weighing for this solution.

10. 0.419 g sulforhodamine B (SRB, Catalog # S1307, Molecular Probes, Inc., Eugene, OR) is added to 0.5 mL 0.2 M HEPES, pH 7.5 and diluted to a final volume of 5 mL with deionized water to yield a 150 mM solution. For blue liposomes, Patent Blue dye should be substituted.

11. Sephadex G-50 (Catalog # G-50-150, Sigma, St. Louis, MO) and a glass chromatography column (such as VWR catalog #

KT420400-1530). The column is packed by first swelling 1 g of Sephadex in 120 mL deionized water in 3 × 50 mL centrifuge tubes overnight, and decanting off water either after centrifugation or settling by gravity then pouring. The volume is then replaced by 1×HSS and decanting procedure repeated. Overall, three 1×HSS buffer exchanges are typically sufficient. The Sephadex mixture is then poured into the chromatography column to a height of approximately 20 cm and allowed to settle while maintaining a flow of 1×HSS through the column for at least 30 min. The top of the column should be level with no gaps or bubbles throughout the remaining height.

12. Dialysis membranes, Specta/Por 2 (Catalog # 132678, Spectrum Laboratories, Inc., Rancho Dominguez, CA).

13. 15- and 50-mL centrifuge tubes (Catalog #: 21008-216 and 21008-242, respectively, VWR).

2.2 Liposome Conjugation

1. Streptavidin (Catalog # S888, Molecular Probes, Inc., Eugene, OR) is diluted to 200 μM with 50 mM potassium phosphate, pH 7.0 and aliquotted into 50 μL portions prior to storage at −20 °C.

2. 1-Ethyl-3-(3-dimethylaminopropyl)carbodiimide hydrochloride (EDC) (Catalog #22980, Thermo Fisher Scientific).

3. 0.1 M 2-(N-morpholino)ethanesulfonic acid (MES) buffer, pH 4.65.

4. Sepharose CL-4B (Sigma-Aldrich #CL4B200) packed as the Sephadex G50 column is above in Subheading 2.1. Sepharose CL-4B is provided as a slurry by the manufacturer, which reduces the preparation time to that required for the HSS buffer exchanges.

2.3 Membrane Preparation

1. TLC applicator (Linomat IV, CAMAG Scientific, Wilmington, NC).

2. Vacuum sealer (Foodsaver, San Francisco, CA).

3. Vacuum oven, capable of 23 and 50 °C.

4. Paper cutter.

5. Fine-tip tweezers (Catalog #25729-081, VWR).

6. Kimwipes, 15″ × 17″ (Catalog # 21905-049, VWR).

7. Nitrocellulose membrane cards (HF090 membrane cards, Catalog # HF09004XSS, EMD Millipore, Cheshire, CT) cut into 20 cm sections.

8. Anti-myoglobin capture antibodies diluted to 1.0 mg/mL with a solution of 0.4 M $NaHCO_3/Na_2CO_3$, pH 9.0 containing 5% (v/v) methanol. Anti-streptavidin control antibodies diluted to 1.5 mg/mL with the same buffer.

9. The blocking reagent is prepared by diluting 100 mL 1% (w/v) casein sodium salt and 3.85 mL 2 M sucrose to a final volume of 1000 mL with HPLC grade water.

10. Sample pads (Unbound glass, Catalog #VFE, Whatman International, Ltd., Florham Park, NJ) cut into 1.9×20 cm sections.

11. Absorbent pads (Cellulose fiber sample pads, Catalog #CFSP223000, EMD Millipore, Billerica, MA) cut into 1.8×20 cm sections.

2.4 Assay Optimization Using Recombinant Myoglobin and Serum

1. Flat bottom, non-binding microtiter plates (Corning, #3610, Corning, NY).

2. Prior to running the assay, recombinant full length myoglobin is diluted in TBS with 0.05% (w/v) Tween 20 and 0.1% (w/v) BSA to concentrations of 100 μg/mL, 10 μg/mL, 1 μg/mL, 100 ng/mL, 10 ng/mL, 1 ng/mL, 100 pg/mL, and 0 pg/mL. 30 μL portions are appropriate.

3. Biotinylated anti-myoglobin antibody is diluted to 10 μg/mL with 0.05% (w/v) Tween 20 and 0.1% (w/v) BSA.

4. Streptavidin-tagged liposomes of known phospholipid concentration.

5. Membranes with capture and control antibodies immobilized cut into 4.5 mm wide strips.

6. Reflectometer ($\lambda = 560$ nm, ESECO Speedmaster, Cushing, OK) or scanner with image analysis software (Optional if more than qualitative results by eye are desired).

3 Methods

3.1 Liposome Preparation

A flowchart of this process is shown in Scheme 1. Note that the times listed in this flowchart, and others in this chapter, reflect overall times for each step, including setup and incubations. For specific times for each step, please refer to the text.

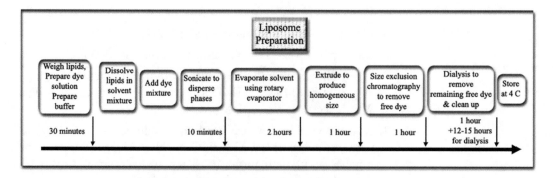

Scheme 1 Flow chart for the preparation of dye-encapsulating liposomes

1. Set rotary evaporator bath and sonicator bath to 45 °C. Fill the condenser on the rotary evaporator with ice and seal condenser top rim to rotary evaporator glass housing with Parafilm.

2. Prepare dye solution as described in materials section using a 15-mL centrifuge tube, cover tube with aluminum foil, and store in the 45 °C sonicator bath.

3. DPPC, DPPG, N-glutaryl DPPE, and cholesterol (40.9:20.1:7.3:51.7 µmol, respectively) are added to a 50-mL round bottom flask (see **Note 1**).

4. A solvent mixture containing 3 mL chloroform and 0.5 mL methanol is added and the mixture sonicated for 1 min in a bath sonicator. We use sonication level 6 with the Aquasonic Model 150D bath sonicator.

5. 2 mL of the 45 °C sulforhodamine B dye solution is added to the lipid mixture during the first 30 s while sonicating for a total of 4 min. Swirl the flask manually during sonication (see **Note 2**).

6. The mixture is then placed onto a rotary evaporator at the highest rotation speed with the bath at 45 °C. The vacuum should be adjusted such that bubbling or foaming of the contents does not occur. Using the Buchi R-114 rotary evaporator, a vacuum of 500 mBar for 20 min, followed by 400 mbar for 20 min is recommended. During evaporation, return unused portion of the SRB solution to the 45 ° C sonicator bath.

7. The mixture is then transiently vortexed preceding and following an additional introduction of 2 mL 45 °C 150 mM SRB. Using a VWR mini-vortexer, we have found that holding the flask at a 45° angle and vortexing at level 4–5 is appropriate, however, be sure to use a stopper on the flask.

8. The mixture is returned to the rotary evaporator for an additional 30 min under slightly higher vacuum than in **step 6** (350 mBar, in our case.) Generally, the evaporation is considered complete 15 min after no further solvent is observed from the condenser. Removal of nearly all organic solvent is required for successful formation of liposomes.

9. While the liposome mixture is on the rotary evaporator, set the extruder support block of the mini-extruder on a hot plate and adjust temperature such that the temperature of the block does not exceed 65 °C.

10. Set-up the mini-extruder as outlined in the manufacturer's instructions. Label 2–50 mL centrifuge tubes for each extrusion size and clamp them in a 45 °C water bath during the extrusion process and before application to the size-exclusion column.

11. The liposomes are then extruded at 60–65 °C 19 times through 1.0 μm pore membranes, followed by 19 times through 0.4 μm pore membranes. The liposomes must be maintained in the 45 °C water bath during and after extrusion.

12. The level of the 1×HSS buffer volume on the Sephadex G-50 column prepared in **step 12** of Subheading 2.1 is reduced to just below the level of the Sephadex. Do not allow the Sephadex to dry out.

13. The liposomes are pipetted onto the top of the column without disturbing the top of the Sephadex. This addition should be done as carefully, but rapidly, as possible using a glass Pasteur pipet in a circular path along the inside of the glass column just above the Sephadex. *Important:* the tubes containing the liposomes need to be kept in 45 °C at all times and the transfer to the column needs to be efficient. If the liposomes cool to room temperature, they will form clumps with external dye on the column and will not separate.

14. Once all of the liposome volume has entered the column, two ~1 mL aliquots of 1×HSS should be pipetted onto the top of the column allowing each one to enter the column before the next addition. This is done to create a separation between the liposomes/free dye and the remaining 1×HSS used for liposome elution.

15. After carefully pipetting another 1–2 mL of 1×HSS on top of the column, allow 1×HSS to flow freely (typically ~4 mL/min) and collect the eluting liposomes in test tubes. The liposomes will elute from the column first, forming a dark magenta band, followed by the free dye yielding a darker and larger elution volume. If the column is run properly, some separation of liposomes and free dye can usually be seen.

16. The liposome containing fractions are then combined based either visually or on their measured optical density at 532 nm. Optical density measurements are made at 532 nm by diluting 5 μL liposomes with 995 μL 1×HSS in a 1.5 mL spectrometer cell, or, more conveniently, 1 μL liposomes in 199 μL 1×HSS in a clear microtiter plate (Corning #3795). Be selective about which liposomes are pooled together: only the most highly concentrated fractions should be together, the intermediates together, and if needed, the low concentration fractions. This is very important if the liposomes are to be subsequently coupled to proteins where a high liposome concentration is necessary (*see* **Note 3**).

17. The combined fractions are placed into dialysis bags and dialyzed overnight against the sucrose-HEPES-saline buffer. Be sure to transfer liposomes from the test tubes to the dialysis bags over a beaker to reclaim any potential spills. Replace buffer and continue dialysis until external dialysate is no longer notably pink.

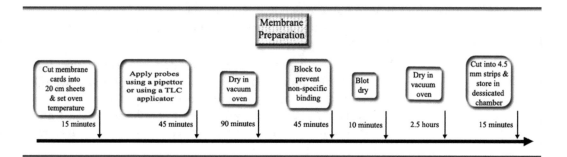

Scheme 2 Flow chart for the preparation of lateral flow membranes with immobilized capture antibodies

18. Clean everything thoroughly, including removing the o-rings from the Teflon holders and the needles and plungers from the syringes of the extruder.

19. Store dialyzed liposomes in 15-mL centrifuge tubes at 4 °C.

3.2 Membrane Preparation

A flowchart of this process is shown in Scheme 2.

1. Prepare blocking reagent (0.1% (w/v) casein, sodium salt, 0.25% (w/v) sucrose) or remove previously made blocking reagent from the refrigerator and allow to reach ambient temperature.

2. Cut the HF090 membrane cards into 20 cm sections (*see* **Note 4**).

3. Label each membrane card section with a line drawn ~0.5 cm from its top on the back side using a permanent marker. The top of the HF090 membrane cards has a single covered adhesive section, whereas the bottom has an adhesive section covered by two separate peel strips.

4. For each 20 cm membrane card section, prepare 50 μL of capture antibodies at 1.0 mg/mL in 0.4 M NaHCO$_3$/Na$_2$CO$_3$, pH 9.0 containing 5% (v/v) methanol. Prepare also 50 μL of control antibodies at 1.5 mg/mL in the same solution.

3.2.1 Automated Application of Capture Antibodies Using a Linomat IV or Similar TLC Plate Applicator

The following instructions are for operation of the CAMAG Linomat IV, but should be adaptable for similar applicators using the respective manufacturer's directions.

5. Set the Limonat IV parameters as follows:

Plate width:	200	Space:	0
Start position:	0	Rate:	12 s/μL
Band width:	190	Volume:	38

6. Fill the syringe slowly with the capture antibody solution mixture to avoid air bubbles, then insert syringe fully into syringe holder.

7. Gradually lower the plunger mechanism by pressing in the front and rear buttons until the silver lever is just above the syringe plunger.

8. Press the GAS button on the front of the Linomat, then lower the silver lever onto the syringe plunger by holding down the ↓ key until you see liquid consistently bubble from the end of your syringe. You may simultaneously hold down the + and ↓ buttons (faster) if there is a fair distance between the silver lever and syringe plunger.

9. Place one of the 20 cm membrane sections so that is lined up with the bottom and side markers on the black surface. Use the supplied flat magnets to hold down the membrane at its top and right sides. Slide the tower to 3.0 cm from the base of the membrane card using the ruled marks.

10. Press Calc, then Run. Repeat steps for remaining membrane sections.

11. Thoroughly wash the syringe with HPLC grade water, then repeat steps with the anti-streptavidin solution to form a control zone 3.5 cm from the base of the membrane card following the same procedure.

3.2.2 Manual Application of Capture Antibodies Using a Pipettor

If a TLC or similar applicator is not available, the capture antibody solution may be applied manually to the lateral flow membranes. Here, a round spot for the capture and control zones would result, versus the line seen in Fig. 3 generated using the Linomat for antibody application. This procedure is useful for optimization purposes, but is more laborious than the TLC applicator procedure due to the need to apply antibodies and tape individual membranes.

1. Cut the membrane sections into 4.5 mm pieces. We've found it beneficial to tape a laminated piece of paper to the paper cutter with lines every 4.5 mm for a guide. Then, align one end of one section with this paper and advance the section, making perpendicular cuts every 4.5 mm.

2. Using tape with the sticky side exposed placed onto a piece of cardboard, lightly tape the top of each membrane to a piece of clean cardboard using tweezers.

3. Apply 1 μL of the capture antibody solution to each membrane 3.0 cm from the bottom of the strip using a microvolume pipettor. The more consistent the antibody application is, the more consistent the LFA results will be. The antibody solution will rapidly seep into the membrane.

4. Deposit the control zone 3.5 cm from the base of the membrane card following the same procedure.

3.2.3 Membrane Drying and Blocking Steps

5. Place the membranes or membrane cards in the vacuum oven at 40 °C and set vacuum to 15″ Hg for 1.5 h. Alternatively, membranes or membrane cards may be stored overnight in a tightly sealed Tupperware container with the bottom filled to 0.5″ with Drierite and covered with layers of large Kimwipes.

6. Remove membranes from vacuum oven or desiccated storage container.

7. Pour blocking reagent into a plastic Tupperware container so that it is <50% full.

8. Using tweezers, gently remove each membrane or membrane card and place in the blocking reagent laminated side down. Be sure to allow each membrane to gradually sink into your blocking reagent, otherwise blocking will be uneven. You can gently help them along by pressing the ends under the solution.

9. Place container on shaker using slow agitation for 30 min. The speed should be set such that the membranes are being covered by solution and moving freely, but not so much that they become bunched up and solution comes over the sides.

10. Using tweezers, remove each membrane or membrane section and lie flat on three layers of large Kimwipes. Place 2–3 layers of Kimwipes on top of membranes, then add a heavy flat object on top of the membranes, such as a heavy lab supplies catalog.

11. Retape membranes back onto the cardboard.

12. Return the taped membranes to the vacuum oven at 25–30 °C and set vacuum to 15″ Hg for 3–4 h.

13. Remove membrane strips or sheets from cardboard and store in vacuum sealed bags at 4 °C or as above in **step 17** at in sealed Tupperware container at ambient temperature.

LFA assembly steps

14. Cut the absorbent pad into 1.8 × 20.0 cm sections corresponding to the number of membrane cards prepared in **step 2**.

15. Gently peel away the film covering the top section of the HF090 card.

16. Apply the absorbent pad such that it overlaps the nitrocellulose by 2 mm (Fig. 6).

Fig. 6 (*Left*) Layout of assembled LFA membrane showing overlaps and positions of sample and absorbent pads and antibody deposition zones. (*Right*) Image of assembled HF090 membrane with VFE unbound glass sample pad and cellulose fiber absorbent pad before insertion into cassette

17. Apply gentle, but even, pressure along the length of the absorbent pad to secure it to the underlying adhesive.

18. If the membranes are to be used for assay optimization purposes only in the absence of whole blood, the bottoms may be cut off. To cut off the bottoms, align the paper cutter such that the lowest 0.5 mm of nitrocellulose will be removed when the cut is made (*see* **Note 5**).

19. The bottom of the HF090 membrane cards can accommodate both a conjugate and a sample pad. For the assays described within, only a sample pad is used. Prior to assembly, this sample pad is cut into 1.9 × 20 cm sections and blocked with 0.05% (w/v) casein, 0.25% (w/v) sucrose for 30 min and dried in a vacuum oven at 21 °C and 15″ Hg overnight.

20. Gently peel away the film covering the bottom sections of the HF090 card.

21. Apply the sample pad such that it overlaps the nitrocellulose by 2 mm.

22. Apply gentle, but even, pressure along the length of the sample pad to secure it to the underlying adhesive. Excessive pressure may crush the delicate sample pads and impede flow of the samples.

23. The assembled membrane sheets may be cut into strips of desired widths prior to storage as described in **step 12**. Our membranes are typically cut into 4.5 mm widths.

3.3 Analysis of Liposomes

If available, liposome size measurements should be made using dynamic light scattering or alternative sub-micron particle size measurement technique. We make liposome size distribution measurements using a DynaPro LSR (Proterion Corporation, Piscataway, NJ) using the Dynamics (version 6.3.01) software program and the Cumulants method of analysis [70, 71]. The Bartlett assays are necessary for the preparation of protein or antibody-conjugated liposomes (*see* **Notes 6** and **7**).

3.3.1 Bartlett Assays for Phospholipid Content

1. Liposome samples (20 µL) in triplicate are dehydrated at 180 °C for 10 min, then mixed and heated with 1.5 mL of 3.33 N H_2SO_4 for 2 h at the same temperature. Standards prepared from potassium phosphate dibasic in deionized water (16, 32, 64, 128, and 256 nmol phosphate per tube in triplicate) are subjected concurrently to the same procedure. Note that inorganic phosphates will interfere with the Bartlett assay and need to be considered if these are used in alternate buffers or diluents used for liposome preparation.

2. 100 µL of 30% hydrogen peroxide is added and the mixture is returned to the oven for 1.5 h. The tubes are permitted to cool

to ambient temperature prior to, and vortexed vigorously following, each addition.

3. Lastly, 4.6 mL of 0.22% ammonium molybdate and 0.2 mL of the Fiske–Subbarow reagent [72] are added.

4. The Fiske-Subbarow reagent is prepared by mixing 40 mL 15% (w/v) sodium bisulfite, 0.2 g sodium sulfite, and 0.1 g 1-amino-4-naptholsulfonic acid at ambient temperature for 30 min then filtering out undissolved solids. This reagent needs to be prepared fresh at least weekly.

5. The tubes are then heated in a boiling water bath for 7 min, then quickly cooled in an ice water bath. The color of the highest standard should be moderate to dark blue after heating. If it is not sufficiently dark, continue heating for several minutes until it becomes this shade of blue.

6. The absorbance at 830 nm is recorded using a Powerwave XL microtiter plate reader (Bio-Tek Instruments, Winooski, VT).

7. The phospholipid content of the liposomes is determined from a calibration curve prepared from the phosphate standards analyzed in each run. The total lipid concentration is calculated by multiplying the phospholipid concentration by the initial ratio of total lipid to phospholipid.

3.4 Liposome Conjugation

The liposome conjugation procedure assumes that the phospholipid concentration, determined as described in Subheading 3.3 or by other means, is known. The procedure described within is has been optimized for the conjugation of streptavidin to liposomes and may be modified to accommodate other proteins by varying the EDC and protein concentrations. It utilizes EDC to form a zero-length linkage between carboxylic acid groups on the liposomes and protein amine groups. Before beginning this procedure, ensure that the protein to be conjugated is purified and does not contain any stabilizing proteins or buffer components with amine groups.

1. Remove EDC from −20 °C storage and allow to warm to ambient temperature while proceeding with the following steps.

2. Add desired volume of liposomes prepared in Subheading 3.1 to a 1.5-mL microcentrifuge tube.

3. Using the known phospholipid concentration, calculate the nmol of total lipid present. Be sure to account for non-phospholipid constituents such as cholesterol in your calculation.

4. Determine the amount of protein needed for conjugation. Typically, 0.05 mol% of the total lipid content is a reasonable amount.

5. Determine the amount of EDC needed for conjugation. 0.3 molar equivalents EDC based on the total lipid content is optimal. Calculate how much volume of a 100 mg/mL solution of EDC is needed.

6. Add the amount of protein calculated in step 23 to the microcentrifuge tube containing the liposomes and gently vortex until homogeneous.

7. Weigh the requisite amount of EDC into a clean, dry microcentrifuge tube.

8. Dilute EDC to a concentration of 100 mg/mL with 0.1 M MES buffer, pH 4.65 immediately before proceeding with **step 9**.

9. Add desired volume of EDC and immediately vortex the solution for approximately 30 s.

10. Incubate the mixture at room temperature in the dark for 30 min.

11. Purify the conjugated liposomes through a SEC column packed with Sepharose CL-4B using 1× HSS as an elution buffer [73].

12. All fractions may be combined together for subsequent phospholipid determination measurements as described in Subheading 3.3.

3.5 Assay Optimization Using Recombinant Myoglobin

The steps within this section are helpful for determining the best conditions for antibody/antigen interaction and signaling. Here, the buffer composition, biotinylated antibody, and liposome concentrations may be varied to attain the desired level of detection. This information can be utilized to streamline the development of the assay in whole blood. If needed, the capture antibody concentration can also be optimized during membrane preparation which can be done easily if the manual application approach is used.

1. Add 10 μL myoglobin solutions to the bottom of wells of a flat-bottom, non-binding microtiter plate (*see* **Note 7**).

2. Insert a 4.5 mm membrane strip with immobilized capture and control antibodies with the base cut off and allow volume to wick up membrane.

3. Once the solution has completely wicked, the membrane strips are transferred to wells containing 10 μL of MESS with 0.1% (w/v) BSA.

4. Once this solution has wicked, membranes are transferred to wells containing a solution with 6 μg/mL biotinylated anti-myoglobin antibodies and liposomes diluted to 250 μM phospholipid.

5. Once this solution has wicked, membranes are then transferred to wells containing 30 μL of MESS buffer.

6. Remove membrane once all solution has wicked and place on a sheet of paper to dry. A dark colored sheet allows for more contrast when results are scanned. Drying takes ∼ 10 min at room temperature.

3.6 Assay Format for the Analysis of Myoglobin in Whole Blood

In our studies, 16 mL of human venous blood was drawn into EDTA vacutainer tubes by venipuncture at the Gannett Health Center at Cornell University. Participants provided informed consent for this study, which was approved by the institutional review board at Cornell University. No participant identifier was associated with each sample.

Important: Human blood should be treated as it is potentially infectious and proper precautions for handling and disposal should be followed, in accordance with your organization's protocols. Obtaining human blood samples also requires informed consent of your subjects, in accordance with your organization's institutional review board.

1. Fully assembled membrane strips with blocked sample pads, immobilized capture and control antibodies, and absorbent pads are inserted into a commercially available cassette. The dimensions of the absorbent pad may be adjusted to ensure that the pressure points of the cassette line up with the sample and absorbent pad overlaps; the sample hole lines up with the sample pad; and the test and control lines are visible.

2. Whole blood is spiked with 0–350 ng/mL myoglobin.

3. 50 μL of the spiked blood is added to 50 μL of an antibody/liposome solution and is gently vortexed for 5 min.

4. 100 μL of the blood mixture is transferred to the sample pad.

5. Allow assay to run for 15 min. A signal proportional to the myoglobin concentration should result (Fig. 7).

3.7 Analysis of Results

The membranes may be read by eye for qualitative results, however, semiquantitative results may be obtained if a scanner, digital camera, or reflectometer is available. After the membranes have fully dried (typically 20–30 min), scan an image of the membranes using a computer and flatbed scanner, then use any available densitometry software to quantify band intensity. Alternatively, if a reflectometer is available, the readings should be taken at this time following the instrument manufacturer's instructions. For optimal reproducibility and sensitivity, the intensity of the bands should be read as soon as the membranes dry as it tends to decrease over time (days to weeks) and exposure to light. While membranes may also be scanned or read using a reflectometer when wet, care must be taken to avoid leaving impressions in the nitrocellulose. Whether read wet or dry, consistency in the timing of measurement is key for optimal reproducibility.

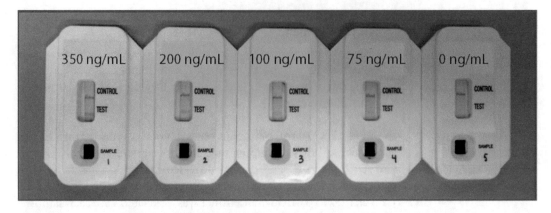

Fig. 7 Whole-blood lateral flow assay at spiked myoglobin concentrations ranging from 350 ng/mL (*left*) to 0 ng/mL. 50 μL blood samples were spiked with various myoglobin concentrations and added to 50 μL of an antibody/liposome solution. This mixture was gently vortexed, and after 5 min, all 100 μL were added to the sample pad. Results were observed after 15 min. A visible signal at the test line could be seen at a spiked myoglobin concentration of 75 ng/mL (*see* **Notes 8** and **9**)

4   Notes

1. For the preparation of liposomes to be tagged to antibodies or other proteins, the N-glutaryl DPPE phospholipid introduces a carboxylic acid group to the lipid bilayer allowing for facile conjugation to primary amine groups using 1-ethyl-3-[3-dimethylaminopropyl]carbodiimide hydrochloride (EDC) [73].

2. Be sure that the solution in the flask is being sonicated moderately vigorously and that the flask is not located in a "dead spot" of the bath. We use sonication level 6 with the Aquasonic Model 150D bath sonicator.

3. If liposomes are instead formed with a cholesterol-modified DNA probe for nucleic acid sequence detection [74], the focus on liposome concentration is less important since no post-formation conjugation needs to take place.

4. Here, HF090 membranes worked out well for this assay as they provided the requisite capillary flow, performance for our analyte, and ability to secure the sample and absorbent pads. However, blank backing cards (#MIBA-010, DCN, Carlsbad, CA) with adhesive for the center portion may be purchased to allow the user to apply their preferred nitrocellulose or other analytical membrane material.

5. By cutting off a small portion of the nitrocellulose, good contact of the nitrocellulose with the fluid during the assay is ensured.

6. For antibody conjugation, the liposomal phospholipid concentration needs to be determined using Bartlett assays [75] (*see* Subheading 3.3). Both antibodies of the matched set should be conjugated to liposomes to determine which serves as the better detection antibody. Note that EDC is very hydroscopic—use care to protect the reagent from moisture.

7. Alternatively, the phospholipid concentration of liposomes can be measured by inductively coupled plasma—atomic emission spectroscopy (ICP-AES). A volume of 20 μL of liposomes is diluted in nitric acid ($c = 0.5$ mol/L) to a total volume of 3 mL. The mixture is vortexed and sonicated thoroughly. Phosphorus standard solutions are prepared with concentrations of 1, 5, 10, 25, 50, and 100 μmol/L phosphate in 0.5 mol/L HNO_3 for calibration. The phosphorus content is then measured via ICP-AES (at the phosphorus specific wavelength of 178.29 nm [76].

8. No signal was observed in this assay with normal levels of circulating myoglobin, but a visible signal was observed if a normal blood sample was spiked with 75 ng/mL myoglobin, which mimics an elevation outside of the normal range. Normal levels of myoglobin in human blood range from 0 to 85 ng/mL, whereas the median level of myoglobin 90 min after an AMI has been reported to be 353 ng/mL [77]. The level at which a signal is observed in a LFA can be altered by optimizing the assay as noted in Subheading 3.5.

9. In any sandwich immunoassay run without a wash step, the risk of a high-dose Hook effect needs to be checked for to avoid underestimating the concentration of the analyte in the sample. This can be done by making dilutions of the sample and repeating the assay to ensure that the signal decreases as expected, rather than increases as would be the case with the Hook Effect.

Acknowledgments

The authors gratefully acknowledge Vincent Sy and Peter Bent of GE Healthcare for their sage advice in blood separation for LFA applications. We also are indebted to Tara Holter, Jocelyn Jing Tan, Kit Meyers, and Barb Leonard for their preliminary work in the development of this assay. The authors acknowledge partial funding through the Multistate Federal Formula Grant "Development of a novel rapid-on-site biosensor for food safety" Project #2012-13-132; the CD4 Initiative, Imperial College, London; and a Cornell Engineering Learning Initiatives Undergraduate Research Award funded by James Moore, '62 to Ricki Korff.

References

1. Helbig JH, Luck PC, Kunz B, Bubert A (2006) Evaluation of the Duopath Legionella lateral flow assay for identification of Legionella pneumophila and Legionella species culture isolates. Appl Environ Microbiol 72(6):4489–4491

2. Koide M, Haranaga S, Higa F, Tateyama M, Yamane N, Fujita J (2007) Comparative evaluation of Duopath Legionella lateral flow assay against the conventional culture method using Legionella pneumophila and Legionella anisa strains. Jpn J Infect Dis 60(4):214–216

3. http://www.kit.nl/biomedical_research/assets/images/Lepto_lateral_flow_protocol.doc

4. Johnston SP, Ballard MM, Beach MJ, Causer L, Wilkins PP (2003) Evaluation of three commercial assays for detection of Giardia and Cryptosporidium organisms in fecal specimens. J Clin Microbiol 41(2):623–626

5. Ketema F, Zeh C, Edelman DC, Saville R, Constantine NT (2001) Assessment of the performance of a rapid, lateral flow assay for the detection of antibodies to HIV. J Acquir Immune Defic Syndr 27(1):63–70

6. Cazacu AC, Demmler GJ, Neuman MA et al (2004) Comparison of a new lateral-flow chromatographic membrane immunoassay to viral culture for rapid detection and differentiation of influenza A and B viruses in respiratory specimens. J Clin Microbiol 42(8):3661–3664

7. Mokkapati VK, Sam Niedbala R, Kardos K et al (2007) Evaluation of UPlink-RSV: prototype rapid antigen test for detection of respiratory syncytial virus infection. Ann N Y Acad Sci 1098:476–485

8. Slinger R, Milk R, Gaboury I, Diaz-Mitoma F (2004) Evaluation of the QuickLab RSV test, a new rapid lateral-flow immunoassay for detection of respiratory syncytial virus antigen. J Clin Microbiol 42(8):3731–3733

9. van Hengel AJ, Capelletti C, Brohee M, Anklam E (2006) Validation of two commercial lateral flow devices for the detection of peanut proteins in cookies: interlaboratory study. J AOAC Int 89(2):462–468

10. http://www.craigmedical.com/salivascreen6_drug_tests.htm

11. http://www.brittneylimited.com/PDF%20Files/FDA_Files/Pregnancy-strip.pdf

12. Tiplady S (2013) Lateral flow and consumer diagnostics. In: Wild D (ed) *The immunoassay handbook*, 4th edn. Elsevier, Oxford, U.K.

13. Jung K, Zachow J, Lein M et al (1999) Rapid detection of elevated prostate-specific antigen levels in blood: performance of various membrane strip tests compared. Urology 53(1):155–160

14. http://www.brittneylimited.com/PDF%20Files/PSA-Prostrate-strip.pdf

15. http://www.nhdiag.com/botulism.shtml

16. Sharma SK, Eblen BS, Bull RL, Burr DH, Whiting RC (2005) Evaluation of lateral-flow clostridium botulinum neurotoxin detection kits for food analysis. Appl Environ Microbiol 71(7):3935–3941

17. Park CH, Kim HJ, Hixon DL, Bubert A (2003) Evaluation of the duopath verotoxin test for detection of shiga toxins in cultures of human stools. J Clin Microbiol 41(6):2650–2653

18. Capps KL, McLaughlin EM, Murray AWA et al (2004) Validation of three rapid screening methods for detection of verotoxin-producing *Escherichia coli* in foods: interlaboratory study. J AOAC Int 87(1):68–77

19. Smits HL, Eapen CK, Sugathan S et al (2001) Lateral-flow assay for rapid serodiagnosis of human leptospirosis. Clin Diagn Lab Immunol 8(1):166–169

20. Kassler WJ, Dillon BA, Haley C, Jones WK, Goldman A (1997) On-site, rapid HIV testing with same-day results and counseling. AIDS 11(8):1045–1051

21. Prevention USCfDCa (1996) Provisional public health service recommendations for chemoprophylaxis after occupational exposure to HIV. MMWR Morb Mortal Wkly Rep 45:468–480

22. Baeumner AJ, Jones C, Wong CY, Price A (2004) A generic sandwich-type biosensor with nanomolar detection limits. Anal Bioanal Chem 378(6):1587–1593

23. Baeumner AJ, Pretz J, Fang S (2004) A universal nucleic acid sequence biosensor with nanomolar detection limits. Anal Chem 76(4):888–894

24. Seal J, Braven H, Wallace P (2006) Point-of-care nucleic acid lateral-flow tests. IVD Technol 8:41–54

25. Tomoda H, Aoki N (2000) Prognostic value of C-reactive protein levels within six hours after the onset of acute myocardial infarction. Am Heart J 140(2):324–328

26. Qureshi A, Gurbuz Y, Niazi JH (2012) Biosensors for cardiac biomarkers detection: a review. Sens Actuators B Chem 171–172:62–76

27. Anderson JL, Adams CD, Antman EM et al (2011) 2011 ACCF/AHA Focused Update

Incorporated Into the ACC/AHA 2007 Guidelines for the Management of Patients With Unstable Angina/Non-ST-Elevation Myocardial Infarction: A Report of the American College of Cardiology Foundation/American Heart Association Task Force on Practice Guidelines. Circulation 123(18):e426–e579

28. Brogan GX Jr, Friedman S, McCuskey C et al (1994) Evaluation of a new rapid quantitative immunoassay for serum myoglobin versus CK-MB for ruling out acute myocardial infarction in the emergency department. Ann Emerg Med 24(4):665–671

29. McCord J, Nowak RM, McCullough PA et al (2001) Ninety-minute exclusion of acute myocardial infarction by use of quantitative point-of-care testing of myoglobin and troponin I. Circulation 104(13):1483–1488

30. Alexander LL (2016) Early prediction of acute myocardial infarction from clinical history, examination, and electrocardiogram in the emergency room. Ann Emerg Med 20 (11):1270

31. Schiff HB, MacSearraigh ET, Kallmeyer JC (1978) Myoglobinuria, rhabdomyolysis and marathon running. Q J Med 47(188):463–472

32. C-Peptide ELISA Kit ab178641. www.abcam.com/C-Peptide-ELISA-Kit-ab178641.pdf. Accessed 23 Mar 2016

33. Brown M (2008) Antibodies: Key to a robust lateral flow immunoassay. In: Wong RC, Tse HY (eds) Lateral flow immunoassay. Humana Press, New York

34. O'Farrell B (2013) Lateral flow immunoasay systems: Evolution from the current state of the art to the next generation of highly sensitive, quantitative rapid assays. In: Wild D (ed) The immunoassay handbook, 4th edn. Elsevier, Oxford

35. O'Farrell B, Bauer J (2006) Developing highly sensitive, more-reproducible lateral-flow assays. Part 1: New approaches to old problems. IVD Technol 10(6):11–15

36. Hans H, Beer EJ, Pflanz K, Klewitz TM (2002) Qualification of cellulose nitrate membranes for lateral-flow assays. IVD Technol 8:35–42

37. Harlow E, Lane D (1988) Antibodies: a laboratory manual. Cold Spring Harbor, Cold Spring Harbor Laboratories

38. Jones K (2001) Membrane immobilization of nucleic acids. Part 2: Probe attachment techniques. IVD Technol pp 1–7

39. Edwards KA, Baeumner AJ (2006) Optimization of DNA-tagged dye-encapsulating liposomes for lateral-flow assays based on sandwich hybridization. Anal Bioanal Chem 386:1335–1343

40. Reinholt SJ, Sonnenfeldt A, Naik A, Frey MW, Baeumner AJ (2014) Developing new materials for paper-based diagnostics using electrospun nanofibers. Anal Bioanal Chem 406 (14):3297–3304

41. Manis AE, Bowman JR, Bowlin GL, Simpson DG (2007) Electrospun nitrocellulose and nylon: design and fabrication of novel high performance platforms for protein blotting applications. J Biol Eng 1:2–2

42. Matlock-Colangelo L, Baeumner AJ (2012) Recent progress in the design of nanofiber-based biosensing devices. Lab Chip 12 (15):2612–2620

43. Mansfield M (2009) Nitrocellulose membranes for lateral flow immunoassays: a technical treatise. In: Wong RC, Tse HY (eds) Lateral flow immunoassay. Humana Press, New York, pp 95–113

44. Bangs LI (1999) Lateral flow tests. Vol Tech Note 303:1–6

45. Millipore. Rapid lateral flow test strips—considerations for product development. http://www.emdmillipore.com/US/en/products/ivd-oem-materials-reagents/lateral-flow-membranes/n6mb.qB.L0YAAAE_gut3.Lxi,nav?bd=1. Accessed 18 Mar 2016

46. Batteiger B, Newhall WJ, Jones RB (1982) The use of Tween 20 as a blocking agent in the immunological detection of proteins transferred to nitrocellulose membranes. J Immunol Methods 55(3):297–307

47. Jones KD (1999) Troubleshooting protein binding in nitrocellulose membranes. Part I: Principles. IVD Technol 5:32–41

48. Van Dam AP, Van den Brink HG, Smeenk RJ (1990) Technical problems concerning the use of immunoblots for the detection of antinuclear antibodies. J Immunol Methods 129 (1):63–70

49. Zampieri S, Ghirardello A, Doria A et al (2000) The use of Tween 20 in immunoblotting assays for the detection of autoantibodies in connective tissue diseases. J Immunol Methods 239 (1–2):1–11

50. Smits HL, Chee HD, Eapen CK et al (2001) Latex based, rapid and easy assay for human leptospirosis in a single test format. Trop Med Int Health 6(2):114–118

51. Carney J, Braven H, Seal J, Whitworth E (2006) Present and future applications of gold in rapid assays. IVD Technol 12:41–49

52. Chan CP, Sum KW, Cheung KY et al (2003) Development of a quantitative lateral-flow assay for rapid detection of fatty acid-binding protein. J Immunol Methods 279 (1–2):91–100

53. Wang S, Zhang C, Wang J, Zhang Y (2005) Development of colloidal gold-based flow-through and lateral-flow immunoassays for the rapid detection of the insecticide carbaryl. Anal Chim Acta 546(2):161–166

54. Klewitz T, Gessler F, Beer H, Pflanz K, Scheper T (2006) Immunochromatographic assay for determination of botulinum neurotoxin type D. Sens Actuat Chem 113(2):582–589

55. Shyu RH, Shyu HF, Liu HW, Tang SS (2002) Colloidal gold-based immunochromatographic assay for detection of ricin. Toxicon 40(3):255–258

56. Greenwald R, Esfandiari J, Lesellier S et al (2003) Improved serodetection of Mycobacterium bovis infection in badgers (*Meles meles*) using multiantigen test formats. Diagn Microbiol Infect Dis 46(3):197–203

57. Birnbaum S, Uden C, Magnusson CG, Nilsson S (1992) Latex-based thin-layer immunoaffinity chromatography for quantitation of protein analytes. Anal Biochem 206(1):168–171

58. Corstjens P, Zuiderwijk M, Brink A et al (2001) Use of up-converting phosphor reporters in lateral-flow assays to detect specific nucleic acid sequences: a rapid, sensitive DNA test to identify human papillomavirus type 16 infection. Clin Chem 47(10):1885–1893

59. LaBorde R, O'Farrell B (2002) Paramagnetic-particle detection in lateral-flow assays. IVD Technol 8(3):36–41

60. Rule GS, Montagna RA, Durst RA (1996) Rapid method for visual identification of specific DNA sequences based on DNA-tagged liposomes. Clin Chem 42(8):1206–1209

61. Szoka F, Olson F, Heath T, Vail W, Mayhew E, Papahadjopoulos D (1980) Preparation of unilamellar liposomes of intermediate size (0.1–0.2 mumol) by a combination of reverse phase evaporation and extrusion through polycarbonate membranes. Biochim Biophys Acta 601(3):559–571

62. Szoka F Jr, Papahadjopoulos D (1978) Procedure for preparation of liposomes with large internal aqueous space and high capture by reverse-phase evaporation. Proc Natl Acad Sci U S A 75(9):4194–4198

63. Edwards KA, Baeumner AJ (2006) Sequential injection analysis system for the sandwich hybridization-based detection of nucleic acids. Anal Chem 78(6):1958–1966

64. Rule GS, Montagna RA, Durst RA (1997) Characteristics of DNA-tagged liposomes allowing their use in capillary-migration,

sandwich-hybridization assays. Anal Biochem 244(2):260–269

65. Rongen HAH, Bult A, vanBennekom WP (1997) Liposomes and immunoassays. J Immunol Methods 204(2):105–133

66. Edwards KA, Baeumner AJ (2013) Periplasmic binding protein-based detection of maltose using liposomes: a new class of biorecognition elements in competitive assays. Anal Chem 85(5):2770–2778

67. Chen CS, Baeumner AJ, Durst RA (2005) Protein G-liposomal nanovesicles as universal reagents for immunoassays. Talanta 67(1):205–211

68. Plant AL, Brizgys MV, Locasio-Brown L, Durst RA (1989) Generic liposome reagent for immunoassays. Anal Biochem 176(2):420–426

69. Edwards KA, Baeumner AJ (2007) DNA-oligonucleotide encapsulating liposomes as a secondary signal amplification means. Anal Chem 79(5):1806–1815

70. Koppel DE (1972) Analysis of macromolecular polydispersity in intensity correlation spectroscopy—method of Cumulants. J Chem Phys 57(11):4814–4820

71. Frisken BJ (2001) Revisiting the method of cumulants for the analysis of dynamic light-scattering data. Appl Opt 40(24):4087–4091

72. Fiske CH, Subbarow Y (1925) The colorimetric determination of phosphorus. J Biol Chem 66(2):375–400

73. Edwards KA, Curtis KL, Sailor J, Baeumner A (2008) Universal liposomes: preparation and usage for the detection of mRNA. Anal Bioanal Chem 391(5):1689–1702

74. Edwards KA, Baeumner AJ (2009) Liposome-enhanced lateral-flow assays for the sandwich-hybridization detection of RNA. Methods Mol Biol 504:185–215

75. Bartlett GR (1959) Phosphorus assay in column chromatography. J Biol Chem 234(3):466–468

76. Fenzl C, Hirsch T, Baeumner A Liposomes with high refractive index encapsulants as tunable signal amplification tools in surface Plasmon resonance spectroscopy. Anal Chem 87(21):11157–11163

77. Sallach SM, Nowak R, Hudson MP et al (2004) A change in serum myoglobin to detect acute myocardial infarction in patients with normal troponin I levels. Am J Cardiol 94(7):864–867

Chapter 26

Development of Dual Quantitative Lateral Flow Immunoassay for the Detection of Mycotoxins

Yuan-Kai Wang, Ya-Xian Yan, and Jian-He Sun

Abstract

Lateral flow immunoassays have been widely used in recent years for detection of toxins, heavy metals, and biomarkers. To improve the efficiency of individual lateral flow immunoassays, multiplex analytical strips play an important role in the detection of several important analytes. In this chapter, development of a dual lateral flow immunoassay is presented for detection of a variety of low molecular weight molecules. Various buffers, additives, and materials are introduced and evaluated. Depending on the analyte to be tested, the technique allows for selection of optimum buffers, additives, and other materials.

Key words Lateral flow, Gold nanoparticles, Fumonisin B1, Deoxynivalenol, Aflatoxin B1

1 Introduction

With the advantages of speed, low cost, and ease of use, lateral flow immunoassays can be developed to detect toxins, biomarkers, drugs, etc., and are especially useful for point-of-care applications [1]. Like other immunoassays, the principle of lateral flow immunoassays includes antigen–antibody and antibody–antibody (such as mouse antibody and goat anti-mouse antibody) binding. Moreover, capillary action is employed to induce fluid flow through the device without need of further mechanical or human intervention. Signal detection is provided via gold nanoparticles or quantum dot fluorescence. For most lateral flow immunoassays minimal training is necessary and results can be observed in 20 min. These assays will help address future needs in area as diverse as environmental monitoring, biomarker assays, and identification of toxins and drugs.

However, qualitative lateral flow immunoassays still lack adequate sensitivity for many applications. Current detection limits range from 1–10 ng/mL. Moreover, this immunoassay provides only "yes or no" qualitative results. The future challenge of

Avraham Rasooly and Ben Prickril (eds.), *Biosensors and Biodetection: Methods and Protocols Volume 1: Optical-Based Detectors*, Methods in Molecular Biology, vol. 1571, DOI 10.1007/978-1-4939-6848-0_26,
© Springer Science+Business Media LLC 2017

Table 1
Comparison between competitive and sandwich lateral flow immunoassay

Differences of strip	Indirect competitive lateral flow immunoassay	Sandwich lateral flow immunoassay
Conjugation pad	Colloidal gold–monoclonal antibody conjugate	Colloidal gold–monoclonal antibody 1# conjugate
Test line	Analyte–protein conjugate	Monoclonal antibody 2#
Detect analytes	Low-weight molecules	Proteins, bacteria, etc.
Result judgment of test line	Red line indicates negative	Red line indicates positive

analytical methods such as this will be to provide more sensitive, quantitative, and accurate results.

Based on the measurement of strip reader device for the test line and control line, quantitative lateral flow immunoassay can be developed to determine certain concentrations of analytes [2]. Also, different analytes can be detected simultaneously in multiplex quantitative lateral flow immunoassay for the high through-put screening, which can save time, cost, and human power.

According to the different analytes, lateral flow immunoassay can be divided into competitive analysis [3] and sandwich analysis [4] (*see* Table 1). Competitive lateral flow immunoassays are employed to detect the low-weight molecules (toxins, heavy metals, drugs, etc.). In most competitive lateral flow assays, the competition is between the migrating target analyte in the sample and the immobilized analyte on the strip (capture antigen) for the binding to the migrating labeled antibody. The degree of inhibition of migrating labeled antibody binding to the capture antigen is proportional to the level of the migrating target analyte in the sample.

To form a strip for lateral flow immunoassay, several essential parts including sample pad, conjugation pad, absorption pad, nitrocellulose (NC) membrane, and a sticky PVC baseplate are needed (*see* Fig. 1). A sticky baseplate is used as the framework, and other pads can be readily pasted onto it. Because the mycotoxins are low molecular weight molecules, competitive lateral flow immunoassay is selected in this study. To assemble the competitive test strip, the antigen conjugates/secondary antibody-coated NC membrane is first pasted onto the sticky side of PVC baseplate. Next, the conjugation pad (0.5 × 0.5 cm) sample pad (2 × 0.5 cm) and absorption pad (3 × 0.5 cm) are pasted to the lower and upper portions of the baseplate, with 1–2 mm overlap to the NC membrane on both sides. For the reactant in each part of the strip, colloidal gold–antibody (such as mouse monoclonal antibody from against seleted

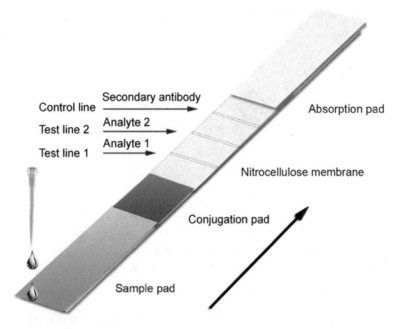

Fig. 1 Schematic diagram of the quantitative lateral flow dual immunoassay

analytes) is added on the conjugation pad and pretreated for dehydration. Analyte–protein conjugate is immobilized in a line onto the NC membrane as the test line. A secondary antibody (such as goat anti-mouse antibody) is immobilized in a line onto the NC membrane as the control line, which can react with the antibody in the colloidal gold solution. When the sample solution is added on the sample pad, the solution will flow on the strip in the direction of arrow under capillary action, which passes through the conjugation pad, NC membrane, and absorption pad in sequence. Normally the colloidal gold and colloidal gold–antibody solutions are red, thus if reaction occurred on the test line or control line, a red color can be observed. To judge the positive or negative samples, if the extract contains no analytes (negative sample), the dehydrated colloidal gold–monoclonal antibody will be resolved and flow with the extract under capillary action. Then the mixture will react with the analyte–protein conjugate to form a complex. Thus the red color can be observed in the test line. The fluid moves forward continuously and reacts with the secondary antibody to form a new complex, and the clear red line can be observed in the control line (*see* Fig. 2). When the extract contains analyte (positive sample), the colloidal gold–monoclonal antibody will first bind with the analyte. Then the complex of the colloidal gold–monoclonal antibody will flow over the strip. Because the antibody has already bound with the analyte, the colloidal gold–monoclonal antibody cannot bind with the analyte–protein conjugate in the test line, and no color will be observed. But the antibody in colloidal gold solution can still

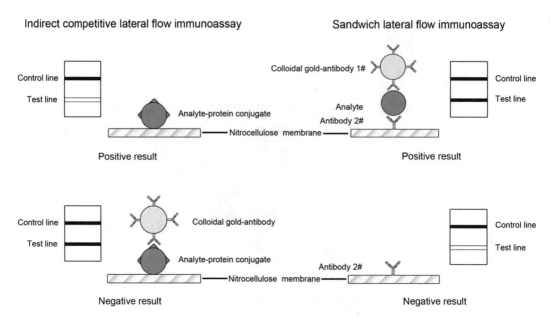

Fig. 2 Results judgment and reactions in test line of indirect competitive and sandwich lateral flow immunoassay

bind with secondary antibody in the control line, and red color will be observed on the control line (*see* Fig. 2).

The sandwich lateral flow immunoassays are mainly used to detect large molecule analytes, such as proteins, bacteria, and biomarkers. For each analyte, the monoclonal antibodies 1# and 2# are the antibodies against the different epitopes of the analyte. Normally, the monoclonal antibody conjugated with colloidal gold nanoparticles named 1#, and the monoclonal antibody is immobilized in a line on the NC membrane as the test line named 2#. The anaylte in the extract of samples will be reacted with colloidal gold–monoclonal antibody 1# conjugate firstly, and continuously reacted with the monoclonal antibody 2# on the test line. Thus a clear red line can be observed, which indicate the positive result (*see* Fig. 2).

To our knowledge, the uses of inappropriate buffers for lateral flow immunoassay will induce instability of the gold nanoparticle-labeled antibodies and affect sensitivity (*see* **Note 1**). The intensities of the test lines, stability and release rate of the mixtures (gold nanoparticle-labeled antibodies and sample extraction), and detection sensitivity are markedly affected by the types of buffer and concentrations of their components. Moreover, in different lateral flow immunoassays (different analytes, antibodies, or materials), the optimum buffers may be different. To optimize the immunoassay, different buffers, additives, surfactants (*see* **Note 2**) would be evaluated. The optimum buffers and concentrations of components are needed to determine for lower detection limits, balance of intensities of the test and control lines, and stability of the gold nanoparticle-labeled antibodies. The buffers in lateral flow immunoassay are mainly PBS and borate buffer (from 2 mM to 100 mM), and the pH values are ranges from 6.5 to 8.2. The surfactant is also an essential component for the conjugation/sample pad pretreating buffer, and it will affect the color of lines and the sensitivity. Different types of surfactants (Tween 20, Triton X-100, Tetronic 1307, etc.) can be evaluated, and selected in the certain lateral flow immunoassay.

2 Materials

1. Tetrachloroauric acid (Sigma Chemical, St. Louis, MO, USA).
2. The monoclonal antibodies against zearalenone (2C9, mAb-ZEN) and fumonisin B1 (6H3, mAb-FB1) are prepared in our laboratory based on the hybridoma approach [5].
3. Bovine serum albumin (BSA, Sangon Biotech, Shanghai, China).
4. Ovalbumin (OVA, Sangon Biotech, Shanghai, China).
5. ZEN-BSA and FB1-OVA conjugates are prepared in our laboratory.
6. Goat anti-mouse antibody (ICL, Portland, OR, USA).
7. Tetronic 1307 (Pragmatic co. ltd, Elkhart, IN, USA).
8. NC membrane (Millipore 180, from Millipore, Bedford, MA, USA).
9. Glass fiber (8964, from Ahlstrom, Helsinki, Finland).
10. Glass fiber (SB08, from Jiening Biotech, Shanghai, China).
11. Strip reagent dispenser (XYZ-3060, from Bio-Dot, Irvine, CA, USA).
12. Strip reader (CHR-110R, from Kaiwood, Taiwan, China).
13. MX2 shaker (Finepcr, Gyeonggi-do, Korea).
14. Incubator (DHP-9082, from Yiheng, Shanghai, China).
15. Centrifuge (5418, from Eppendorf, Hamburg, Germany).
16. Spectrophotometer (UV-1800, from Shimadzu, Tokyo, Japan).
17. Transmission electron microscope system (JEM 2100, from JEOL, Tokyo, Japan).

Other reagents such as sodium citrate, potassium carbonate, sodium chloride, sodium phosphate, sodium dihydrogen phosphate, trehalose, sucrose, sodium azide, potassium chloride, monopotassium phosphate, Tween 20, methanol are purchased from Sinopharm, Shanghai, China.

3   Methods

3.1   Preparation of Colloidal Gold Nanoparticles (25 nm)

The flow chart of methods in this study is shown on Fig. 3. Firstly, 100 mL tetrachloroauric acid solution (0.01%, w/v) is stirred and heated until boiling (*see* **Note 3**) in conical flask. A volume of 0.7 mL 1% sodium citrate (w/v) is then added to the solution quickly [6] (*see* **Note 4**). In the next 2 min, the color of mixture is changed from transparent to dark, then clear wine red. Decreasing the power to gentle heating for 5 min, then the mixture is cooled down to room temperature and stored at 4 °C. The diameter of nanoparticles is determined by the wavelength scanning and transmission electron microscope (*see* Fig. 4). The maximum absorbing wavelength is 525 nm for the 25 nm colloidal gold nanoparticles, based on the equation relating the gold nanoparticle size X (nm) and the maximum absorption wavelength Y (nm): $Y = 0.4271X + 514.56$ [7]. In transmission electron microscopy the mean diameter of gold nanoparticles is 24.8 ± 3.2 nm as measured by averaging a random selection of 100 particles.

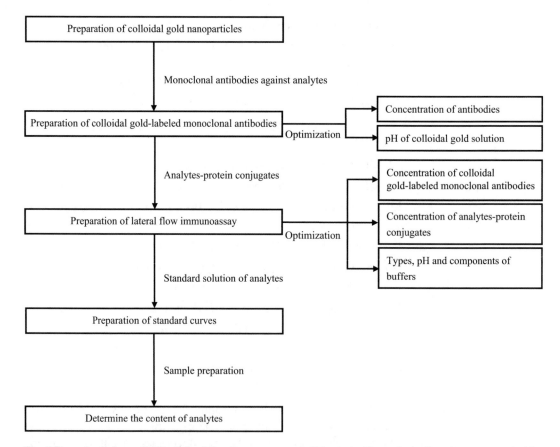

Fig. 3 Flow chart of quantitative lateral flow immunoassay in this study. The optimization steps are needed for each lateral flow immunoassay

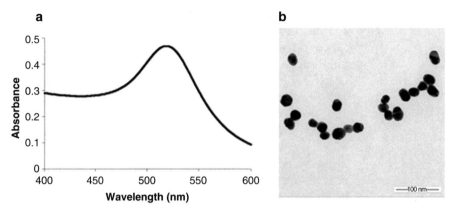

Fig. 4 Measurement of 25 nm colloidal gold nanoparticles by wavelength scanning (**a**) and transmission electron microscope (**b**)

3.2 Preparation of Colloidal Gold-Labeled Monoclonal Antibodies

The monoclonal antibody mAb-ZEN is evaluated and showed no cross-reactivity with FB1, deoxynivalenol, or aflatoxin B1, which usually coexist in samples. The monoclonal antibody mAb-FB1 is also evaluated and showed no cross-reactivity with ZEN, deoxynivalenol, and aflatoxin B1. To optimize the binding between colloidal gold particles and antibodies, colloidal gold solutions at different pH and antibody concentrations are evaluated as follows (*see* **Note 5**). The pH values of colloidal gold solutions (1 mL) are adjusted by 0.2 M potassium carbonate to 6.0, 6.5, 7.0, 7.5, 8.0, 8.5, and 9.0. 0.1 mg/mL of antibody in 100 μL 2 mM borate buffer (pH 7.4) is added to each colloidal gold solution. After vortexing for 10 min, let the mixture sit for 10 min and room temperature. Then 100 μL of 10% sodium chloride are added to each tube. After vortexing for 10 min, the mixtures are incubated at room temperature for 30 min. After adding sodium chloride, the coagulation of colloidal gold–monoclonal antibody will be observed in the improper pH values. Finally, the solutions are determined by wavelength scanning respectively, and the highest optical density of certain pH value solution in 525 nm is selected as the optimum pH value.

Different concentrations of antibodies are also optimized. The optimum pH which evaluated in the previous method is employed. Different concentrations of antibodies mAb-ZEN (0.5, 1, 2, 3, 4, 5, and 6 μg/mL) or mAb-FB1 (3, 4, 5, 6, 7, 8, and 9 μg/mL) in 100 μL 2 mM borate buffer (pH 7.4) are prepared and added to each colloidal gold solution. After vortexing for 10 min, let the mixture sit for 10 min and room temperature. Then 100 μL of 10% sodium chloride are added to each tube. After vortexing for 10 min, the mixtures are incubated at room temperature for 30 min. At last, the solutions are determined by wavelength scanning respectively and the highest optical density in 525 nm, and relative lower concentration (to improve the sensitivity) of certain antibody solution is selected as the optimum antibody concentration.

To prepare the colloidal gold-labeled monoclonal antibodies, 1.0 mL of 2 mM borate buffer (pH 7.4) containing mAb-ZEN (3 μg/mL) or mAb-FB1 (7 μg/mL) is added slowly with 1-min duration to 10 mL colloidal gold solution (pH 6.5 for mAb-ZEN or pH 7.0 for mAb-FB1). After gentle stirring for 30 min, 1.0 mL 10% BSA (w/v) is added to the mixture, which is again stirred for another 30 min. The mixture is then centrifuged at $1500 \times g$ for 20 min. The supernatant sample is collected and centrifuged at $8000 \times g$ for 30 min. Finally, the colloidal liquid is washed three times by adding 10 mL of 2 mM borate buffer (pH 7.4) and centrifuging at $6000 \, g$ for 30 min. The resulting colloidal gold-labeled antibodies are then stored in 2 mM borate buffer (pH 7.4) containing 6% trehalose (w/v), 4% sucrose (w/v), 1% BSA (w/v), and 0.05% sodium azide (w/v) at 4 °C. To evaluate the quality of colloidal gold-labeled monoclonal antibodies, 2 μL 0.1 mg/mL goat anti-mouse antibody is added on the NC membrane firstly. After dried at room temperature, 2 μL of colloidal gold-labeled monoclonal antibodies are added on the NC membrane which located lower than the goat anti-mouse antibody respectively. At last, 50 μL ddH$_2$O is added on the low side of the NC membrane. Red dot is observed at the goat anti-mouse antibody point, which indicated the well quality of colloidal gold-labeled monoclonal antibodies.

3.3 Preparation of Lateral Flow Immunoassay

Mycotoxin-protein conjugates (ZEN-BSA and FB1-OVA) are prepared firstly. To improve the sensitivity of detection, the concentrations of coated antigens (0.05 mg/mL for ZEN-BSA, and 0.2 mg/mL for FB1-OVA) are optimized by checkerboard titration test which the details are showed as follows (*see* Fig. 5).

Different concentrations of ZEN-BSA (0.025, 0.05, 0.1, 0.2 mg/mL) or FB1-OVA (0.05, 0.1, 0.2, 0.4 mg/mL) in 50 mM phosphate buffer saline (PBS, pH 7.4) with 7% methanol (w/w) are prepared. Then the dispensed volume is set to 1.0 μL/cm, which means that 1.0 μL of solution is dispensed onto each centimeter of NC membrane. Each of the above solutions is dispensed onto the NC membrane to form the test lines in each strip. The conjugation pad made of glass fiber (8964) is dipped in and pretreated by 10 mM PBS (pH 7.4) containing 6% trehalose (w/v), 1% BSA (w/v), 0.5% Tetronic 1307 (w/v), and 0.05% sodium azide (w/v), and stored at 4 °C after freeze dehydration. The pad is cut into 0.5 × 0.5 cm sections, and 10 μL different dilution ratio of colloid gold labeled antibody (diluted by 20 mM borate buffer, pH 8.2, containing 6% trehalose, 1% BSA, and 0.05% sodium azide) is added and dehydrated under vacuum (*see* **Note 6**). The sample pad made of glass fiber (SB08) is dipped in and pretreated by 50 mM borate buffer (pH 7.4) containing 5% trehalose, 1% BSA, 0.1% Tetronic 1307, and stored at 4 °C after freeze dehydration. Finally, 150 μL ddH$_2$O is added on the sample pad.

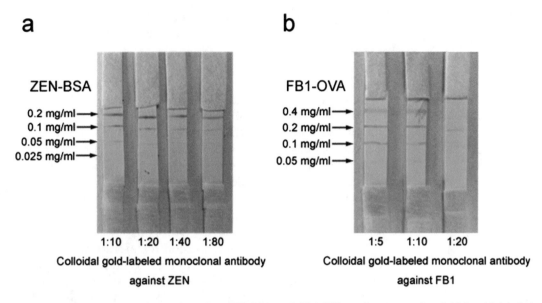

Fig. 5 Different concentrations of coating ZEN-BSA and FB1-OVA conjugates and colloidal gold labeled antibody in lateral flow immunoassay for ZEN (**a**) and FB1 (**b**)

The fluid will pass through the sample pad, conjugation pad, NC membrane, and absorption pad under capillary action. Clear red lines (test lines) can be observed if the concentration of coated antigen and colloidal gold-labeled monoclonal antibody are enough, and selected as the optimum coating concentration.

Based on the optimization of individual test strip, dual lateral flow strip is developed subsequently. ZEN-BSA (0.05 mg/mL) and FB1-OVA (0.2 mg/mL) are dispensed onto the NC membrane as two test lines (*see* Fig. 1). The goat anti-mouse antibody (0.07 mg/mL) is also dispensed onto the NC membrane (1.0 μL/cm) to form the control line, positioned at 0.5 cm above the Test 2 line. Mixtures of the two colloidal gold-labeled antibodies (mAb-ZEN and mAb-FB1) are prepared at different ratios (1:9, 1.5:8.5, 2:8, 2.5:8.5, and 3:7, v/v) to acquire similar product intensities in the Test 2 and Test 1 lines. A low concentration of colloidal gold-labeled antibodies results in higher sensitivity, but narrows the detection range. Thus, the use of an optimum dilution ratio could balance the sensitivity and detection range in the lateral flow immunoassay. The optimum ratio of the gold nanoparticles-mAb-ZEN and gold nanoparticles-mAb-FB1 is 1.5:8.5 (v/v). Furthermore, different dilution ratios (1:1.5, 1:2, 1:2.5, 1:3, 1:3.5, and 1:4, v/v) of these mixtures are assessed using the 20 mM borate buffer (pH 8.2) containing 6% trehalose (w/v), 1% BSA (w/v), and 0.05% sodium azide (w/v). The optimum dilution ratio of the colloidal gold-labeled antibodies mixture is 1:2.5 (v/v) in the colloidal gold–monoclonal antibody dilution buffer. Then 10 μL mixtue is added on the pretreated conjugation pad and dehydrated under vacuum. The assembly of dual lateral flow immunoassay is similar with the individual method which introduced previously.

3.4 Test Procedure

Twofold dilution series of ZEN (from 25 ng/mL) or FB1 (from 500 ng/mL) are prepared in 50 mM PBST. 250 μL standard solutions are tested to determine the detection limit and ranges of the strip. During reaction in the test strip, ZEN or FB1 in the mixture would compete for binding with the specific antibodies gold nanoparticles–monoclonal antibody for ZEN or gold nanoparticles–monoclonal antibody for FB1 against the coating antigen conjugates, thus altering the reaction intensity on the test lines (Test 1 or Test 2 lines, respectively, *see* Fig. 6).

The intensity of the test line and control lines are captured with the strip reader after 20 min of reaction. Concentrations of ZEN or FB1 in the samples are quantified from the dose–response curves (band intensity vs concentrations of ZEN or FB1 in the standard solutions) which are run simultaneously in triplicate. The strip reader results from different concentrations of ZEN and FB1 are used to determine the calibration curves of ZEN and FB1 (*see* Fig. 7, by Origin 6.0, from OriginLab, Northampton, MA, USA). Relative standard deviations for different samples are determined and normally need to be lower than 15%. After extraction

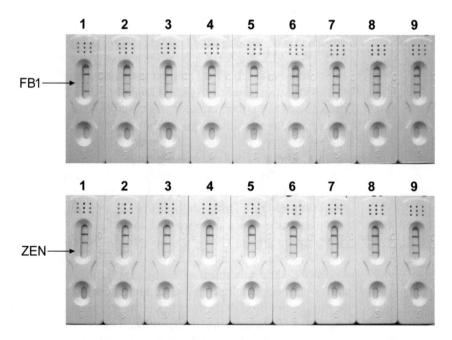

Fig. 6 Lateral flow dual immunoassay strips for fumonisin B1 (*top*) and zearalenone (*bottom*). The concentrations of FB1 from 1 to 9 (*top*): 500, 250, 125, 62.5, 31.3, 15.6, 7.81, 3.91 and 0 ng/mL, and those of ZEN from 1 to 9 (*bottom*): 25, 12.5, 6.25, 3.13, 1.56, 0.781, 0.391, 0.195 and 0 ng/mL

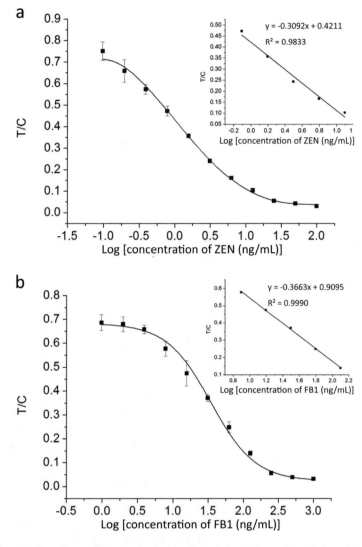

Fig. 7 Calibration curves of zearalenone (panel **a**) and fumonisin B1 (panel **b**) in the lateral flow dual immunoassay. The X-axes are the log concentrations of ZEN or FB1. The Y-axes are the ratio of the relative optical density of the test line to the control line, a ratio that represents the degree of competitive inhibition. The detection range is 0.94–7.52 ng/mL for ZEN and 9.34–100.45 ng/mL for FB1. The error bars indicate the standard deviation

from the spiked and natural samples, the dilution ratios (1:2.5, 1:5, 1:10, 1:15, and 1:20, v/v) of the sample extracts are also optimized in order to decrease the methanol content in the extraction buffer and the influence of proteins or other components in the sample matrixes. The spiked and natural samples are determined for ZEN or FB1 by quantitative lateral flow immunoassay and LC-MS/MS. The recovery rates of the spiked samples are compared and the correlation between lateral flow immunoassay and LC-MS/MS results is investigated using linear regression (SPSS software, IBM, New York, USA).

4 Notes

1. PBS could result in the appearance of deposits in gold nano-particle–antibody conjugation. Borate buffer could reduce line intensities when used for coating of the antigens or antibody. Borate buffer is more suitable than PBS for use in conjugation, dilution and storage of gold nanoparticle-labeled antibodies. Trehalose at 6% (w/v) is better than sucrose for stability of the gold nanoparticle-labeled antibodies during storage. BSA is essential for gold nanoparticle-labeled antibody storage and dilution.

2. To optimize lateral flow immunoassay performance, PBS and borate buffers with four different molar concentrations (2, 10, 20, and 50 mM) and two different pH values (7.4 and 8.2) are assessed. The buffers for antigen or antibody coating and sample pad pretreating as well as for conjugation, storage, and dilution of gold nanoparticle–antibodies are investigated. Types and concentrations of components in the buffers, i.e., BSA (0, 0.5%, and 1%, w/v), sucrose (0, 2%, 4%, 6%, 8%, and 10%, w/v), and trehalose (0, 2%, 4%, 6%, 8%, and 10%, w/v), are also investigated by assessing the detection limits of the test strip and the stability of gold nanoparticles-labeled antibodies.

3. The quality of solvent (ddH_2O) is essential for the preparation of gold nanoparticles.

4. The volume of 1% sodium citrate can be adjusted to prepare different diameters of gold nanoparticles. Normally, the more the volume of 1% sodium citrate, the smaller the nanoparticles prepared.

5. For the different diameters of gold nanoparticles and antibodies, the optimum pH of gold nanoparticles and concentrations of antibodies are different, but normally from 6.0 to 9.0. Thus a series of dilutions from 6.0–9.0 (such as 6.0, 6.5, 7.0, 7.5, 8.0, 8.5, 9.0) can be used for other different diameters and antibody optimizations.

6. Not every colloidal gold-labeled monoclonal antibody works well after the dehydration (freeze or evaporation). The researchers can add the liquid colloidal gold-labeled monoclonal antibody solution on the sample pad directly. But the sample pad is needed to be pretreated by 10 mM PBS (pH 7.4) containing 6% trehalose (w/v), 1% BSA (w/v), 0.5% Tetronic 1307 (w/v), and 0.05% sodium azide (w/v), and stored at 4 °C after freeze dehydration.

References

1. Quesada-González D, Merkoçi A (2015) Nanoparticle-based lateral flow biosensors. Biosens Bioelectron 73:47–63

2. Feng S, Caire R, Cortazar B et al (2014) Immunochromatographic diagnostic test analysis using Google Glass. ACS Nano 8(3): 3069–3079

3. Kolosova AY, Sibanda L, Dumoulin F et al (2008) Lateral-flow colloidal gold-based immunoassay for the rapid detection of deoxynivalenol with two indicator ranges. Anal Chim Acta 616 (2):235–244

4. Warren AD, Kwong GA, Wood DK et al (2014) Point-of-care diagnostics for noncommunicable diseases using synthetic urinary biomarkers and paper microfluidics. Proc Natl Acad Sci U S A 111(10):3671–3676

5. Wang YK, Wang J, Wang YC et al (2011) Preparation of anti-zearalenone monoclonal antibodies and development of an indirect competitive ELISA for zearalenone. Microbiology China 38(12):1793–1800

6. Frens G (1973) Preparation of gold dispersions of varying particle size: controlled nucleation for the regulation of the particle size in monodisperse gold suspensions. Nat Phys Sci 241:20–22

7. Chen X, Liu S (2004) Colloidal gold labeling immunoassay and its application in rapid detection of small molecule. Pharm Biotechnol 11:278–280

INDEX

Avraham Rasooly and Ben Prickril (eds.), *Biosensors and Biodetection: Methods and Protocols Volume 1:
Optical-Based Detectors*, Methods in Molecular Biology, vol. 1571, DOI 10.1007/978-1-4939-6848-0,
© Springer Science+Business Media LLC 2017

Printed in the United States
By Bookmasters